jacaranda *plus*

eBook *plus*

Next generation teaching and learning

Access all formats of your online Jacaranda resources in three easy steps!

To access your resources:

1. ▸ go to **www.jacplus.com.au**
2. ▸ log in to your existing account, or create a new account
3. ▸ enter your unique registration code(s).

Note
- Only one JacPLUS account is required to register all your Jacaranda digital products.
- By registering the code(s) within your JacPLUS bookshelf, you are agreeing to purchase the resource(s). Please view the terms and conditions when registering.

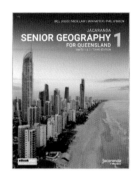

REGISTRATION CODE

Electronic versions of this title are available online; these include eBookPLUS and PDFs. Your unique registration codes for this title are:

QSGNY4WDUR9
RWR5NH4ETW7
RK7SUGU3DPT

Each code above provides access for one user to the eBookPLUS and PDFs.

NEED HELP?
If you would like to discuss specific digital licensing options or request digital trials, or if you require any other assistance, email support@jacplus.com.au or telephone 1800 JAC PLUS (1800 522 7587).

jacaranda
A Wiley Brand

Jacaranda Senior Geography 1 for Queensland Units 1 & 2 Third Edition

JACARANDA
SENIOR GEOGRAPHY 1
FOR QUEENSLAND
UNITS 1 & 2 | THIRD EDITION

jacaranda
A Wiley Brand

JACARANDA
SENIOR GEOGRAPHY 1
FOR QUEENSLAND
UNITS 1 & 2 | THIRD EDITION

BILL DODD
MICK LAW
IAIN MEYER
PHIL O'BRIEN

jacaranda
A Wiley Brand

Third edition published 2019 by
John Wiley & Sons Australia, Ltd
42 McDougall Street, Milton, Qld 4064

First edition published 2000
Second edition published 2008

Typeset in 11/14 pt TimesLTStd

© Bill Dodd, Mick Law, Iain Meyer, Phil O'Brien 2000, 2008, 2019

The moral rights of the authors have been asserted.

ISBN: 978-0-7303-6378-1

Reproduction and communication for educational purposes
The Australian *Copyright Act 1968* (the Act) allows a maximum of one chapter or 10% of the pages of this work, whichever is the greater, to be reproduced and/or communicated by any educational institution for its educational purposes provided that the educational institution (or the body that administers it) has given a remuneration notice to Copyright Agency Limited (CAL).

Reproduction and communication for other purposes
Except as permitted under the Act (for example, a fair dealing for the purposes of study, research, criticism or review), no part of this book may be reproduced, stored in a retrieval system, communicated or transmitted in any form or by any means without prior written permission. All inquiries should be made to the publisher.

Trademarks
Jacaranda, the JacPLUS logo, the learnON, assessON and studyON logos, Wiley and the Wiley logo, and any related trade dress are trademarks or registered trademarks of John Wiley & Sons Inc. and/or its affiliates in the United States, Australia and in other countries, and may not be used without written permission. All other trademarks are the property of their respective owners.

Front and back cover images: © Visual Collective / Shutterstock

Cartography by Spatial Vision Innovations Pty Ltd, Melbourne, www.spatialvision.com.au, MAPgraphics Pty Ltd Brisbane and the Wiley Art Studio

Illustrated by Harry Slaghekke (pages 210, 211, 212, 213), various artists and Wiley Composition Services

Typeset in India by diacriTech

Printed in Singapore by
Markono Print Media Pte Ltd

All activities have been written with the safety of both teacher and student in mind. Some, however, involve physical activity or the use of equipment or tools. **All due care should be taken when performing such activities**. Neither the publisher nor the authors can accept responsibility for any injury that may be sustained when completing activities described in this textbook.

 A catalogue record for this book is available from the National Library of Australia

10 9 8 7 6 5 4

CONTENTS

About this resource ... vii
About eBookPLUS ... ix
Acknowledgements ... x

UNIT 1   LIVING IN HAZARD ZONES   1

1 Natural hazards   3

  1.1   Overview ... 3
  1.2   Natural hazards ... 4
  1.3   Types of natural hazards ... 6
  1.4   Natural hazards in Australia ... 8
  1.5   The impact of hazards ... 11
  1.6   Assessing and responding to hazards ... 15
  1.7   Atmospheric hazards ... 20
  1.8   Geological hazards ... 40
  1.9   Geomorphic hazards ... 63
  1.10   Review ... 70

2 Ecological hazards   72

  2.1   Overview ... 72
  2.2   Ecological hazards ... 73
  2.3   Plant and animal invasions ... 80
  2.4   Pollutants ... 94
  2.5   Marine hazard zones ... 97
  2.6   Atmospheric pollutants ... 107
  2.7   Lithospheric pollutants ... 119
  2.8   Infectious and vector-borne diseases ... 125

UNIT 2   THE CHALLENGES OF CREATING SUSTAINABLE PLACES   137

3 Challenges for Australian places   139

  3.1   Overview ... 139
  3.2   Places in Australia ... 140
  3.3   Hierarchy of settlements ... 143
  3.4   Where are Australian places? ... 156
  3.5   Factors affecting Australia's population distribution ... 163

3.6	Challenges facing remote and rural Australia	175
3.7	FiFo communities in central Queensland	179
3.8	Challenges facing urban Australia	182
3.9	West End, Brisbane	195
3.10	Responding to the challenges facing places in Australia	203
3.11	Preparing for your field study	207

4 Challenges for megacities — 215

4.1	Overview	215
4.2	Global patterns of urbanisation	216
4.3	Air quality in the megacity	226
4.4	Health and sanitation management	233
4.5	Vulnerability to natural hazards	246
4.6	Heat islands	251
4.7	Food security	257
4.8	Waste management	261
4.9	Economic opportunities and challenges	267
4.10	Sustainable development in megacities	272

Glossary — 284
Index — 290

ABOUT THIS RESOURCE

Jacaranda Senior Geography for Queensland 1 (*Units 1 & 2*) *Third Edition* is tailored to address the intent and structure of the new Senior Geography syllabus, and to inspire students' sense of geographical curiosity. The *Jacaranda Senior Geography for Queensland* series provides easy-to-follow text and is supported by a bank of resources for both teachers and students. At Jacaranda we believe that every student should experience success and build confidence, while those who want to be challenged are supported as they progress to more difficult concepts and questions.

Building a sense of inquiry with strong geographical knowledge and skills

Chapter openers begin with key questions and activities to encourage students to consider their existing knowledge, and to begin forming questions about a topic. These activities also provide diagnostic opportunities for teachers.

Every chapter includes a range of case studies and examples to build a strong understanding of the impacts of geographical processes and patterns.

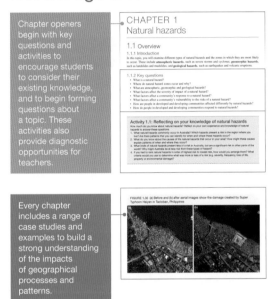

Each subtopic includes carefully graded questions.

Activities are aligned with Marzano and Kendall's taxonomy of cognitive process — retrieval, comprehension, analysis and knowledge utilisation.

A wide range of diagrams, models, data tables, maps and images build students' understanding and skills in interpreting various data formats.

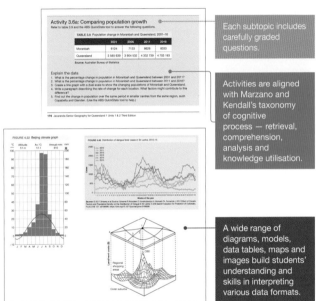

Preparing students for internal and external assessment success

Two complete, unseen data analysis activities with short- and extended-response questions are provided for each of the four chapters. These can be modified to suit your class, and are available only in the teacher eGuidePLUS. These tasks have been carefully designed to build students' confidence and ability with unseen data analysis tasks, and in constructing written responses to a wide range of data types. Sample responses are also provided for each activity.

An extensive glossary of terms is provided in print and as a hover-over feature in the eBookPLUS.

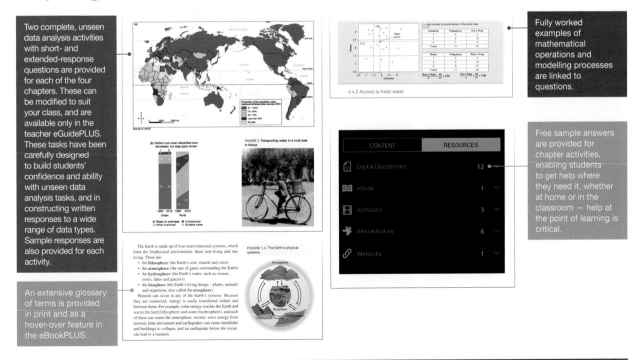

Fully worked examples of mathematical operations and modelling processes are linked to questions.

Free sample answers are provided for chapter activities, enabling students to get help where they need it, whether at home or in the classroom — help at the point of learning is critical.

eBookPLUS features

Digital documents: Downloadable activities to support skill development, and maps to ensure students are confident completing mapping exercises digitally and on paper.

Digital documents: Downloadable glossary to assist in knowledge development and exam preparation.

Digital documents: Downloadable PDF of the entire chapter of the print text.

Interactivities and Video eLessons: Bonus, step-by-step SkillBuilder instructional videos and multimedia activities consolidate students' core geographical skills. These are placed at the point of learning to enhance understanding and establish strong connections between knowledge and practical skills.

Weblinks: Direct access to an extensive range of additional data sources and information through links placed at the point of learning.

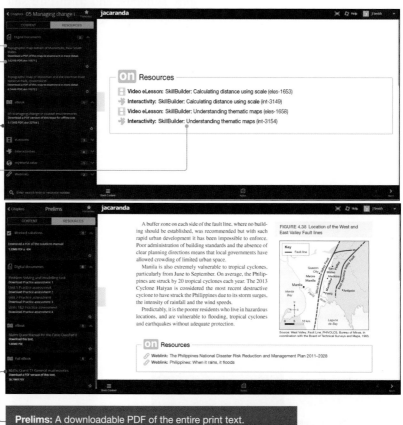

Prelims: A downloadable PDF of the entire print text.

Additional resources for teachers available in the eGuidePLUS

Teacher digital documents: Access to two quarantined assessment activities for each topic. Short answer and extended response tasks are provided with sample answers. Activities and sample answers are downloadable in Word format to allow teachers to customise as they need.

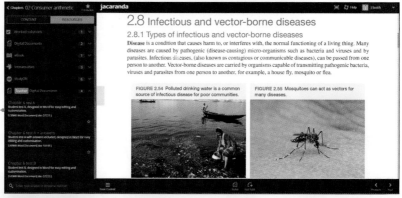

Teacher digital documents: Teaching notes and work programs are provided to assist with classroom planning.

Teacher weblinks: Teachers have access to an extensive list of case study resources, each with annotations to assist with course mapping and planning.

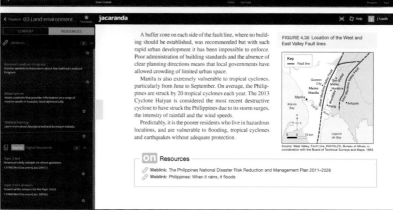

About eBookPLUS

jacaranda*plus*

This book features eBookPLUS: an electronic version of the entire textbook and supporting digital resources. It is available for you online at the JacarandaPLUS website (www.jacplus.com.au).

Using JacarandaPLUS
To access your eBookPLUS resources, simply log on to www.jacplus.com.au using your existing JacarandaPLUS login and enter the registration code. If you are new to JacarandaPLUS, follow the three easy steps below.

Step 1. Create a user account
The first time you use the JacarandaPLUS system, you will need to create a user account. Go to the JacarandaPLUS home page (www.jacplus.com.au), click on the button to create a new account and follow the instructions on screen. You can then use your nominated email address and password to log in to the JacarandaPLUS system.

Step 2. Enter your registration code
Once you have logged in, enter your unique registration code for this book, which is printed on the inside front cover of your textbook. The title of your textbook will appear in your bookshelf. Click on the link to open your eBookPLUS.

Step 3. Access your eBookPLUS resources
Your eBookPLUS and supporting resources are provided in a chapter-by-chapter format. Simply select the desired chapter from the table of contents. Digital resources are accessed within each chapter via the resources tab.

Once you have created your account, you can use the same email address and password in the future to register any JacarandaPLUS titles you own.

Using eBookPLUS references
eBookPLUS logos are used throughout the printed books to inform you that a digital resource is available to complement the content you are studying.

Searchlight IDs (e.g. **INT-0001**) give you instant access to digital resources. Once you are logged in, simply enter the Searchlight ID for that resource and it will open immediately.

Minimum requirements
JacarandaPLUS requires you to use a supported internet browser and version, otherwise you will not be able to access your resources or view all features and upgrades. The complete list of JacPLUS minimum system requirements can be found at http://jacplus.desk.com.

Troubleshooting
- Go to www.jacplus.com.au and click on the Help link.
- Visit the JacarandaPLUS Support Centre at http://jacplus.desk.com to access a range of step-by-step user guides, ask questions or search for information.
- Contact John Wiley & Sons Australia, Ltd. Email: support@jacplus.com.au Phone: 1800 JAC PLUS (1800 522 7587)

ACKNOWLEDGEMENTS

The authors and publisher would like to thank the following copyright holders, organisations and individuals for their assistance and for permission to reproduce copyright material in this book.

Images

• Above Photography: **144** (middle), **148** (e); **148** (c)/Mike Swaine • ABS: **38** (b), **200**; **142**/Based on ABS data; **146, 169**/Based on Australian Bureau of Statistics data; **157, 163** (left), **163** (right)/Australian Bureau of Statistics 2016; **194**/Source Brisbane Inner City SA4 305; **197** (middle)/Source: 2016 Census QuickStats West End Brisbane - Qld Code SSC33063 SSC • AIATSIS: **156**/This map attempts to represent the language, social or nation groups of Aboriginal Australia. It shows only the general locations of larger groupings of people which may include clans, dialects or individual languages in a group. It used published resources from 1988-1994 and is not intended to be exact, nor the boundaries fixed. It is not suitable for native title or other land claims. David R Horton (creator), © AIATSIS, 1996. No reproduction without permission. To purchase a print version visit: www.aiatsis.ashop.com.au • Alamy Australia Pty Ltd: **85** (top)/Suzanne Long; **101**/dbimages; **125** (right)/SURAPOL USANAKUL; **125** (left)/ZUMA Press, Inc.; **148** (b)/aeropix; **148** (d)/William Robinson; **179**/Christian Uhrig / redbrickstock.com; **181** (left)/Genevieve Vallee; **181** (right)/Philip Quirk; **191** (top right)/Greg Balfour Evans; **201** (top)/Rex Allen; **219** (bottom)/Liu Xiaoyang / China Images • Alamy Stock Photo: **73** (bottom)/REUTERS; **120**/Andrew McConnell; **133** (bottom left)/Jake Lyell; **257**/Citizen of the Planet • Allmetsat.com: **263** (bottom)/Source: www.allmetsat.com • Atlantic Oceanographic & Meteorological Laboratory: **36**/NOAA • Australian Bureau of Agriculture and Resource Economics and Sciences: **151**/© Commonwealth of Australia Australian Bureau of Agriculture and Resource Economics and Sciences Redrawn by Spatial Vision. Used under a CC Attribution BY 4.0 International Licence. • Australian Bureau of Statistics: **147** (bottom)/ABS Census of Population and Housing, 2016 • Australscope : **122** (bottom left), **122** (bottom right) • Barron Catchment Care: **105** (bottom) • Bill Dodd: **5, 22** (top), **23**; **55**/Natural Earth Data • Brisbane City Council: **188**/© Brisbane City Council 2018 • Bureau of Meteorology: **10** (right), **34**/Commonwealth of Australia 2018,; **10** (left)/Reproduced by permission of Bureau of Meteorology, © 2018 Commonwealth of Australia.; **32, 33**/Reproduced by permission of Bureau of Meteorology, © 2018 Commonwealth of Australia • Bureau of Mines, in coordination with the Board of Technical Surveys and Maps, 1963: **251**/West Valley_Fault Line_PHIVOLCS, Bureau of Mines, in coordination with the Board of Technical Surveys and Maps, 1963 • Bushfire & Natural Hazards CRC: **20**/Used with permission from the Bushfire and Natural Hazards CRC • Canadian Center of Science and Education CCSE: **252** (bottom)/Sharifi, E., & Lehmann, S. 2014. Comparative Analysis of Surface Urban Heat Island Effect in Central Sydney Journal of Sustainable Development, 73, 23-34. doi: 10.5539/jsd.v7n3p23 • Centre for Research on the Epidemiology of Disasters CRED: **7**/Guha-Sapir D, Hoyois Ph., Below. R. Annual Disaster Statistical Review 2016: The Numbers and Trends. Brussels: CRED; 2016 • China Daily Information Co: **240**/adapted from: Water scarcity in Beijing and countermeasures to solve the problem at river basins scale, Lixia Wang, Jixi Gao, Changxin Zou, Yan Wang, Naifeng Lin, Nanjing Institute of Environmental Sciences, Ministry of Environmental Protection, Nanjing 2 • Compare Infobase Limited: **254**/Copyright © Compare Infobase Ltd • Copyright Clearance Center: **239**/Adapted from Journal of the American Water Resources and the IOP Conference Series: Earth and Environmental Science. • Creative Commons: **86** (bottom), **105** (top)/Department of Environment, Land, Water and Planning; **110**/Institute for Health Metrics and Evaluation; **112** (top)/© Copyright, Commonwealth of Australia Bureau of Infrastructure, Transport and Regional Economics BITRE, 2014, Urban public transport: updated trends, Information Sheet 59, BITRE, Canberra; **113** (bottom)/ACRE/000003/001W, "LANDSDAT imagery produced by Australian Centre for Remote Sensing ACRES AUSLIG, www.auslig.gov.au, Topic 2; **216**/Reproduced under a CC BY 4.0 licence https://creativecommons.org/licenses/by/4.0/ c The World Bank Group from The United Nations Population Divisions World Urbanization Prospects.; **272**/Reproduced under CC BY-4.0 licence https://creativecommons.org/licenses/by/4.0/ © 2018 The World Bank Group, World Bank Staff estimates based on United Nations, World Urbanization Prospects. • CSIRO: **89**, **89** (bottom)/Dr Peter R. Brown • Department of Agriculture and Fisheries: **92**/For the latest version of the Fire Ant biosecurity zone map please refer to www.daf.qld.gov.au • Department of Environment and Energy: **80-81**/© Commonwealth of Australia 2018. • Department of Infrastructure & Transport, Commonwealth: **174** (top left)/© Commonwealth of Australia 2013Source: Bureau of Infrastructure, Transport and Regional Economics BITRE, 2013, Population growth, jobs growth and commuting flows in South East Queensland, Report 134, Canberra ACT. • Department of Natural Resources, Mines and Energy: **195**/Natural Resources, Mines and Energy, Queensland Government, various maps and spatial data sets, licensed under Creative Commons Attribution 4.0 sourced on 31 May 2018 • Department of the Environment and Energy: **162**/© Commonwealth of Australia 2018. Source: State of the Environment 2011 Committee. Australia state of the environment 2011.Independent report to the Australian Government Minister for Sustainability, Environment, Water, Population and Communities. Canberra • Digital Finance Analytics: **190**/© Digital Finance Analytics 2018 • Elitza Germanov/Marine Megafauna Foundation: **100** (top) • Elsevier: **219** (top)/Alirol E, Getaz L, Stoll B, Chappuis F, Loutan L, Urbanisation and infectious diseases in a globalised world, Lancet Infect Dis, 112, 131–241. Reprinted under STM guidelines. • Geo-Mexico: **116** • Geoscience Australia: **8**, **78** (top)/© Commonwealth of Australia Geoscience Australia 2018. Redrawn by Spatial Vision; **12**/Source: Adapted from Risk and Impact, Geoscience Australia 2018 http://www.ga.gov.au/scientific-topics/hazards/risk-and-impact • Geoscience Australia, Openstreetmap: **174** (top right)/© Commonwealth of Australia Geoscience Australia 2018; © OpenStreetMap contributors • Getty Images: **37** (a), **37** (b)/DigitalGlobe; **113** (top)/JORGE UZON/AFPJORGE UZON/AFP; **123** (bottom)/Kelly Cheng; **133** (bottom right)/ToniFlap; **241** (bottom left)/© Sergio Dorantes/Corbis/VCG; **261**/Tharaka Basnayaka/NurPhoto; **267**/xijian; **270**/Nelson Ching / Bloomberg • Getty Images Australia: **37** (top)/Kevin Frayer; **205** (top)/Planet Observer • Gillian Needham: **47** (bottom)/c Gillian Needham • Grant Singleton: **88** • Greek Orthodox Community of St George, Brisbane C/- Paniyiri Greek Festival : **196** (bottom right), **196** (bottom left), **196** (bottom middle)/Andrew Porfyri for Paniyiri Greek Festival. Reprinted with permission. • Hindawi: **255** (top)/© 2014 Bablu Kumar et al. Source: Bablu Kumar, Kopal Verma, and Umesh Kulshrestha, "Deposition and Mineralogical Characteristics of Atmospheric Dust in relation to Land Use and Land Cover Change in Delhi India," Geography Journal, vol. 2014, Article ID 325 • id consulting: **160**/This material was compiled and presented by .id, the population experts. www.id.com.au. This material is a derivative of ABS Data that can be accessed from the website of the Australian Bureau of Statistics at www.abs.gov.au, and which data can be license • Institute for Health Metrics and Evaluation: **129** (top)/Institute for Health Metrics and Evaluation IHME. GBD Compare. Seattle, WA: IHME, University of Washington, 2017. Available from http://vizhub.healthdata.org/gbd-compare. Accessed 2018. • iStockphoto: **95** (top)/Daniel Stein • John Wiley & Sons, Inc.: **83**/"Mapping the global state of invasive alien species: patterns of invasion and policy responses" by Anna J. Turbelin, Bruce D. Malamud and Robert A. Francis, Global Ecology and Biogeography, Global Ecol. Biogeogr. 2017 26, 78–92, Fig 1 • Manju Mohan : **256** (bottom)/Assessment of Urban Heat Island Effect in Megacity Delhi by Manju Mohan, 11 April 2013 <https://www.slideshare.net/Delhi2050/urban-heat-islands> • MAPgraphics: **9**/Map by MAPgraphics Pty Ltd, Brisbane; **24**, **58** (bottom), **96**, **114**; **123** (top)/MAPgraphics, with data from the U.S Geological Survey; **154**, **231**, **235**, **244**/MAPgraphics Pty Ltd, Brisbane • MDPI: **247** (left), **247** (right)/Zhang H, Ma W, Wang X. Rapid Urbanization and Implications for Flood Risk Management in Hinterland of the Pearl River Delta, China: The Foshan Study. Sensors Basel, Switzerland. 2008;84:2223-2239. • Melbourne Water: **248** (top)/Photo courtesy of Melbourne Water • Mick Law: **212** (bottom), **213** (bottom) • Microsoft Corporation: **199**, **200** (top)/Screenshots reprinted by permission from Microsoft Corporation • NASA: **25**/NOAA GOES Project • NASA Earth Observatory: **26** • NASA/Visible earth: **59** (right); **166**, **166**/Reto Stöckli, NASA Earth Observatory • Natalie Osborne: **201**/Right to the City • National Weather Service: **252** (top right)/National Weather Service • Natural Earth: **120** (top)/Natural Earth Data www.naturalearthdata.com • Natural Earth Data <www.naturalearthdata.com>: **98** (top)/Natural Earth Map by Spatial Vision; **100** (bottom)/Natural Earth Map • Nature Publishing Group: **102** (left), **102** (right)/L. Lebreton, B. Slat, F. Ferrari, B. Sainte-Rose, J. Aitken, R. Marthouse, S. Hajbane, S. Cunsolo, A. Schwarz, A. Levivier, K. Noble,

P. Debeljak, H. Maral, R. Schoeneich-Argent, R. Brambini & J. Reisser, Evidence that the Great Pacific Garbage Patch is r • NDRRMC Office of Civil Defense, Philippines: **66** • News Regional Media : **177** (top)/Photo News Corp / Brenda Strong • Newspix: **28**/Bruce Long • Omni Resources: **115** • Open Street Maps: **212** (top)/© State of Queensland 2018. Source: Qld Globe & QImagery • Peter Bellingham: **148** (a)/Peter Bellingham Photography • Philippine Statistics Authority: **38** (a) • Photolibrary: **234**/Imagestate RM/The Print Collector • PLoS ONE: **228**/Luo Y, Chen H, Zhu Q, Peng C, Yang G, Yang Y, et al. 2014 Relationship between Air Pollutants and Economic Development of the Provincial Capital Cities in China during the Past Decade. PLoS ONE 98: e104013. https://doi.org/10.1371/journal.pone.0104013; **263** (top)/© 2017 Sirisena et al. Source: Sirisena P, Noordeen F, Kurukulasuriya H, Romesh TA, Fernando L 2017 Effect of Climatic Factors and Population Density on the Distribution of Dengue in Sri Lanka: A GIS Based Evaluation for Prediction of Outbreaks. PLoS ONE 1 • Public Domain: **97**, **177** (bottom) • Public Domain out of copyright: **191** (top left)/Paddington Kindergarten viewed from the playground, Paddington, Brisbane, ca. 1949. Brisbane John Oxley Library, State Library of Queensland; Anon 2008; **203**/Garden Cities of Tomorrow by Ebenezer Howard 1902; **266** (middle) • Public Domain/ Natural Earth: **122** (top)/Made with Natural Earth Data www.naturalearthdata.com / ETOPO1 / MAPgraphics and U.S. Geological Survey. • Queensland Globe: **143**, **144** (bottom), **145** (top), **147** (top left), **147** (top right), **149**, **150**, **152**, **159** (a), **159** (b), **159** (c), **159** (d), **159** (e), **174** (a), **174** (b), **174** (c), **174** (d), **182** (top), **191** (bottom)/© State of Queensland 2018. • Queensland Globe, QImagery: **182** (bottom)/© State of Queensland 2018 Source: QImagery Brisbane 1997, QAP5562 Frame 25 • Queensland Treasury: **169**/© The State of Queensland Treasury 2018. Source : ABS 2033.0.55.001 Census of Population and Housing: Socio-Economic Indexes for Areas SEIFA, Australia, 2016; GBRMPA; Google; ZENRIN; **173**/© The State of Queensland Treasury 2011; Source: Queensland Government population projections to 2031: local government areas, 2011 edition, Office of Economic and Statistical Research • Queensland Treasury, ABS: **158**/c The State of Queensland 2009 • ReliefWeb: **242** (bottom)/Based on OCHA/ReliefWebSource: OCHA Indonesia Jakarta 2014 <https://reliefweb.int/sites/reliefweb.int/files/resources/Update%20on%20Jakarta%20Flood%20as%20of%2021Jan2014-R.pdf> • Right to the City Brisbane : **202**/Giselle Penny, Luke van de Vorst & Marilena Hewitt of Right to the City. Reprinted with permission. • Sandra Duncanson: **56** • Science Photo Library: **86** (top)/WAYNE LAWLER; **91**/JAMES H. ROBINSON • Shutterstock: **1**/Martin Valigursky; **3**/fboudrias; **3**/THPStock; **16** (b)/Anne Greenwood; **16** (c)/Igor Corovic; **16** (a)/paintings; **17**/Stephen Denness; **18**/Myszka; **19**/Madjacktaylor1970; **21**/Vladi333; **40**/GENNADY TEPLITSKIY; **45**, **48** (bottom)/NigelSpiers; **47** (top)/Breck P. Kent; **49**/Adwo; **51**/Sara_Escobar; **54**/ahmad zikri; **59** (left)/Gandolfo Cannatella; **61**/Nina Janesikova/Shutterstock THPStock; **61**/zjuzjaka; **67** (a)/N8Allen; **67** (bottom)/Tuangtong Soraprasert; **67** (b)/woe; **72**/jukurae; **73** (top)/TotemArt; **74** (bottom)/Ilya Images; **74** (top)/Sergey Uryadnikov; **75**/Fotos593; **78** (bottom)/Taras Vyshnya; **79**/Hypervision Creative; **81** (bottom)/Belozorova Elena; **81** (top)/Houshmand Rabbani; **85** (bottom)/Aeed Bird; **85** (centre)/DuxX; **87**/ND700; **88**/Edwin Godinho; **95** (bottom)/IanRedding; **95** (centre)/Mikadun; **98** (bottom)/Ethan Daniels; **98** (centre)/Rich Carey; **109**/Anticiclo; **120** (bottom)/Kwasi Kyei Mensah; **137**/Jannick Clausen; **139**/Melanie Marriott; **144** (top)/cornfield; **144** (middle)/VEK Australia; **145**/Steven Bostock; **176**/Lienka; **205** (bottom)/ChameleonsEye; **206** (bottom)/anucha maneechote; **206** (middle)/Claudine Van Massenhove; **206** (top left)/Gordon Bell; **206** (bottom left), **206** (top right)/Scott Kenneth Brodie; **206** (bottom right)/Tupungato; **215**/onemu; **218**; **224** (top)/Celso Diniz; **224** (bottom)/flocu; **225**/bodom; **268**/Arsirya; **278**/Catalin Lazar; **282**/Gyuszko-Photo • Spatial Vision: **42**/Redrawn by Spatial Vision based on information from the Smithsonian National Museum of Natural History; **58** (top), **167**, **245** (top right), **269**; **62**/"© OpenStreetMap contributors".; **167**/Redrawn by Spatial Vision based on the information from the Nature Conservancy and GIS Data; **238**/Maximilian Dörrbecker Chumwa • Springer Nature: **111** (bottom)/Environment: Mexico's scientist in chief, Jeff Tollefson, Published online 20 October 2010; **242** (top)/© Springer Science+Business Media B.V. 2011 Abidin, H.Z., Andreas, H., Gumilar, I. et al. Nat Hazards 2011 59: 1753. https://doi.org/10.1007/s11069-011-9866-9 • State Library of Queensland: **196** (top)/Brisbane John Oxley Library, State Library of Queensland ; 2003 • State Library of Tasmania: **229**/Van Ryne, I. The city of Batavia in the island of Java and capital of all the Dutch factories & settlements in the East Indies, London : Robt. Sayer, ca. 1740. Digitised item from: Allport Library and Museum of Fine Arts, Tasmanian Archive and Heritage Off • The World Bank: **230**/Redrawn by Spatial Vision based on World Bank material Indonesia's Urban Development Towards Inclusive and Sustainable Economic Growth" by Taimur Samad, 19 September 2012 • UK Environment Agency , Ordnance Survey : **250**/Contains OS data © Crown copyright and database right 2018; Contains public sector information licensed under the Open Government Licence v3.0 • UN-Habitat United Nations Human Settlements Programme: **275**, **281**/Copyright © United Nations Human Settlements Programme, 2016, World Cities Report 2016 • United Nations: **218** (middle)/United Nations, Department of Economic and Social Affairs, Population Division 2014, Our urbanizing world, Population Facts No. 2014/3; **220**/United Nations, Department of Economic and Social Affairs, Population Division 2015. World Urbanization Prospects: The 2014 Revision, ST/ESA/SER.A/366; **274**/Source: From; **280**/From Sustainable Development Goals, 2015 United Nations • United Nations University Pres: **124**/N.F. Glazovsky, The Aral Sea Basin, from • United States Environmental Protection Agency : **253**/US Environmental Protection Agency Source: Learn About Heat Islands <https://www.epa.gov/heat-islands/learn-about-heat-islands> • US Library of Congress: **217** (right)/Library of Congress, Prints & Photographs Division, [LC-DIG-ds-00182]; **217** (left)/Library of Congress, Prints & Photographs Division, [LC-DIG-pga-02708] • Victorian Eco-Innovation Lab, The University of Melbourne: **260** (bottom)/Sheridan, J., Larsen, K. and Carey, R. 2015 Melbourne's foodbowl: Now and at seven million. Victorian Eco-Innovation Lab, The University of Melbourne • Wendell Cox: **233** (left), **233** (right), **241** (top right)/New Geography, The Evolving Urban Form: Jakarta Jabotabek, by Wendell Cox 05/31/2011, <http://www.newgeography.com/content/002255-the-evolving-urban-form-jakarta-jabotabek> • Wikimedia Commons: **35**/By Nilfanion [Public domain], via Wikimedia Commons; **48** (top)/By Mikenorton [CC BY-SA 3.0 https://creativecommons.org/licenses/by-sa/3.0 from Wikimedia Commons • Wiley art: **111** (top) • Wiley art, Natural Earth Data <www.naturalearthdata.com>: **39**/Natural Earth Data • World Health Organisation: **110**/"Reprinted from http://www.who.int/sustainable-development/news-events/breath-life/air-pollution-by-numbers.jpg"; **127**/© WHO 2018; **128**, **129** (bottom)/© 2018 WHO http://www.who.int/healthinfo/global_burden_disease/estimates/en/index1.html; **133** (top)/Reprinted from "World Malaria Report 2015 - Map - Projected Changes in Malaria incidence rates, by country, 2000-2015. http://www.who.int/gho/malaria/malaria_003.jpg?ua=1

Text

• ABS: **157**, **193**, **198** (middle); **159** (top), **160**,**161**,**162**/Source: 3101.0 - Australian Demographic Statistics, Sep 2017; **166**/Source: 2016 Census QuickStats Mount Isa Code SED30057 SED; **197**/Source: 2016 Census QuickStats West End Brisbane - Qld Code SSC33063 SSC; **198** (top)/Source: 2016 Census QuickStats West End Brisbane - QldCode SSC33063 SSC; **198** (bottom)/Source: 2016 Census QuickStats, West End Brisbane - Qld Code SSC33063 SSC • American Meteorological Society: **255** (bottom), **256** (top) • Bureau of Meteorology: **29-30**, **32**/Reproduced by permission of Bureau of Meteorology, © 2018 Commonwealth of Australia • Centre for Research on the Epidemiology of Disasters CRED: **71**/EM-DAT: The Emergency Events Database – Université catholique de Louvain UCL – CRED, – www.emdat.be, Brussels, Belgium <http://cred.be/sites/default/files/CRED_Disaster_Mortality.pdf> • Creative Commons: **50-52**/Morten Wendelbo Lecturer, Bush School of Government and Public Service; Research Fellow, Scowcroft Institute of International Affairs; and, Policy Sciences Lecturer, Texas A&M University Libraries, Texas A&M University, published in The Conversation; **80-81**/© Commonwealth of AustraliaCreative Commons License; **82**/McLeod, R. 2004 Counting the Cost: Impact of Invasive Animals in Australia 2004. Cooperative Research Centre for Pest Animal Control. Canberra.; **83** (bottom)/Department of Environment, Land, Water and Planning; **112** (bottom)/Australian Bureau of Statistics; **118**/© The State of Queensland Department of Environment and Heritage Protection 2012–2018; **275-276**/This article was originally published on < • Department of State Development QLD: **190**/© The State of Queensland Department of State Development, Manufacturing, Infrastructure and Planning 2018 • GNS Science: **54-55** • NASA Earth Observatory: **152** • National

Dengue Control Unit - Ministry of Health - Sri Lanka: **262**, **264**/© 2014 National Dengue Control Unit - Ministry of Health - Sri Lanka • New Zealand Parliamentary Library: **49**/Parliamentary Library of New Zealand, 'Social effects of the Canterbury earthquakes', 8 October 2014. Used under CC BY 3.0 licence • Public Domain: **99**; **108**/© 2008-2016 World Air Quality / United States Environmental Protection Agency; **117**/© 2008-2016 World Air Quality • Public Domain out of copyright: **60-61**, **203** • Queensland Treasury: **179**/© The State of Queensland Treasury 2018 Source: QGSO, Resource Regions Population Reports • Trove: **200**/1930 THE BRISBANE RIVER., The Brisbane Courier Qld. : 1864 - 1933, 22 March, p. 10. , viewed 21 Jun 2018, http://nla.gov.au/nla.news-article21506287 • United Nations: **221** (bottom), **221** (top), **222**, **223**/From World Urbanization Prospects, 2014 Revision, by Department of Economic and Social Affairs, Population Division. © United Nations 2015 Reprinted with the permission of the United Nations • University of Sydney: **170** (bottom)/The University of Sydney, School of Architecture, Design and Planning • USGS: **44**/Abridged from The Severity of an Earthquake, USGS General Interest Publication 1989-288-913 • Victorian Eco-Innovation Lab, The University of Melbourne: **260** (top)/Sheridan, J., Larsen, K. and Carey, R. 2015 Melbourne's foodbowl: Now and at seven million. Victorian Eco-Innovation Lab, The University of Melbourne • Wet Tropics Healthy Waterways Partnership: **106** (bottom), **106** (top), **107**/Wet Tropics Report Card 2017 Results- Reporting on data July 2015 to June 2016. Wet Tropics Healthy Waterways Partnership, Cairns. • Wiley Created: **258-259**/Epidemiology Unit Ministry of Health, Sri Lanka, and Geetha Mayadunne and K. Romeshun, Sri Lankan Journal of Applied Statistics, Vol 14-1. • World Health Organisation: **126**/© WHO 2018 http://www.who.int/sustainable-development/news-events/breath-life/air-pollution-by-numbers.jpg

Every effort has been made to trace the ownership of copyright material. Information that will enable the publisher to rectify any error or omission in subsequent reprints will be welcome. In such cases, please contact the Permissions Section of John Wiley & Sons Australia, Ltd.

UNIT 1
LIVING IN HAZARD ZONES

Life on Earth is full of risk. Natural and ecological hazards – such as earthquakes, cyclones, diseases and pollution – have the potential to damage and permanently change the environment, cause harm to people and other living things, and affect our way of life and wellbeing. By understanding the nature of hazards and their impacts, we can make decisions and take actions to reduce our vulnerability and minimise the risks we face.

CHAPTER 1 Natural hazards (Unit 1, Topic 1) 3

CHAPTER 2 Ecological hazards (Unit 1, Topic 2) 72

CHAPTER 1
Natural hazards

1.1 Overview

1.1.1 Introduction

In this topic, you will examine different types of natural hazards and the zones in which they are most likely to occur. These include **atmospheric hazards**, such as severe storms and cyclones; **geomorphic hazards**, such as landslides and mudslides; and **geological hazards**, such as earthquakes and volcanic eruptions.

You will also learn about the processes and patterns of natural hazards, and why they are sources of risk. By analysing data and information you will assess why some hazards seem more common, predictable or frequent, while others occur seemingly randomly.

Finally, you will apply your understanding of natural hazards to examine their potential impacts and how different communities might be able to minimise the damaging effects they have on people, property and the environment.

FIGURE 1.1 Volcano Fuego in Antigua, Guatemala, 2018

1.1.2 Key questions
- What is a natural hazard?
- Where do natural hazard zones occur and why?
- What are atmospheric, geomorphic and geological hazards?
- What factors affect the severity of impact of a natural hazard?
- What factors affect a community's response to a natural hazard?
- What factors affect a community's vulnerability to the risks of a natural hazard?
- How are people in developed and developing communities affected differently by natural hazards?
- How do people in developed and developing communities respond to natural hazards?

Activity 1.1: Reflecting on your knowledge of natural hazards

How much do you know about natural hazards? Reflect on your own experience and knowledge of natural hazards to answer these questions.

1. What natural hazards commonly occur in Australia? Which hazards present a risk in the region where you live? Are there patterns that you can identify for when and where these hazards occur?
2. What do you know about the causes of the natural hazards that occur in your area? How might these causes explain patterns of when and where they occur?
3. What kinds of natural hazards present less of a risk in Australia, but are a significant risk in other parts of the world? Why might Australia be at less risk from these types of hazard?
4. If you had to rank natural hazards in order of highest risk to lowest risk, how would you arrange them? What criteria would you use to determine what was more or less of a risk (e.g. severity, frequency, loss of life, property or environmental damage)?

1.2 Natural hazards

1.2.1 Hazards and disasters

A **natural hazard** is any extreme geophysical event that has the potential to cause harm to people, other living things, property and the environment. They can occur in the Earth's crust, on the surface of the Earth or in the atmosphere, and are created by powerful forces that generate high levels of destructive energy. Because of the dynamic nature of the Earth, natural hazards occur almost everywhere and affect all parts of the **biophysical environment**: natural, managed and built.

A natural event becomes a hazard when its **magnitude** (size), speed of onset, duration or frequency create serious risk to people and have the potential to result in considerable damage. These hazards create **risk** (exposure to some form of dangerous situation). When individuals and communities are at risk, they have to assess how to manage and lessen the effects of that risk for their communities and the local area.

Natural hazards can cause death or injury to people as well as damage buildings, property, infrastructure, crops and farmland. When a hazard is responsible for many deaths, loss of homes and will cost huge sums of money for repairs and compensation, it is called a **natural disaster**. Natural disasters may involve extensive disruption that requires a long-term recovery plan. For example, the 'Black Saturday' bushfires in Victoria (2009) became Australia's worst fire disaster when 173 people lost their lives and more than 400 people were injured. In 2017, Cyclone Debbie caused a small number of fatalities in Queensland and New South Wales, mostly due to floodwaters. The cyclone's destructive winds and heavy rainfall also caused economic damage, with an estimated cost of more than $1.5 billion in Queensland alone, including economic damage of approximately $150 million to the mining industry and about $150 million to the sugar industry.

1.2.2 Hazard zones

The term **hazard zone** is used to identify areas at risk of being affected by a specific hazard or hazards, and to indicate which areas are at greater or lesser risk. For example, areas that have been flooded or are at high risk of flooding due to the location of drains will be identified and outlined in a local council's urban flood map. This kind of hazard zone map would be used in urban land-use planning or by insurance companies when they are calculating premiums for clients.

Hazards and risks can also change over time. For example, hazard zones are declared around active volcanoes when the risk of eruption is assessed to be greater. When Mt Agung erupted in Bali, Indonesia, in 2017, people were evacuated from the area surrounding the base of the volcano to a distance of 12 km – any closer was considered a high-risk zone for ash and rock fallout. Because these volcanoes have a history of releasing toxic fumes and **pyroclastic clouds** (hot clouds of gas and debris from the volcano), the distance needed to be substantial. There was also the risk of lahars (mudflows) forming because of the heavy tropical rainfall.

1.2.3 Risk management

When humans are faced with the risks of natural hazards, they engage in **risk management** strategies, or ways to prevent or mitigate the risk or its effects based on the known consequences of the hazard.

Risk management includes a number of different elements. **Prevention** is about stopping a hazard from occurring. **Mitigation** is about reducing or eliminating a hazard's force or level of impact if it does happen. **Preparedness** refers to actions taken by communities so they can maintain an ability to respond to, and recover from, natural hazards if they occur. This involves strategies such as planning, community education, information management, communications and developing warning systems.

Activity 1.2: Assessing hazards

Examine figure 1.2 carefully to answer the following questions.

FIGURE 1.2 Road covered by floodwater

Source: Bill Dodd

Explain and analyse a ground level image

1. Describe the features of the natural hazard in figure 1.2.
2. Identify the factors that might have contributed to this hazard occurring. Consider the biophysical (living and non-living things), managed (human-controlled), and constructed (built) features of the environment.
3. Look closely at figure 1.2. Based on the visual evidence, what kind of event might have created this hazard?

Apply your understanding and propose a response

4. Might this hazard happen again? What steps could be taken to eliminate or control the risk permanently?
5. What possible scenarios might unfold if a person with children in their car attempted to drive through the water?

Synthesise the information

6. If you managed road safety for the local council in this area, what actions would you take to mitigate the risk this hazard poses in both the short-term and the long-term? Propose two strategies you could employ and justify why they would help to reduce the risk.

1.3 Types of natural hazards

Because of the way they form and how and where they happen, natural hazards are grouped into categories. Some hazards fit into several categories because they are caused by a combination of processes.

Atmospheric hazards occur in the atmosphere. These include severe storms, tropical cyclones (typhoons and hurricanes), tornadoes, blizzards, wind storms and drought (these may also be referred to as climatological – related to weather).

Geological hazards are natural events that occur in the Earth's crust, such as a volcanic eruption, earthquake or tsunami.

Geomorphic hazards are events on the Earth's surface, such as avalanches, landslides, mudslides (mass wasting). These may be triggered by natural or human processes, for example, by an earthquake or torrential rain, or by land clearing on mountain slopes.

Climatological hazards occur due to the climatic conditions of an area, such as bushfires, droughts and heatwaves. In each of these examples, the hazard is made worse by severe deficiencies of water over a prolonged period.

Hydrological hazards are extreme events with a significant water component, such as flash flooding due to storms, cyclones, ice melt or storm surges and tsunamis (which are hydrological hazards, caused by geological processes).

1.3.1 The systems approach

Examining the locations, processes and effects of natural hazards is best done using a systems approach. A **system** is any network of objects, places, events or organisms that work together as a whole. If a change occurs in one part of the system, that change will often affect the balance and operation of the whole system.

There are two basic types of systems, closed and open, depending on the exchange of energy and matter. **Closed systems** have boundaries, which allow the system to exchange energy with its surroundings, but not matter. An example of a closed system is an oven, which allows energy to enter and keep things warm, but does not allow heat to escape and warm the room. An **open system** is one that can exchange both matter and energy across its boundaries. A river catchment is an example of an open system because both matter, especially water and energy (in the form of sunlight and long-wavelength radiation), enter and leave the system.

The Earth is made up of four interconnected systems, which form the biophysical environment: three non-living and one living. These are:
- the **lithosphere** (the Earth's core, mantle and crust)
- the **atmosphere** (the mix of gases surrounding the Earth)
- the **hydrosphere** (the Earth's water, such as oceans, rivers, lakes and glaciers)
- the **biosphere** (the Earth's living things – plants, animals and organisms, also called the **ecosphere**).

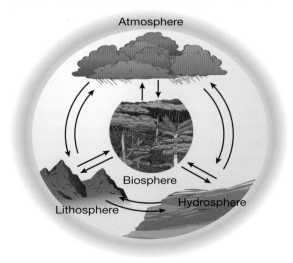

FIGURE 1.3 The Earth's physical systems

Hazards can occur in any of the Earth's systems. Because they are connected, energy is easily transferred within and between them. For example, solar energy reaches the Earth and warms the land (lithosphere) and water (hydrosphere), and each of these can warm the atmosphere; seismic wave energy from tectonic plate movement and earthquakes can cause landslides and buildings to collapse, and an earthquake below the ocean can lead to a tsunami.

Activity 1.3: Impact of hazards

Examine figure 1.4 carefully to answer the following questions.

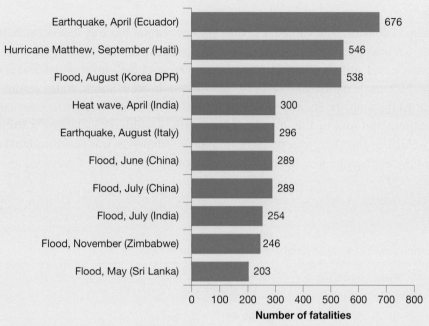

FIGURE 1.4 Natural disasters (2016) with the highest number of fatalities

Disaster	Fatalities
Earthquake, April (Ecuador)	676
Hurricane Matthew, September (Haiti)	546
Flood, August (Korea DPR)	538
Heat wave, April (India)	300
Earthquake, August (Italy)	296
Flood, June (China)	289
Flood, July (China)	289
Flood, July (India)	254
Flood, November (Zimbabwe)	246
Flood, May (Sri Lanka)	203

Source: Guha-Sapir D, Hoyois Ph., Below. R. Annual Disaster Statistical Review 2016: The Numbers and Trends. Brussels: CRED; 2016

Explain the fatality data

1. Based on the data in figure 1.4, identify what type of hazard caused the most fatalities in 2016.
2. Identify the three non-living systems of the Earth where the natural hazards listed in figure 1.4 occurred.
3. Group the hazards according to the systems in which they occurred. Which system was responsible for most deaths?

Analyse the data and apply your knowledge of hazard zones

4. Research the natural disasters listed in figure 1.4 online. Using spatial technology or a print map of the world, shade the areas affected by these disasters, using colour to show the different types of hazard.
5. Based on your map data, do certain types of hazards seem to occur in specific regions or areas? Write a short paragraph to describe the geographic patterns you can identify.
6. Create a table showing the natural disasters that occurred around the world during the last year. Sort the data into a table showing the types of hazard, the location of the event and the number of fatalities. Does the data support your answer to question 5? Explain in a paragraph whether your assessment is supported by recent catastrophic hazards.

 Resources

- **Weblink:** Reliefweb Updates: Current disaster responses
- **Weblink:** EM-DAT: The International Disaster Database

1.4 Natural hazards in Australia

Australia is considered a naturally hazardous country because its risk level is relatively high, especially for atmospheric hazards. However, it is a very large continent with a small population. Consequently, few natural hazards in Australia become catastrophic natural disasters on the scale of the 2004 earthquake and tsunami in the Indian Ocean, which killed more than 270 000 people, or the earthquake that killed approximately 160 000 people in Haiti in 2010.

In Australia there is considerable variation in the types of natural hazards that occur between and within states (see figures 1.5 and 1.6). This influences the way people perceive natural hazards. Factors such as knowledge, experience and attitude all affect people's judgement of their level of risk and ability to cope. For example, until the Newcastle earthquake of 1989, few people in New South Wales would have considered earthquakes a risk. At the time, the Building Code of Australia, which is designed to safeguard people against major structural failure and loss of life, classified Newcastle's buildings as having a low earthquake risk. Consequently, specific building design for protection against earthquakes was not considered necessary. Given this, the impact of the earthquake was significant because of the low levels of preparedness and preparation for such an event.

FIGURE 1.5 Distribution of Australia's geomorphic and geological hazards

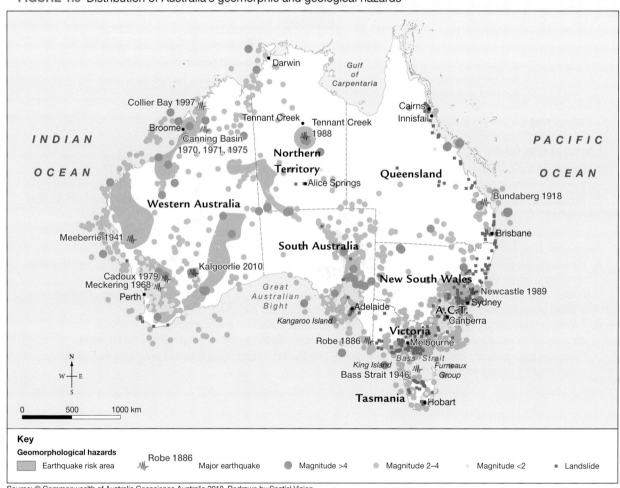

Source: © Commonwealth of Australia Geoscience Australia 2018. Redrawn by Spatial Vision.

FIGURE 1.6 Distribution of selected atmospheric, climatological and hydrological hazards in Australia

Source: © Commonwealth of Australia Geoscience Australia 2018. Redrawn by Spatial Vision.

Resources

Weblink: Geoscience Australia
Interactivity: Australia's natural hazards and disasters (int-5281)

Activity 1.4: Hazard zones in Australia

Examine figures 1.5, 1.6, 1.7 and 1.8 to answer the following questions about hazard zones in Australia.

Explain the data

1. List the natural hazards that are identified in the Australian maps.
2. Which of these hazards do you think might be the most closely linked to:
   (a) long periods of low rainfall?
   (b) movement of the lithosphere?
   (c) very high temperatures?

CHAPTER 1 Natural hazards 9

3. Which parts of Australia are most adversely affected by bushfires? During what time of the year?
4. Which states experienced a severe or extreme heatwave during January–February 2017?
5. Which states of Australia have recorded seismic tremors and earthquakes above 4 on the Richter Scale?

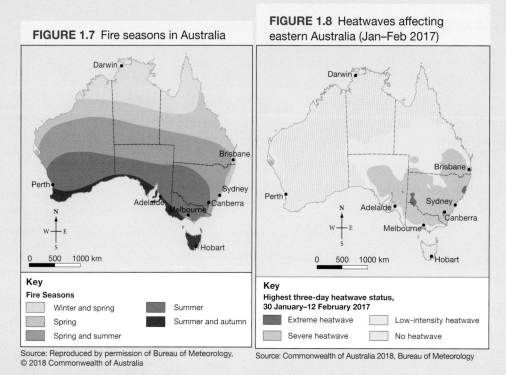

FIGURE 1.7 Fire seasons in Australia

FIGURE 1.8 Heatwaves affecting eastern Australia (Jan–Feb 2017)

Source: Reproduced by permission of Bureau of Meteorology, © 2018 Commonwealth of Australia

Source: Commonwealth of Australia 2018, Bureau of Meteorology

6. Based on the maps above, and using an interactive mapping tool that allows annotation, create a map of Australian places that have experienced a major earthquake and are in a hazard zone for tropical cyclones. Calculate the approximate distance from where you live to the closest location you have marked on your map.

Analyse the data and apply your knowledge

7. Considering both the fire seasons and the areas affected by the heatwave in 2017, which areas do you think were the most at risk from bushfires during the summer of 2017?
8. Locate the place where you live on each map. What natural hazards have been most disruptive to your area? Does this data support your experience of hazard patterns in your area? What challenges might these hazards create for people in your area?
9. Where do you consider to be the most hazardous or least hazardous places to live in in Australia? Use information from each of the maps to arrive at an answer.
10. Based on these maps, what new challenges might you face if you moved from Darwin to a bushland property in central Victoria? What steps could you take to minimise the risk to your family?

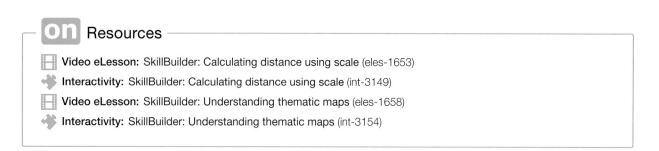

Resources

Video eLesson: SkillBuilder: Calculating distance using scale (eles-1653)

Interactivity: SkillBuilder: Calculating distance using scale (int-3149)

Video eLesson: SkillBuilder: Understanding thematic maps (eles-1658)

Interactivity: SkillBuilder: Understanding thematic maps (int-3154)

1.5 The impact of hazards

1.5.1 Variables affecting impact

The impact of a hazard on a specific area and its people, both in the long- and short-term, depend on a range of factors.

Cause

What is the origin of the hazard? For example, a landslide might be triggered by deforestation of slopes (human causes) whereas a flood might be caused by torrential rain after a storm or tropical cyclone (natural causes). Other hazards may be triggered by a combination of causes.

Frequency

How often does it happen? Some hazards are seasonal, such as bushfires or cyclones, while others can occur at any time or without warning, such as earthquakes or a tsunami. If hazards occur with greater frequency, this leaves less time for rebuilding and risk management strategies to be put in place. If hazards occur infrequently, people may not be well-prepared for an event to occur.

Duration

How long does it last? A severe storm may only last for an hour or so, while a drought could go on for months or years. Coping with the impact of an event over a long period of time will stretch the available economic and social resources, and might mean that people affected need to leave the area permanently, or that the land may no longer be safe for human habitation. A long-running natural disaster will also affect the wellbeing of the people in the area.

Speed of onset

How quickly does it appear and was there time for any warning or response? For example, flash flooding can occur quickly without giving people time to move to safety or prepare, such as the 2011 flooding in the Lockyer Valley. A volcano may begin emitting smoke or gases in the days or weeks before an eruption, giving people time to evacuate.

Predictability

Is this kind of event foreseeable or does it occur unexpectedly? Is it a random occurrence or a regular seasonal event? A hurricane can be monitored and tracked to allow authorities to warn people in its path, but a significant earthquake might occur in an area where there has been little or no recent seismic activity.

Prevention, preparedness and adaptation

How much control do people have over the impacts and outcomes? Can they prevent a hazard from occurring or prepare and adapt to increase their chances of surviving a hazard? The threat in some bushfire hazard zones can be mitigated with controlled burns and careful land management, but this does not always lower the risk when other factors occur, such as very high temperatures, winds and arson. People might prepare for the risk of cyclones by shuttering or boarding up windows as a cyclone approaches, choosing to move away during high risk periods or building cyclone shelters. Authorities might construct sea walls to prevent storm surges flooding coastal communities.

Damage potential and magnitude

How large or intense was the hazard? Is there the potential for loss of life or large-scale damage to infrastructure and the environment? The size or magnitude of a hazard may not equate to greater damage. For example, a weaker cyclone hitting a heavily populated urban area may cause more damage to buildings and infrastructure than a stronger cyclone that makes landfall in an unpopulated area.

Ability to respond

Can people respond quickly, or must they wait for assistance due to safety concerns or inaccessibility? A landslide in a remote mountain area may cut off all access for emergency crews or make existing access dangerous because of unstable ground.

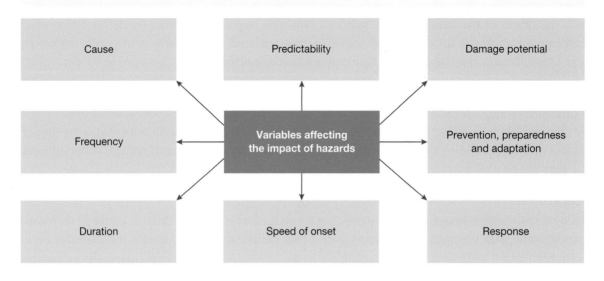

FIGURE 1.9 Factors affecting the impact of hazards

Activity 1.5a: Variables affecting impact

Using the information you have learned about hazard impacts, answer the following questions.

Explain and describe how the impact of hazards can vary

1. Explain which of the factors listed in figure 1.9 will be beyond the control of a community in the path of a category 5 cyclone.
2. Which variable refers to how quickly a hazard occurs? Give an example.
3. Explain the difference between the variables *prevention, preparedness and adaptation* and *predictability*. Give examples to support your explanation.
4. Describe and give examples to explain how each of the factors listed in figure 1.9 might affect the impact of a cyclone on a major city, such as Brisbane.

1.5.2 Factors affecting vulnerability

Even though hazards can be quite different in their structures, the way they form and the way they disrupt an area, there are still some common factors that influence people's vulnerability to their impact.

Physical factors

Factors in the physical environment, such as the weather, the season (summer or winter) or terrain, can affect how people cope in the short-term when a hazard occurs. When Cyclone Debbie struck north Queensland in April 2017, relentless torrential rain made it difficult for people to put temporary covers on unroofed and damaged houses several days after the initial gales. Flooded roads also made it impossible for emergency workers such as the police to reach those in need. An example of the terrain worsening the severity of impact is when the 7.5 magnitude earthquake struck the central highlands of Papua New Guinea in February 2018 – people were still waiting for assistance over a week later because roads were blocked by landslides.

Economic factors

Preparedness, mitigation, prevention and adaptation strategies can be expensive to implement, so a community's level of economic development can affect the impact of a natural hazard. Countries with limited financial resources also have a greater chance of fatalities from hazards because they lack the money to provide the required emergency aid quickly. This emergency aid might include well-resourced emergency response teams, medical supplies and healthcare workers, shelter for survivors, and fresh food or water. When hazards such as earthquakes occur in less developed countries, affected residents must often wait for overseas aid. For example, after the Papua New Guinea earthquake in 2018, much of the rescue and recovery work was organised by an oil drilling company that was in the area because the government did not have the resources to do so.

Social and political factors

After a natural disaster, the social structures of a community and country also become an important part of the recovery. An immediate and positive response from internal government agencies (political and military), such as declaring a state of emergency, helps to ensure that the rescue and recovery process runs smoothly and efficiently. Aid efforts are also affected by the ways in which government bodies, community organisations and the media mobilise non-affected people to help sufferers, for example with generous donations of medical aid, fresh water and food, building supplies, clothing and money. Following a natural disaster, morale also needs to be high and positive for rescue and rebuilding efforts to continue, especially when victims have lost loved ones or are left homeless.

Climate change

In the past, the Earth's atmospheric and ocean systems were regarded as stable. Very little change was evident and natural hazards only occurred from time to time. Today, the potential for adjustments to the Earth's natural systems has accelerated and its impacts have become magnified. Increased levels of greenhouse gases, such as carbon dioxide (CO_2) and methane (CH_4), in the atmosphere are shown to be contributing to a higher frequency of rare (extreme) weather events and even climate change.

It is evident that natural hazards are now having more impact on populations in terms of fatalities, injuries and property damage than in previous times. However, experts will not always agree on the underlying causes. Is the increased impact due to bigger populations and therefore more people are now exposed to risk? Is it because people are having to live in more exposed places such as hillsides, flood plains or beside volcanoes? Or is it because of changes to climates caused by global warming?

Most scientists believe that increases in CO_2 levels and subsequent general warming of the atmosphere (global warming) contribute to climate change and consequently affect natural hazards. The most obvious indicators are:
- rising global temperatures due to the Earth not releasing heat
- more frequent and extreme droughts
- more frequent and damaging wildfires
- more severe and destructive tropical cyclones and hurricanes
- more frequent and destructive tornadoes
- rapid melting of glaciers, sea ice and ice-caps
- melting of permafrost in tundra regions
- gradual sea level rising that adversely affects estuaries and low-lying coastal plains
- weakening of the polar vortex causing prolonged icy periods and intrusions of warm air into parts of the northern hemisphere.

The onset of climatic changes and an expanding human–nature interface is also making communities vulnerable to bushfires (wildfires), which remain our most lethal natural hazard, particularly in southern states.

> **Resources**
>
> 🔗 **Weblink:** Vulnerability to extreme weather events

1.5.3 Primary, secondary and tertiary impacts

Examining the processes and effects of natural hazards involves looking at the systems that those hazards belong to – a system is a dynamic unit with inputs, processes and outputs. Most importantly, if a change occurs in one part of the system, it will affect the whole system.

Some of these impacts will happen immediately. For example, if there is an earthquake the ground will shake, items will fall from shelves and buildings may topple. If a severe storm hits, houses may be unroofed, power could be lost and there will be local flooding. These are called **primary impacts** because they are immediate and happen first. They are also most likely to cause death or injury. It is these events emergency services (police, medical and fire) will attend to as soon as possible, depending on safety and access.

In the hours or days after a disaster, other issues will become apparent. Some people may need medical treatment or to attend to injured animals, food and water could run out, houses might not be safe to occupy, power and sewerage might not work, communication could be cut, roads could be closed, people might not be able to get to work, schools could be closed, shops and banks may not open, and transport systems can shut down and so on. These are **secondary impacts** and may continue for some time until repairs are made or help arrives.

There are also **tertiary impacts**, which are long-term. After some time, businesses or industries may be forced to close or relocate if the cost to restore them is unviable or their buildings are not able to be repaired. The tertiary impacts for individuals might include a range of physical, social or economic affects. Houses may not be allowed to be rebuilt on some sites forcing people to move, outbreaks of diseases might occur, people may become afraid and relocate, or insurance premiums become very high and unaffordable. All these factors may affect people's desire to remain in a hazard-prone area or the ways they adapt to life in the hazard zone if they are unable or unwilling to leave.

Activity 1.5b: Explain and analyse the effects of a cyclone

The severity of impacts of a hazard are often measured in terms of the environmental, economic and social effects. After a cyclone has passed over, it is a normal response for people to want to find out how friends, neighbours and others in their community survived. The following list includes some of the effects of Tropical Cyclone Yasi when it made landfall in Queensland in 2011.

- About 150 homes destroyed or uninhabitable
- Trees shredded including in National Park areas
- Wildlife killed or left without habitat and food
- Powerlines down over large area
- Schools closed
- Emergency staff (SES) unable to help people
- Roads flooded and bridges washed away
- Beaches and marinas destroyed
- Medical staff unable to get to hospitals
- Tourism sites and resorts closed for months
- 85 per cent of Queensland's banana crop destroyed
- Much of the sugar cane crop ruined
- Phone and communication towers damaged
- People unable to get to work
- Businesses closed down and workers made redundant
- Water treatment plants damaged
- Sewerage infrastructure damaged
- Supermarkets and shops unable to be resupplied

Analyse information about the effects of a cyclone

1. Create a table to categorise the impacts into groups according to whether it would be an environmental, economic or social factor. Use a table like the one below.

Environmental effects	Economic effects	Social effects

2. Using these ideas, write a paragraph about each type of effect, giving reasons for why they might have occurred and outlining some of the secondary and tertiary effects that may have occurred as a result. For example: <u>Environmental effects</u>. *Destructive winds over 180 km/h shredded and uprooted trees in the coastal national parks, forcing cassowaries out of the rainforests onto roadways in search of food. This also resulted in a number being struck by cars or chased by domestic dogs. Powerful storm surges inundated coastal areas with salt water resulting in ...*

1.6 Assessing and responding to hazards

1.6.1 Risk assessment

To manage a risk and reduce the possibility of harm, planners assess likely hazards and the potential worst-case scenario. A tool known as a risk assessment allows organisers to conduct a step-by-step analysis of these problems. Once the hazards have been identified, actions can be put in place to improve a community's preparedness, prevent and/or mitigate risk and help to build adaptation strategies so that people can remain safe. This process is called **risk management**.

A risk assessment for a natural hazard involves the following steps.

1. Understand the possible effects of the hazard, largely based on previous events and data.
2. Examine the physical features and topographic area where the hazard may occur.
3. Appreciate the type and distribution of human features, infrastructure and logistics in the area.
4. Consider the demographic profile of the area and ability of residents to respond to challenging situations.
5. Understand the role and availability of media, communications, emergency services and support teams that may be accessed.

A risk may be mitigated by a reduction in the size of any of the three main variables: (a) type of hazard, (b) elements exposed (for example, people and buildings) and (c) vulnerability. Consider figure 1.10 based on Crichton's Risk Triangle.

According to David Crichton, people have little or no control over a hazard's type, frequency and magnitude, but they can do something about exposure and vulnerability, particularly in places where improved technologies and communications allow residents to be better informed. **Exposure** refers to the things likely to be affected by a hazard, including people, crops, livestock, buildings and infrastructure. It also refers to intangible assets with economic value, such as work or communications. For example, if a tropical cyclone was heading towards a settlement, the residents and homes would be exposed to substantial risk.

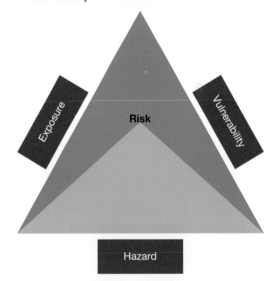

FIGURE 1.10 Reducing exposure and vulnerability to a hazard

Risk management triangle

Source: Adapted from Risk and Impact, Geoscience Australia 2018 http://www.ga.gov.au/scientific-topics/hazards/risk-and-impact

Vulnerability is a term used to measure the degree of risk according to its location, amount of preparedness, and counter-response resources available.

In the risk management model (right), the large blue area assumes the initial risk with each of the three variables contributing equally to the risk. However, if steps are taken to reduce both exposure (for example, by not living too close to a volcano) and vulnerability (for example, closely monitoring and recording volcanoes for signs of activity), the smaller, green triangle shows that the overall risk has been lowered.

Researching events of the past also gives people an increased probability of being able to forecast what might happen in the future. People can't prevent natural hazards from happening, but communities can reduce the risk and manage the effects and response. In most cases, it is often proactive effort and readiness that prevent a natural hazard from becoming a disaster. This is best achieved through education, technologies and funding.

Estimating a level of risk about a specific event (for example, an approaching cyclone) is difficult because precise locations and/or time of impact cannot always be known. However, risk models combining information about past events, including frequency and intensity, are now being developed to help experts predict possible hazard scenarios.

Crichton's model can be summarised as:

$$\text{RISK} = \text{HAZARD TYPE} \times \text{EXPOSURE} \times \text{VULNERABILITY}$$

These models are now used as a guide for emergency services to prepare for a range of effects, including response strategies and damage estimates. The scale and frequency of various natural disasters now form an integral part of any risk assessment and equivalent insurance considerations.

Equally important is that efforts are made to ensure critical infrastructure sites such as power stations, water treatment systems, sewage disposal plants and telecommunication networks are less vulnerable to natural hazards. When minimal disruption occurs to infrastructure, recovery is more rapid and less costly.

1.6.2 Managing impact

While scientific technology, such as satellites and weather instruments, enables us to measure changes to the Earth's surface, oceans and atmosphere, it is the accurate analysis of data and clear communication to the public that are most crucial for shaping the level of impact. Precise and current information passed on to emergency personnel, community decision makers and the media also means that people have time to prepare, respond or evacuate before the hazard occurs. Governments, councils and emergency teams need to be aware of the extent of the hazard zone, the event's potential severity and degree of impact if they're to prepare the best response in terms of safety and mitigation of damage. This process is often referred to as disaster management.

FIGURE 1.11 Emergency response teams help reduce vulnerability to a hazard

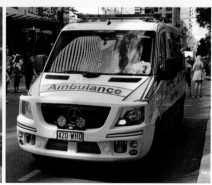

> **Resources**
>
> 🔗 **Weblink:** Risk and impact
> 🔗 **Weblink:** Disaster management in Queensland

1.6.3 Black Saturday bushfires

Australia is a very dry continent and is often prone to bush and grass fires. Fires are extremely hazardous events that occur mainly in the southern half of the continent. Some of the worst affected areas are in Victoria, Tasmania, southern New South Wales and South Australia, and the south east of Western Australia. When bushfires are raging out of control, they are referred to as wildfires.

Bushfires, or wildfires, are a climatological hazard because of their connection to hot, dry periods of weather, but they are frequently started by humans or infrastructure failure. Common causes are careless campers or smokers, arson, vehicle accidents or power line damage. Lightning strikes also start fires in some places.

Fire experts recommend that residents in fire-prone areas have a safety evacuation plan in the event of a bushfire. People must choose to either remain in their home to be protected from radiant heat produced by the fire or decide to evacuate.

Most buildings catch fire due to wind-carried embers landing on roofs and in eaves, so it is possible that physically fit, well-prepared people can fight a slow starting fire if they have protective clothing from the ambient heat, suitable pumps and hoses, and a plentiful water supply. People who choose to leave their homes should do so before the threat of fire is close. Leaving when a fire is close presents a significant risk because of the low visibility and breathing problems caused by smoke, the danger of ambient heat and embers, the unpredictability of changing winds, and the likelihood of fallen trees and other road obstructions that make getting to safety difficult. These conditions are also very frightening and confusing, leading to people making potentially disastrous decisions from fear and lack of experience.

In southern states, some wildfires become out of control. The size and ferocity of high-intensity crown fires (fires burning in the crowns of trees or canopy of the forest) combined with thick choking smoke and powerful winds make water bombing and back-burning impossible, and puts firefighters on the ground at risk of being killed.

FIGURE 1.12 Bushfire devastation in South Gippsland, Victoria, after the 2009 Black Saturday bushfires

This scenario unfolded in Victoria in February 2009 during what are now known as the Black Saturday fires, which devastated parts of Victoria including the Kinglake and Marysville areas, about 100 km north-east of Melbourne, and in central Gippsland to Melbourne's east. The fires burnt out over 450 000 hectares of forest, farmland and national park, killing 173 people and injuring 414 others. At least 3500 buildings were destroyed, including more than 2000 homes. More than 19 000 Country Fire Authority (CFA) personnel fought the fires.

What factors influenced the Black Saturday bushfires?

The Black Saturday bushfires and the extent of their severity were influenced by a number of factors, including:
- a series of days of 40+°C, with temperatures reaching 46°C on 7 February
- strong surface wind that gusted up to 100 km/h
- a change of wind direction late in the day that pushed the fire fronts into new areas
- crowning fires in heavily forested areas that could not be controlled by ground-based fire crews

- strong convection columns (columns of rising hot air over the fires) that took burning bark high into the air and created spot fires kilometres ahead of the fire fronts
- horizontal convective rolls that fanned flames with strong and unpredictable winds, and an 'undular bore' (the wave of cold front that increases the strength of winds on the ground)
- the heavily treed and hilly terrain, which made access difficult for fire crews.

The subsequent Royal Commission into the Black Saturday fires revealed overall costs were conservatively AU$4.4 billion (25 per cent of this was insurance claims), 13 per cent of homes were uninsured, while the RSPCA estimated that more than one million domestic animals and wildlife died.

How did the fires change response strategy?

Victoria's bushfire safety policy at the time was 'Prepare, Stay and Defend or Leave Early' – colloquially known as the 'Stay or Go' policy. The policy and the causes, response to and impacts of the fires were all scrutinised in the 2009 Victorian Bushfires Royal Commission.

Some of the questions raised about the 'Stay or Go' policy during the Royal Commission included:
- When should alerts be communicated to the public and what information should they contain?
- Did residents have sufficient understanding of the risks and their vulnerability to make an informed decision about whether they could successfully defend their property?
- Did residents have sufficient understanding of the physical and mental demands of protecting a property from a passing wildfire?
- What if conditions changed and the fire was larger than first anticipated?
- What were the risks of mass evacuations and people being cut off or trapped on roadsides?
- Did the plan adequately consider variations in bush density, access and topography?
- Did the plan adequately consider vulnerable groups, such as children or the elderly, who are not physically strong enough to help combat fires? (Almost half of those who died in the fires were younger than 12, over 70, disabled or chronically ill.)

The Royal Commission recommended 67 changes to Victoria's bushfire safety policy and the state adopted the national 'Prepare. Act. Survive' bushfire response framework. The Commission also recommended changes to policy regarding responding to bushfires, reducing the number of bushfires, reducing the damage caused by bushfires, and building on current knowledge of fire impact and activity specific to Victorian bushfire risk factors. Many of these changes take into account the need for greater shared community responsibility and local council input to ensure policies and plans suit or are tailored to suit the conditions and geography of a specific area, local area-specific education, and the need for different responses and advice based on the risk-factors present on any given day.

FIGURE 1.13 Homes destroyed in the Black Saturday fires

Some of the other recommendations included:
- building or creating more community refuges in high-risk areas to ensure people who were unable to defend their properties, but were too late to leave the area safely, had somewhere safe to shelter
- encouraging vulnerable people to leave the area earlier, particularly considering relocating on high risk days before the threat of fire is present in the area
- requiring that the CFA and Department of Sustainability have a full incident management team and accredited controller in place, and that aerial water-bombing craft and personnel are on standby by 10 am on days of extreme fire danger
- changes to the electricity infrastructure to minimise the risk if fires spark, such as using underground cabling in high-risk areas

- reviewing native bush-clearing permits, building codes and planning rules in high-risk areas to allow for greater consideration of fire risk mitigation
- reviewing prescribed burning programs to manage biodiversity and bushfire risk mitigation
- establishing a national bushfire research centre.

Resources

- **Weblink:** 2009 Victorian Bushfires Royal Commission report
- **Weblink:** Fire danger index calculations
- **Weblink:** Unusual weather events identified during 2009 Black Saturday bushfires

Activity 1.6: Managing hazards

Using the information you have learned about managing hazards, answer the following questions.

Explain and describe how hazards are managed

1. Explain how technology can be used to mitigate the impacts of a bushfire hazard. Give three specific examples to support your explanation.
2. Explain how risk can be mitigated by lowering *exposure* and *vulnerability* to bushfires.
3. Describe the purpose of a risk assessment and the types of information they typically include.

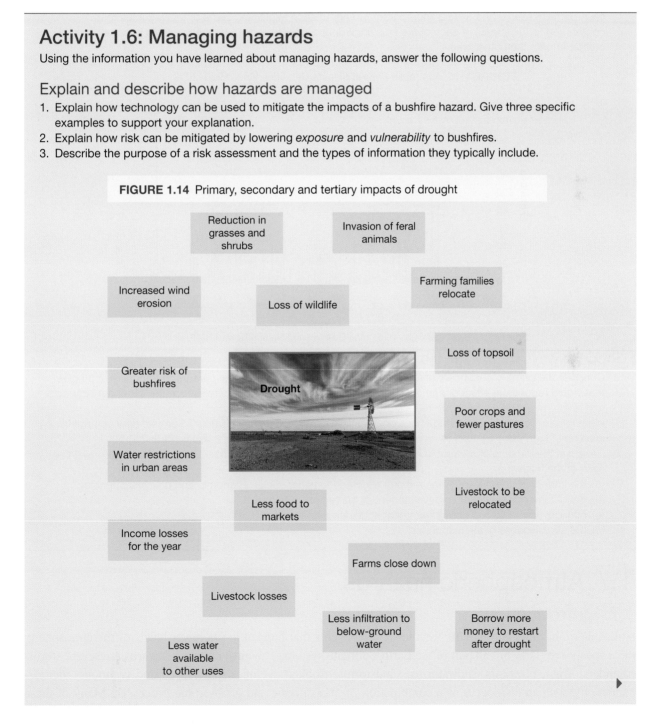

FIGURE 1.14 Primary, secondary and tertiary impacts of drought

Analyse a column graph

4. Examine the impacts of drought in Australia in figure 1.14 and determine which ones are primary, secondary or tertiary. Organise them in a table like the one below.

Primary effects	Secondary effects	Tertiary effects

5. Examine figure 1.15 and locate the number of fatalities for the years in which each of these natural disasters occurred: Tasmanian bushfires (1967), Ash Wednesday fires (1983), Black Saturday fires (2009). Research each of these events and write a short paragraph about each explaining some of the variables that contributed to the impact of each of the fires, including suggestions for why these events may have had a greater impact than other fires.

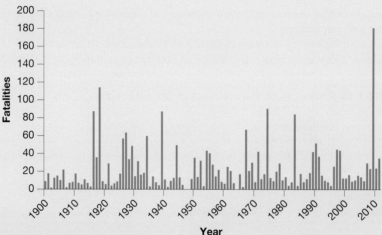

FIGURE 1.15 Australian natural disaster fatalities (fire), 1900–2010

Source: Used with permission from the Bushfire and Natural Hazards CRC

Propose ways to manage the risk

6. Consider the primary, secondary and tertiary impacts of drought. Write one paragraph to outline strategies that would reduce the effect of primary impacts, one to describe a strategy to reduce the secondary impacts, and one to describe a strategy to reduce the tertiary impacts.
7. What impact could climate change have on bushfires and droughts in the future? Suggest how communities in areas of high risk might act to reduce their exposure and vulnerability to these hazards?
8. Research the impacts of drought in Australia and in the Sahel region of Africa. Write a paragraph explaining each of the following:
   (a) the impact on the people of the regions affected (consider social, economic and physical impacts)
   (b) the differences in economic development of the regions affected
   (c) how governments and non-government agencies responded
   (d) the effectiveness of the responses.

1.7 Atmospheric hazards

1.7.1 Introduction

Atmospheric hazards are extreme weather-related events that happen in the lower levels of the atmosphere: the troposphere. They are all part of the Earth's climate system. The most common hazards are severe storms (thunderstorms), blizzards, snowstorms, sandstorms, tropical cyclones (hurricanes and typhoons) and tornadoes. Floods can also occur as a combination of atmospheric and geomorphic processes. Many of these hazards arrive and occur quickly during a short period, for example over a few hours or a few days. Other longer-term atmospheric hazards that are linked to natural cycles or the climate are dry spells and drought. Some atmospheric hazards also occur from human causes, such as air pollution from dust, chemical vapours, industrial fumes, fogs and smog. Because these have an adverse toxicological effect on living things and can impede breathing, they are considered hazards.

1.7.2 Processes that create atmospheric hazards

The atmosphere is a clear layer of gases surrounding the planet. It keeps animals and plants alive and protects them from extreme cold. Within the atmosphere are numerous circulations of air, energy and water – a complex system powered by energy from the sun. The atmosphere around the Earth is 78 per cent nitrogen (N_2), 21 per cent oxygen (O_2) and 0.9 per cent argon (A). Carbon dioxide (CO_2) and other gases including, water vapour (H_2O), ozone (O_3), methane (CH_4), nitrous oxide (N_2O) and hydrogen (H_2) make up the remainder. Despite the relatively constant composition of the atmosphere, across the globe there are factors that create differences in the air's physical properties that generate the conditions to create atmospheric hazards.

FIGURE 1.16 A hurricane viewed from space

One of these factors is the imbalance in the amount of solar radiation that reaches the surface at different latitudes. Because the atmosphere is transparent, the sun cannot heat the atmosphere directly. Instead, it heats the Earth, which heats the air above it. The roundness of the planet means that different parts of the Earth receive different levels of solar radiation. Low-latitude areas close to the equator receive more solar energy per unit area than high-latitude areas – the equator receives more than twice the solar energy over a year than the poles.

Another factor that affects the heat variation in the atmosphere is the distribution of land and sea on the planet. Most of the world is covered by water, not land. Because water takes longer to heat than land, once it's warm, it holds heat longer and cools more slowly. The seas and oceans act like huge heat reservoirs, maintaining relatively constant temperatures, unlike land masses. Consequently, there is a net surplus of radiation between 35°S and 40°N. The reason this occurs at different latitudes in each hemisphere is partly because there is more ocean and less landmass in the southern hemisphere.

This global imbalance of heat energy in the atmosphere is largely corrected by horizontal transfers of energy. The general circulation of air accounts for 80 per cent of the horizontal energy transfer. Ocean currents complete the remaining 20 per cent. Warm ocean and air currents transport warm water to the higher latitudes, which are cooler. The Earth rotates from west to east, which causes a deflection of these flows (the **Coriolis effect**). It is also responsible for deflecting ocean currents, which are large threads of warm or cold water that circulate in the oceans. In the southern hemisphere, currents moving away from the equator are warm and circulate in an anticlockwise direction, whereas in the northern hemisphere they flow in a clockwise direction.

Because the gases that make up the atmosphere have weight, they exert pressure. At any point above the Earth's surface there is a column of air exerting pressure. When air is heated it expands, loses weight and exerts less pressure. Therefore, if temperatures vary from place to place over the Earth, it is not surprising that associated atmospheric pressure also varies.

The subtropical highs that affect Australia most of the year are made up of subsiding air that has moved from the equator; however, low-pressure systems behave very differently. Instead of air subsiding and then diverging near the Earth's surface, it converges and spirals upward. The rising air is relatively unstable and these low-pressure systems are usually associated with unsettled, cloudy, wet and windy weather. The strength of the wind depends on the **pressure gradient**. This is the difference between the pressure at the centre of the system and that of the surrounding air. If the difference is large, the pressure gradient is steep and the resultant winds are strong.

On synoptic charts, strong winds are indicated by tightly packed isobars, just as a steep gradient on a topographic map is shown by tightly packed contour lines. The winds generated by the pressure gradient usually flow parallel to the isobars. The pattern of the isobars also allows us to identify **troughs** of low pressure and ridges of high pressure. Troughs can be recognised by a distinctive V-shaped pattern of isobars similar to the contour pattern of valleys on a topographic map.

> **on Resources**
>
> **Video eLesson:** SkillBuilder: Reading a weather map (eles-1637)
>
> **Interactivity:** SkillBuilder: Reading a weather map (int-3133)

1.7.3 Thunderstorms

One of the most common types of atmospheric hazard is the thunderstorm. They occur all around the world, with up to 2000 happening at any one time. Thunderstorms form when warm moist air rises high into the atmosphere due to relatively hot weather and unstable air. Water vapour condenses forming huge cumulonimbus clouds.

In warmer climates, thunderstorm cells are related to low air pressure, where strong winds and updrafts carry moisture up to 20 km into the sky. This rapid thermal updraft allows huge volumes of water to remain suspended in the sky until it eventually falls as rain. Thunderstorms may only last for an hour or so but can cause havoc when they do occur due to their enormous release of energy.

Thunderstorms also develop in cooler areas. When a mass of cold air, along a cold front, forces warm air to rise, large cumulonimbus clouds form and a thunderstorm eventuates.

The main hazards associated with severe thunderstorms are torrential rain and associated flooding, hail, destructive winds and lightning strikes.

Types of thunderstorms

There are three main types of thunderstorm, each with its own distinctive features. These are:
- a single-cell storm – this type is limited to a single heavy downpour, then breaks up quickly as cool downdraughts of wind smother the original warm air. A single-cell storm may only last an hour or so and seldom produces severe weather.
- the multicell thunderstorm – this type is most common and larger, often consisting of successive storms in sequence. Because it is larger and stronger, a multicell thunderstorm produces severe weather with heavy rain, hail and wind gusts.

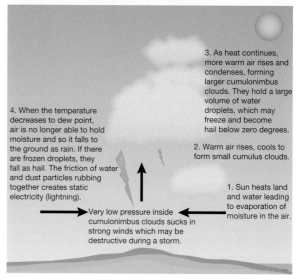

FIGURE 1.17 Formation of thunderstorms by warm convection currents (tropics)

Source: Bill Dodd

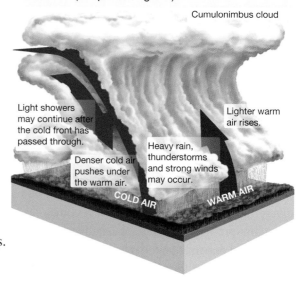

FIGURE 1.18 Formation of thunderstorms from a cold front (temperate regions)

- the supercell – this is a very large and dangerous storm with a continuous powerful updraught that seems to control the surrounding atmosphere. It has a dominant cloud shape that reaches high into the troposphere and a dark, threatening appearance. A supercell may last for many hours and is capable of very heavy rain, severe hail and destructive winds.

FIGURE 1.19 A storm cell moves across the Brisbane suburbs

Source: Bill Dodd

Activity 1.7a: Impacts of thunderstorms

Using the information you have learned about thunderstorms, answer the following questions.

Explain the potential impacts of thunderstorms

1. Most parts of Australia experience severe thunderstorms. Explain how the risk to people and property from thunderstorms can be lessened by proposing three ways that vulnerability and exposure can be reduced in urban or suburban environments.
2. Compare these to three ways that vulnerability and exposure might be reduced in rural or remote environments.
3. Explain how economic factors or underlying health issues can affect an individual's vulnerability to thunderstorm events. Consider factors such as a susceptibility to thunderstorm asthma, housing security and physical mobility.
4. What strategies could authorities put in place to mitigate the risk for vulnerable members of their community in Australia?

Resources

Interactivity: How a thunderstorm works (int-5615)

1.7.4 Tropical cyclones

Tropical cyclones are very large storms (100–2000 km in diameter) that bring heavy, driving rain and destructive winds to coastal and inland regions in tropical and sub-tropical parts of the world. Those that form over the Atlantic Ocean or eastern Pacific Ocean are called hurricanes while those that form in the western Pacific Ocean and travel north are called typhoons. Regardless of what they are called, they all form the same way.

Tropical cyclones usually form in the **inter-tropical convergence zone** (ITCZ), an area of low atmospheric pressure around the equator, because they require specific conditions to form: warmer sea temperatures, rising warm air, humidity and the right levels of **wind shear** (rapid change in the velocity or direction of the wind).

Once water temperatures exceed 26.5°C and the surrounding air pressure falls below 990 hectopascals, low-pressure cells can develop into larger tropical storms, mostly between the 5° and 30° latitudes. These newly formed cells draw in more warm moist air from the ocean surface and increase significantly in size. As the huge storm clouds extend high into the troposphere, these systems take on their characteristic circular shape with an 'eye' in the middle. Winds around the cyclone become gale force and can reach speeds in excess of 280 km/h, but the eye remains calm. Depending on the category of cyclone, the eye may be anywhere between 40 and 100 km wide.

FIGURE 1.20 World distribution of tropical cyclones by names used in different regions

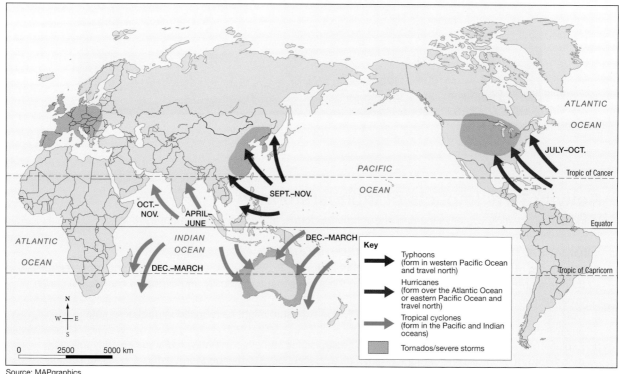

Source: MAPgraphics

The formation of cyclones is also affected by the Coriolis effect, the force that deflects winds clockwise in the southern hemisphere and anti-clockwise in the northern hemisphere. The effect of this force becomes more intense the further you move away from the equator, and is one of the key factors in shaping cyclones into their characteristic circular formation. Close to the equator, the Coriolis effect is weak. As a result, cyclones rarely form here.

Cyclones are well-known for their erratic movement, particularly as they approach land, and may persist for several days offshore, but once they cross onto land they lose their energy source (the rising warm ocean air) and become a rain depression (low pressure area).

Interestingly, it is incredibly rare – and some would argue impossible – for tropical cyclones to form in the south Atlantic Ocean between South America and Africa because the sea surface temperatures are too cold and the wind shear is too strong to allow storms to develop. However, in 2004, a strong storm made landfall in Brazil that many consider to be the first ever southern Atlantic tropical cyclone.

FIGURE 1.21 How a cyclone forms

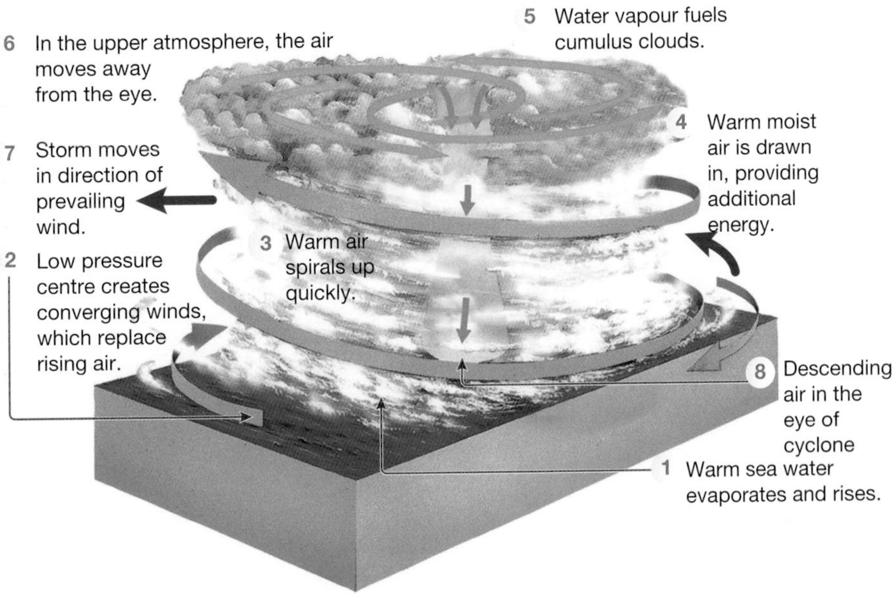

5 Water vapour fuels cumulus clouds.
6 In the upper atmosphere, the air moves away from the eye.
4 Warm moist air is drawn in, providing additional energy.
7 Storm moves in direction of prevailing wind.
3 Warm air spirals up quickly.
2 Low pressure centre creates converging winds, which replace rising air.
8 Descending air in the eye of cyclone
1 Warm sea water evaporates and rises.

Resources

- Weblink: Tropical cyclone intensity
- Weblink: Coriolis effect
- Weblink: Rare south Atlantic tropical cyclone

Cyclone categories

When tropical cyclones make landfall, they can be very destructive, bringing gale force winds. Different scales are used to measure the intensity of these winds around the world.

In the USA hurricanes are measured with the **Saffir-Simpson Hurricane Wind Scale**, which rates the sustained speed of wind on a scale from 1 to 5, and provides examples of the kinds of damage expected for winds of that level. The Australian Bureau of Meteorology categorises the intensity of tropical cyclones using its own five-point scale, which also takes into account the atmospheric pressure, shown in table 1.1. Maximum sustained wind speed is determined by the peak mean wind speed, measured 10 metres above the surface of flat land or open water. In Australia this mean is measured over a 10-minute period, but in the USA a one-minute mean is used. Tropical cyclones are also measured in terms of their wind gust strength. In Australia, this is measured as the average speed of wind over a three-second period.

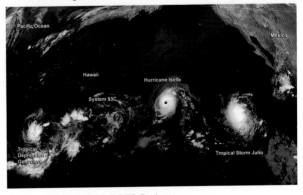

FIGURE 1.22 Satellite imagery showing tropical cyclone, depression and storm activity in the Pacific Ocean, August 2014

Pacific Ocean
Mexico
Hawaii
Hurricane Iselle
System 93C
Tropical Depression Genevieve
Tropical Storm Julio

Source: NASA/NOAA GOES Project

In addition to measuring their intensity, tropical cyclones are also mapped for their paths. This not only helps to determine where cyclones most commonly occur for risk assessment purposes, so communities in the likely path of a cyclone can be warned and given time to prepare, but it also helps to show whether patterns in the intensity and frequency change over time.

Figure 1.23 shows the paths and intensity of tropical cyclones for more than 150 years until September 2006, based on the records of the US National Hurricane Center and the Joint Typhoon Warning Centre. This map rates tropical cyclones using the Saffir-Simpson Scale.

FIGURE 1.23 Paths and intensity of tropical cyclones, tropical depressions and tropical storms

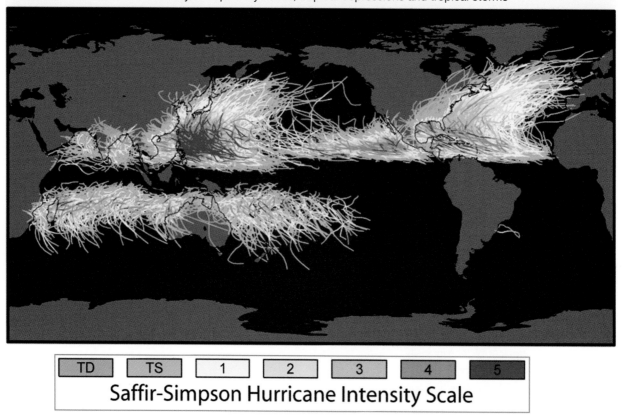

Source: NASA Earth Observatory

TABLE 1.1 Australian Bureau of Meteorology tropical cyclone category system

Category	Australian category name	Strongest wind gust (km/h)	Average maximum wind speed (km/h)	Central pressure (hPa)	Effects	Saffir-Simpson Scale comparison (km/h)
1	Tropical cyclone	90–124	63–90	>985	• Negligible house damage • Damage to crops and trees	119–153
2	Tropical cyclone	125–164	90–125	985–970	• Minor house damage • Risk of power failure • Heavy damage to some crops	154–177
3	Severe tropical cyclone	165–224	125–165	970–955	• Some structural and roof damage • Likely power failure	178–208
4	Severe tropical cyclone	225–279	165–225	955–930	• Significant structural damage and roofing loss • Widespread power loss • Dangerous airborne debris	209–251
5	Severe tropical cyclone	>280	>225	<930	• Extreme danger • Widespread destruction	>252

Note: Storms with average maximum wind speeds between 52 and 54 km/h are referred to as tropical depressions, and tropical lows are storms with average maximum wind speeds between 56 and 61 km/h.

Storm surges

In addition to the immediate damage caused by the strong winds and rain of a cyclone or severe storm, strong winds can also generate very large waves.

Waves are caused by the friction of wind blowing over the sea's surface. The wind tugs at the surface of the water, causing the wave shape to move. This is like shaking a length of rope on the ground. The wave travels along the rope but the rope remains in the same place. Wind speed, duration and the distance over which it blows, commonly known as the fetch, determine the height of waves. During tropical cyclones these large waves are capable of eroding beaches and damaging coastal buildings and facilities.

This damage can be compounded by storm surges, which are increases in the height of sea level above the normal tide level. The resultant water level is known as the storm-tide level. Since atmospheric pressure is the weight of air above the Earth's surface, intense low-pressure systems, such as tropical cyclones, cause localised upward bulging of the sea surface. It is estimated that there is a rise in water level of approximately 10 cm for every 10 hectopascals of difference between the central pressure of the cyclone and the surrounding pressure (the central pressure being lower). This is known as the inverted barometric effect.

However, the height of the storm surge is not solely dependent on the intensity of the cyclone. In fact, a large component of any surge is the effect of strong onshore winds pushing water against the coast.

When wind blows over the sea's surface, it sets up a current in the water as a result of the shear stress on the surface of the water particles. When this current reaches the coast, the water tends to build up against the land. This is known as wind set-up. Consequently, the angle at which the cyclone approaches the coast and the configuration of the coastline itself are important. Shallow, sloping sea beds and the presence of bays and estuaries (which can funnel a surge into a confined area) contribute to abnormally high sea levels. Storm surges are typically between 60 and 80 km in diameter and up to 2.5 m high. Therefore, they can pose a serious threat to people living in low-lying coastal areas and estuaries.

Activity 1.7b: Tropical cyclones

Using the information you have learned about tropical cyclones, answer the following questions.

Comprehend and explain tropical cyclones and identify their patterns

1. Categorise the type of hazards that are tracked on the world map in figure 1.23.
2. Explain the difference between a tropical depression and tropical storm.
3. Based on figure 1.23, which ocean develops the most Category 5 cyclones on the Saffir-Simpson Scale? Based on your understanding of how cyclones form, what atmospheric patterns or features would you expect to find in this area?
4. Between which lines of latitude do most tropical cyclones occur?
5. What categories on the Saffir-Simpson Scale are most common across northern Australia? What category would this type of cyclone be on the Australian Bureau of Meteorology's tropical cyclone category system?
6. Explain the geographical processes that result in the distribution of tropical cyclones shown in figure 1.23.

Analyse the data and apply your understanding

7. Which area of ocean does not seem to develop tropical cyclones, even though it is in the tropics? Suggest reasons why cyclones do not develop in this region, and predict how rising sea temperatures associated with climate change might affect this pattern.
8. If you were thinking of moving to Darwin and building a house, what category rating on the Saffir-Simpson Scale would you expect the builder to construct your home in order to withstand cyclones?
9. Family friends are arriving from Canada to visit you, and ask if early March is a good time to take a caravan holiday in northern Queensland. What advice would you give them?
10. Analyse and compare the Bureau of Meteorology's maps of the average number of tropical cyclones in El Niño years, La Niña years and neutral years using the weblinks in the Resources tab. Suggest how these maps could help authorities to develop cyclone preparedness strategies.

11. Imagine your sister was moving to work in Taiwan for two years. Because you are studying natural hazards in geography, she emails you to ask whether you think it is likely she will experience a cyclone in that time and whether it might be dangerous. Reply to your sister explaining what she could experience.

Resources

Interactivity: How a cyclone forms (int-5299)
Weblink: Cyclone intensity on the Saffir-Simpson Scale
Weblink: Australian Bureau of Meteorology tropical cyclone maps

Climate change

Meteorologists have determined that climate change has contributed to the way tropical cyclones develop and behave. Firstly, tropical cyclones need warm ocean water and quite cool upper atmospheric conditions to form. If the air continues to warm, the difference (gradient) between surface temperatures and upper level temperatures will be reduced, so fewer cyclones may form. Secondly, increased surface temperatures over the ocean combined with higher levels of CO_2 (more CO_2 allows air to hold more moisture than it once could), provide cyclones with a much larger energy source, making them larger and more destructive.

Recent trends observed with tropical cyclones are:
- much higher volumes of rainfall near the centre when they make landfall
- an increase in high-category destructive storms
- slower movement due to weakening of the circulation forces that drive movement. (In northern Australia, cyclones now move almost 20 per cent slower than they did 70 years ago, making them more destructive to settlements and causing more flooding.)

1.7.5 Responding to atmospheric hazards

Improvements in public awareness and communication have enabled communities to be better prepared for storms, cyclones and related flooding than in the past. This is evident from the decreasing number of fatalities and injuries, and despite such hazards becoming more powerful. However, an increasing population and more widespread settlements along the Australian coast have exposed community and government facilities to potentially greater economic damage.

Local authorities and emergency services play an important role in hazard response. They release information to help people prepare for and cope with the impact of atmospheric hazards in their area. In Australia, this information about cyclones comes as a Tropical Cyclone Warning Advice — either a tropical cyclone watch (24–48 hours before the onset) or tropical cyclone warning (onset within 24 hours). This advice relates important information including the area at risk (including a map), the intensity of the cyclone (using the Bureau of Meteorology five-level scale), the movement of the cyclone, the range and maximum strength of wind gusts expected, and advice about what action people should take to mitigate the effects of the cyclone.

FIGURE 1.24 Tropical Cyclone Larry hits Innisfail, March 2006

1.7.6 Preparedness strategies

Housing engineering standards have improved and can now better withstand the wind gusts of cyclones, but most places are not able to sustain the rapid flow of huge volumes of water from the torrential rainfall, storm surges and flooding. Strategies to prepare for cyclones and severe storms include the following.

- Installing underground powerlines, which generally reduce the extent of outages and power failure in the event of a cyclone.
- Sea walls and shore-line sand buffers of at least 150 m in coastal developments help to reduce the impact of tidal surges during cyclones.
- Many urban areas have reclaimed or changed their natural creek, river and wetland systems through development, for example, turning their natural creeks into cement drains. Removing these and creating a wetlands network can capture and slow the water flow during a flood.
- Drinking water sources can be polluted by floodwaters after a cyclone. Very large buildings, such as shopping centres and industrial sheds, can be designed to capture water and have storage tanks for future use. This interception may seem small, but it can reduce stormwater flow in gutters and street drains.
- Cyclones and storms produce extremely high rainfall in a short period of time. Increasing available green space in urban areas allows for greater infiltration of rainfall. Large ovals and outside sports areas can be designed to have run-off flow into suburban wetlands and ponds rather than all water ending up in stormwater drains.
- The destruction caused by cyclones often forces people to seek shelter or live away from home for long periods of time until the damage can be cleared and their homes made safe again. Multi-level car parks and large sports venues can be designed as cyclone shelters and used as temporary storage facilities during emergency periods.

Individual cyclone preparedness and response

The following checklist was prepared by Emergency Management Australia and Australian state emergency services to help people prepare for and respond to a cyclone.

Before the cyclone season

- Check with your local council or your building control authority to see if your home has been built to cyclone standards.
- Check that the walls, roof and eaves of your home are secure.
- Trim treetops and branches well clear of your home (get council permission).
- Preferably fit shutters, or at least metal screens, to all glass areas.
- Clear your property of loose material that could blow about and possibly cause injury or damage during extreme winds.
- In case of a storm surge/tide warning, or other flooding, know your nearest safe high ground and the safest access route to it.
- Prepare an emergency kit containing:
  1. a portable battery radio, torch and spare batteries
  2. water containers, dried or canned food and a can opener
  3. matches, fuel lamp, portable stove, cooking gear, eating utensils
  4. a first aid kit and manual, masking tape for windows and waterproof bags.
- Keep a list of emergency phone numbers on display.
- Check neighbours, especially if recent arrivals, to make sure they are prepared.

When a cyclone watch is issued

- Re-check your property for any loose material and tie down (or fill with water) all large, relatively light items such as boats and rubbish bins.
- Fill vehicles' fuel tanks. Check your emergency kit and fill water containers.
- Ensure household members know which is the strongest part of the house and what to do in the event of a cyclone warning or an evacuation.

- Tune to your local radio/TV/mobile device for further information and warnings.
- Check that neighbours are aware of the situation and are preparing.

When a cyclone warning is issued

Depending on official advice provided by your local authorities as the event evolves, the following actions may be warranted.
- If requested by local authorities, collect children from school or childcare centre and go home.
- Park vehicles under solid shelter (handbrake on and in gear).
- Put wooden or plastic outdoor furniture in your pool or inside with other loose items.
- Close shutters or board-up or heavily tape all windows. Draw curtains and lock doors.
- Pack an evacuation kit of warm clothes, essential medications, baby formula, nappies, valuables, important papers, photos and mementos in waterproof bags to be taken with your emergency kit. Large/heavy valuables could be protected in a strong cupboard.
- Remain indoors (with your pets). Stay tuned to your local radio/TV for further information.

On warning of local evacuation

Based on predicted wind speeds and storm surge heights, evacuation may be necessary. Official advice will be given on local radio/TV/mobile regarding safe routes and when to move.
- Wear strong shoes (not thongs) and tough clothing for protection.
- Lock doors; turn off power, gas and water; take your evacuation and emergency kits.
- If evacuating inland (out of town), take pets and leave early to avoid heavy traffic, flooding and wind hazards.
- If evacuating to a public shelter or higher location, follow police and State/Territory Emergency Services directions.
- If going to a public shelter, take bedding needs and books or games for children.
- Leave pets protected and with food and water.

When the cyclone strikes
- Disconnect all electrical appliances. Listen to your battery radio for updates.
- Stay inside and shelter (well clear of windows) in the strongest part of the building, i.e. cellar, internal hallway or bathroom. Keep evacuation and emergency kits with you.
- If the building starts to break up, protect yourself with mattresses, rugs or blankets under a strong table or bench or hold onto a solid fixture, e.g. a water pipe.
- Beware the calm 'eye'. If the wind drops, don't assume the cyclone is over; violent winds will soon resume from another direction. Wait for the official 'all clear'.
- If driving, stop (handbrake on and in gear), but well away from the sea and clear of trees, power lines and streams. Stay in the vehicle.

After the cyclone
- Don't go outside until officially advised it is safe.
- Check for gas leaks. Don't use electric appliances if wet.
- Listen to local radio for official warnings and advice.
- If you have to evacuate, or did so earlier, don't return until advised. Use a recommended route and don't rush.
- Beware of damaged power lines, bridges, buildings, trees, and don't enter floodwaters.
- Heed all warnings and don't go sightseeing. Check/help neighbours instead.
- Don't make unnecessary telephone calls.

Source: Reproduced by permission of Bureau of Meteorology, © 2018 Commonwealth of Australia

Activity 1.7c: Responding to tropical cyclones

Using the information you have learned about preparation and response strategies, answer the following questions.

Comprehend and explain ways to minimise personal risk

1. List four important things to do around your house before the cyclone season starts and explain why these actions are important in helping to mitigate the effects of the cyclone.
2. List five essential items to have in your household emergency kit and explain which effects of a cyclone they might help to mitigate.
3. Once a cyclone warning has been issued for your locality, you need to make up an 'evacuation kit'. Explain why this is important, giving examples of essential items that should be included and why.
4. If you believe your house may not be strong enough to cope with gale force winds, where should you evacuate to? What should you do with pets?
5. If you were to remain in your home when a cyclone hits, which parts of the building are most likely to be the safest? Explain why.
6. Give reasons why it is important to have a battery-powered radio in the house.
7. Why is it important not to go sightseeing after a cyclone has passed?

Analyse the data and apply your understanding

8. Imagine your home has been cut off by rising floodwaters and emergency workers are unable to restore lost power for at least a week. You are not sure about water quality and the sewerage plant is not working due to power loss. Make a list of things you would do to keep your family and two pet dogs safe until help arrives.
9. Create a cyclone safety poster to display in hotel rooms in Queensland that instructs overseas and interstate visitors about what they should do after a cyclone passes and why these actions are important. Be specific in your advice and think carefully about the factors that might contribute to a visitor's lack of awareness of the risks present after a cyclone has passed.
10. Re-read the cyclone emergency preparation checklist. Which aspects of this list might be problematic or difficult to follow in developing countries? Which might be problematic or difficult to follow in very remote communities? Propose actions that might help overcome the difficulties of implementing these preparedness strategies in these communities.

1.7.7 Tropical cyclones in northern Australia

Typically, the cyclone season in northern Australia runs from November to April. This may vary by a few weeks if the Pacific regions are experiencing a **La Niña**.

According to the Australian Bureau of Meteorology, 10 to 13 tropical cyclones develop on average each season. It is probable that at least one tropical cyclone will cross the Australian coast each season. Most come in from the Pacific Ocean, while a smaller number form in the 'monsoon trough' above the Gulf of Carpentaria and the Northern Territory. They all tend to move in a westerly direction.

 Resources

Weblink: El Niño Southern Oscillation (ENSO)

Activity 1.7d: Reading synoptic maps and weather warnings

Consider the synoptic map and read the weather warning put out by Bureau of Meteorology on Wednesday, 14 March 2018, to answer the following questions on next page.

Ex-tropical cyclone Linda is expected to produce dangerous surf and abnormally high tides along the southern Queensland coast during today and Thursday.

Weather situation:

At 4pm AEST Wednesday, ex-tropical cyclone Linda was located in the Coral Sea about 450 kilometres east to northeast of Fraser Island, moving southwest at 17 kilometres per hour. Ex-Tropical Cyclone Linda, which transitioned into a vigorous subtropical low earlier this morning, is expected to continue its southwest track for the remainder of today before shifting on a more southerly track on Thursday. The system is expected to remain offshore of the southern Queensland coast.

Strong to gale force winds over offshore waters across the southern flank of the low are expected to produce large east to south easterly swells along exposed parts of the southern Queensland coast for the remainder of today and into Thursday. This will combine with high tides to cause hazardous conditions within the warning area.

Dangerous surf conditions with possible beach erosion are expected along the east coast of Fraser Island this afternoon and evening. These hazards should extend southward towards the Sunshine Coast and Gold Coast early Thursday morning, then ease rapidly late afternoon and evening.

On Thursday morning, tides may exceed the highest tide of the year, with inundation of low-lying areas possible on the high tide. Locations which may be affected include Noosa, Maroochydore, Caloundra, Coolangatta and the eastern side of Moreton, Stradbroke, and Fraser Islands.

Queensland Fire and Emergency Services advises that people should:

- Surf Life Saving Australia recommends that you stay out of the water and stay well away from surf-exposed areas.
- Check your property regularly for erosion or inundation by sea water, and if necessary raise goods and electrical items.
- If near the coastline, stay well away from the water's edge.
- Never drive, walk or ride through flood waters. If it's flooded, forget it.
- Keep clear of creeks and storm drains.
- For emergency assistance contact the SES on 132 500.

FIGURE 1.25 Synoptic map and cyclone warning, 14 March 2018

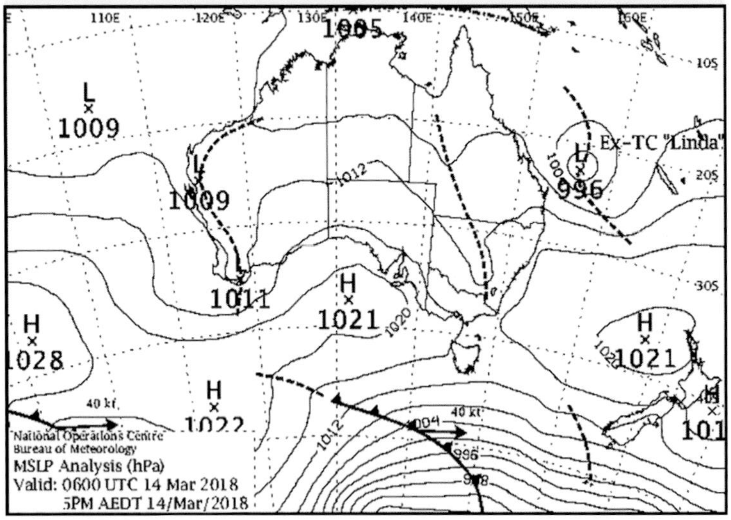

Source: Reproduced by permission of Bureau of Meteorology, © 2018 Commonwealth of Australia

Comprehend and explain cyclone warnings

1. Where was ex-Tropical Cyclone Linda located at 4.00 pm on 14 March 2018? Based on the description and image provided, and by consulting other maps if needed, determine the longitude and latitude. What was its central pressure?

2. List three ways the system was affecting coastal communities in south-east Queensland.
3. Describe the dangerous conditions that were occurring at beaches south of Fraser Island at that time.
4. Define the terms *abnormally high tide* and *inundation of low-lying areas*.
5. Explain how gale force winds (63–87 km/h) could contribute to the highest tide of the year.
6. Mark three other areas of low pressure on the map in addition to ex-Tropical Cyclone Linda. Where is air pressure highest?

Analyse the data to identify the challenges
7. If ex-Tropical Cyclone Linda continued moving and weakening as the BOM suggested, predict where it could be on a map 48 hours after this warning was issued. What challenges would this present for emergency services?

Suggest risk management strategies
8. If you were in charge of closing access to beaches along the Sunshine Coast at this time, what course of action would you take? Create a plan for when you would close access and when you would re-open beaches for swimming again. Justify your choices with evidence from the data provided in the Bureau of Meteorology map and warning.

Tropical Cyclone Debbie 2017

When it hit, Tropical Cyclone Debbie was the deadliest cyclone to hit Australia since Cyclone Tracy in 1974 and the most dangerous to make landfall in Queensland since Cyclone Yasi in 2011. Starting as a tropical low in the Coral Sea, the system gradually strengthened to Category 2. As Tropical Cyclone Debbie moved closer to the coast between Townsville and Mackay, it intensified to a Category 4. After hovering off-shore for nearly 24 hours, Debbie made landfall near Airlie Beach, bringing gusts of wind in excess of 250 km/h and torrential rain. Nearby, Proserpine received 210 mm of rainfall in one hour.

Debbie continued south towards Brisbane, bringing with it torrential rain and widespread flooding of most coastal rivers and creek catchments. It then moved through to the urban areas of south-east Queensland and the Northern Rivers districts of New South Wales. Overall, Tropical Cyclone Debbie caused damage amounting to about AU$2.4 billion. Tragically, fourteen people also died, most as victims of flooding.

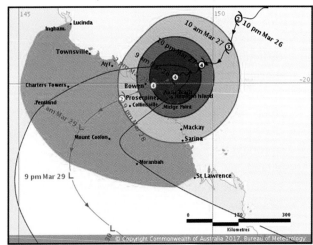

FIGURE 1.26 The path of Tropical Cyclone Debbie

Source: Reproduced by permission of Bureau of Meteorology, © 2018 Commonwealth of Australia

Throughout Queensland, 62 weather stations received record-breaking rainfall including Plane Creek Sugar Mill, south of Mackay, which received 1300 mm of rain across the month of March 2017 – more rain than Brisbane usually receives in a year.

Other effects of Tropical Cyclone Debbie included:
- torrential rainfall along the far northern coast and in coastal river catchments
- severe flooding along coastal areas and in river valleys
- severe damage to farm crops, particularly sugar cane, vegetables and fruits
- widespread power outages and damage to the electrical grid infrastructure
- sewerage plants forced to close due to inundation
- severe damage to homes, buildings and other structures (power poles)
- roads and rail infrastructure damaged and closed
- coastal airports closed for short periods
- people forced to evacuate homes and stay in shelters for some time
- food and medical shortages in isolated townships
- people unable to work (loss of income)
- schools closed and hospitals difficult to access
- marinas, port facilities and island resorts badly damaged, including many tourist/boat operations
- damage to parks and streets from uprooted trees
- inland coal mining operations closed due to open cut pits being filled with water and machinery damage
- beaches severely eroded and coastal areas contaminated with salt water
- livestock drowned or lost.

FIGURE 1.27 Queensland total rainfall, March 2017

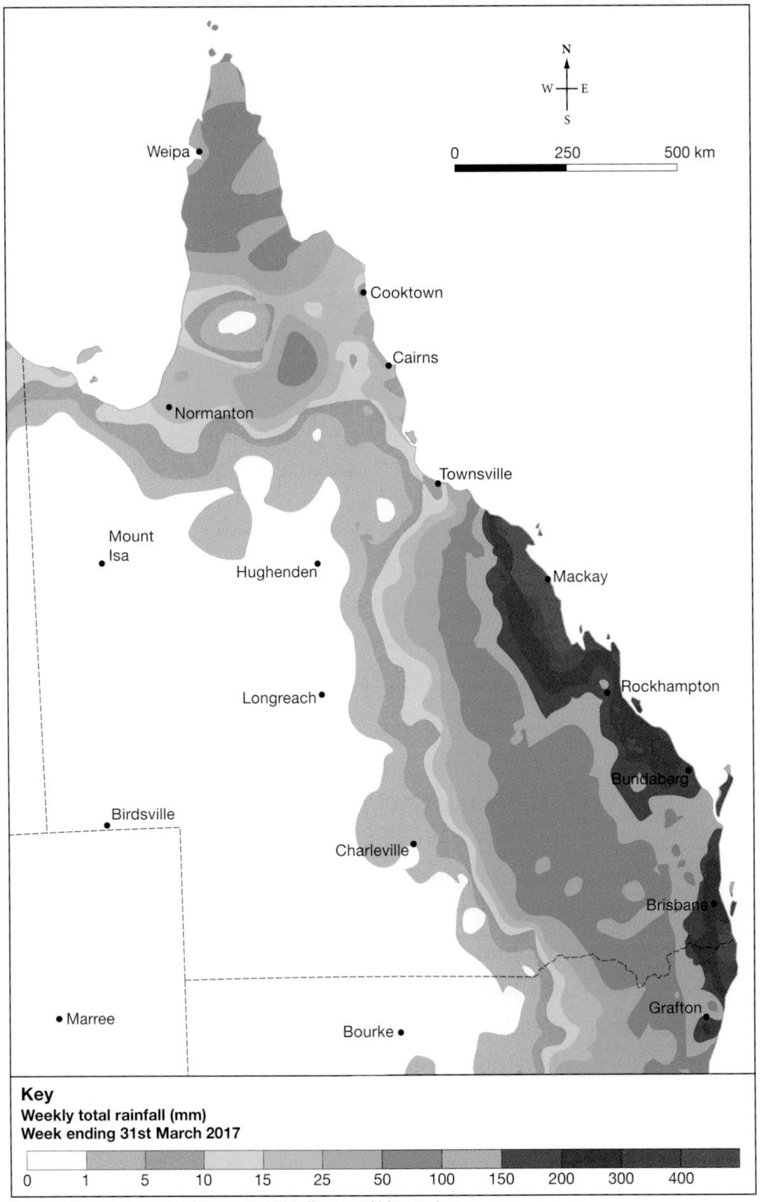

Source: Commonwealth of Australia 2018, Bureau of Meteorology

Activity 1.7e: Tropical Cyclone Debbie

Using the information you have learned about Tropical Cyclone Debbie, answer the following questions.

Explain Tropical Cyclone Debbie's impact

1. Based on the Bureau of Meteorology scale, what category was Cyclone Debbie when it made landfall?
2. How much rain fell in the Mackay area during the week ending 31 March 2017?
3. Which three areas of Queensland experienced the heaviest rainfalls during March 2017?
4. Gold Coast – Brisbane and Mackay received much more rain than the large area between the two cities. Explain why this was the case.

Suggest risk management strategies

5. Imagine you were a tourist at the Whitsunday Islands and could not get back to the mainland as Tropical Cyclone Debbie approached. Suggest and justify strategies that might help to keep you safe.
6. To what extent would good cyclone preparedness have mitigated many of the effects of Tropical Cyclone Debbie? Examine the list of preparedness and response strategies on pages 29–30. Which three strategies do you think would have most significantly reduced risk? Justify your response by explaining how those strategies would reduce your vulnerability and/or exposure to the hazard.
7. If you were in charge of recovery operations after Tropical Cyclone Debbie had crossed the coast, where would you start? What would you do to maximise public safety with water supply, electricity, sewerage and roads? How would you communicate this information to the public if power and communications networks were down?

Resources

- **Weblink:** Bureau of Meteorology tropical cyclones information and warnings
- **Digital doc:** The impact of Tropical Cyclone Debbie (doc-29158)

1.7.8 Typhoons in the western Pacific

Typhoons form the same way as tropical cyclones and have the same damaging effects. The term 'typhoon' is a regional name used to describe severe tropical storms that occur in the western Pacific and Asia. In central America (around the Caribbean Sea) and the eastern Pacific, these same storms are called hurricanes. Even though they have similar features, there are often significant differences in the way they affect people.

FIGURE 1.28 Maximum sustained wind speeds of typhoons in South-East Asia (1980–2005) tracked at six-hour intervals

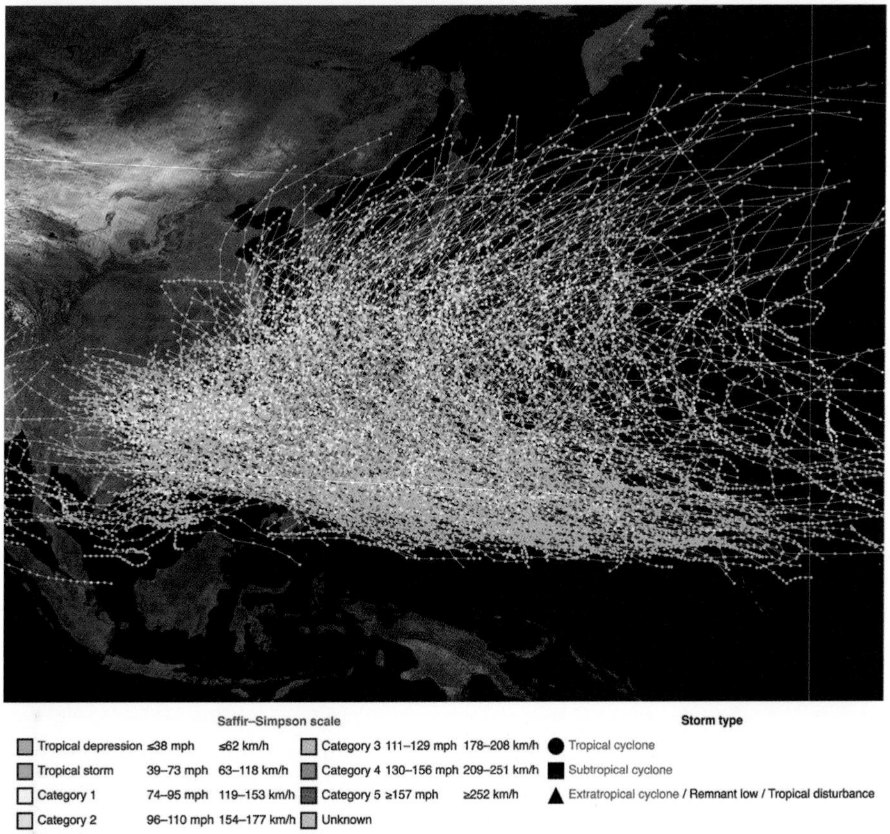

Saffir–Simpson scale / Storm type

- Tropical depression ≤38 mph ≤62 km/h
- Tropical storm 39–73 mph 63–118 km/h
- Category 1 74–95 mph 119–153 km/h
- Category 2 96–110 mph 154–177 km/h
- Category 3 111–129 mph 178–208 km/h
- Category 4 130–156 mph 209–251 km/h
- Category 5 ≥157 mph ≥252 km/h
- Unknown
- ● Tropical cyclone
- ■ Subtropical cyclone
- ▲ Extratropical cyclone / Remnant low / Tropical disturbance

Note: The blue tracks in figure 1.28 show tropical depressions, with increasing darkness of yellow and orange lines representing the increasing intensity of typhoons using the Saffir-Simpson Scale.

Because of the very large expanse of warm, tropical ocean in the western Pacific and the weather patterns, typhoons tend to be more frequent than tropical cyclones and hurricanes. Consider the data in table 1.2 collected between 1981 and 2016 by the Hurricane Research Division of the National Oceanic and Atmospheric Administration in the USA.

The high population and low levels of development in many areas of South-East Asia also mean that the degree of *exposure* and levels of *vulnerability* are greater than many countries where tropical cyclones and hurricanes occur.

TABLE 1.2 Annual frequency of tropical storms and cyclones by location, 1981–2016*

Basin	Tropical Storm or stronger (greater than 17 m/s sustained winds)			Hurricane/Typhoon/Severe Tropical Cyclone (greater than 33 m/s sustained winds)		
	Most	Least	Average	Most	Least	Average
Atlantic	28	4	12.1	15	2	6.4
NE/Central Pacific	28	8	16.6	16	3	8.9
NW Pacific	39	14	26	26	5	16.5
N Indian	10	2	4.8	5	0	1.5
SW Indian	14	4	9.3	8	1	5
Aus SE Indian	16	3	7.5	8	1	3.6
Aus SW Pacific	20	4	9.9	12	1	5.2
Globally	102	69	86	59	34	46.9

Note: * 1981–82 to 2015–16 cyclone season for the southern hemisphere
Source: AOML/NOAA

Over the past 30 years, typhoons from the western Pacific have become more powerful and scientists have predicted they will continue to get stronger due to ocean warming. When coastal water is warmer than usual, it tends to revitalise typhoons before they hit land. Scientists suggest the destructive power of typhoons has intensified by 50 per cent in the past 40 years due to warming seas. They also warn that global warming will continue to make giant storms even stronger in the future, posing a greater threat to heavily populated areas of the Philippines, Vietnam, China, Japan, North Korea and South Korea.

Typhoon Hato was a destructive typhoon that inflicted havoc on communities between the Philippines and Hong Kong (China) in August and September 2017. Also known as Tropical Storm Isang in the Philippines, it caused the deaths of at least 26 people and left a damage trail of approximately US$6.82 billion. With wind speeds of over 180 km/h and generating 10 m waves, Hato destroyed fish farms and boats in the South China Sea, tore down trees and unroofed buildings, flooded urban areas and rice fields, closed ferry services and airports, closed schools and businesses, and stopped power generation, which caused blackouts and forced hospitals to use backup generators. The weather bureau in Hong Kong gave it the highest typhoon intensity category seen for five years.

In November 2013, one of the world's largest ever storms, Super Typhoon Haiyan, unleashed its fury on the island nation of the Philippines, then proceeded westward across the South China Sea, devastating coastal regions of northern Vietnam. Also known in the Philippines as Super Typhoon Yolanda, it originated in the western Pacific, generating highly destructive wind gusts in excess of 300 km/h. Haiyan made landfall near Tacloban City — south of the capital, Manila — destroying almost everything in its path.

Warm air and ocean temperatures ensured Typhoon Haiyan remained powerful as it moved towards Vietnam. Here, coastal and delta regions were exposed to storm surge waves higher than 5 m. Despite weakening as it crossed the coast, the typhoon still had the power to kill or injure many people.

Haiyan smashed houses and buildings, uprooted trees, knocked out water, electricity and transport infrastructure, flooded farms and fishing ports, and created havoc for residents over an area almost as large as Australia. Its devastation was extraordinary, destroying over 70 per cent of houses and infrastructure in its path and killing an estimated 6300 people. In total, about 11 million people were affected. Many people were forced to evacuate, having lost family members, homes, possessions and crops. The economic damage was estimated to be more than US$4.5 billion.

FIGURE 1.29 Debris from Super Typhoon Haiyan, Tacloban, Philippines

FIGURE 1.30 (a) Before and (b) after aerial images show the damage created by Super Typhoon Haiyan in Tacloban, Philippines

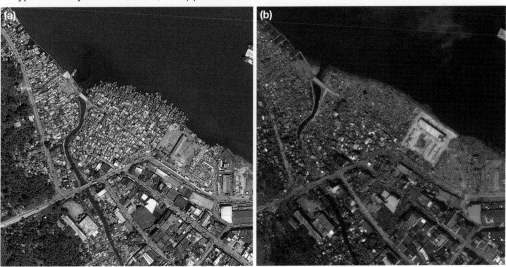

While countries around the world rallied to support survivors in the Philippines and Vietnam with aid, the level of despair and suffering after these events was incredible. Widespread panic and looting from desperate, hungry survivors made the distribution of essential aid difficult.

Resources

- **Video eLesson:** SkillBuilder: Understanding satellite images (eles-1643)
- **Video eLesson:** SkillBuilder: Interpreting an aerial photo (eles-1654)
- **Interactivity:** SkillBuilder: Interpreting an aerial photo (int-3150)
- **Video eLesson:** SkillBuilder: Comparing aerial photographs to investigate spatial change over time (eles-1750)
- **Interactivity:** SkillBuilder: Comparing aerial photographs to investigate spatial change over time (int-3368)

Activity 1.7f: Typhoons in the western Pacific

Using the information you have learned about typhoons in the western Pacific Ocean, answer the following questions.

Explain typhoons and comprehend their impact

1. What names are used to describe tropical storms in
   (a) Asia?
   (b) northern Australia
   (c) central America?
2. Which is more destructive: tropical cyclones, typhoons or hurricanes? Justify your answer with evidence or examples.
3. Which type occurs more frequently? Explain the features or processes that interact to create this pattern.
4. Using the data in table 1.2, create a bar graph to display the average number of tropical cyclones that occur in each region per year.
5. Research and, as precisely as possible, mark in the paths taken by both Typhoon Hato and Super Typhoon Haiyan on a map. Insert two or three information boxes where extensive damage was done. For example, Tacloban, Philippines. You could use the map below, or complete this task digitally. (A print-friendly version of this map has been included in the Resources tab.)

Analyse the data to identify the challenges

6. Examine figures 1.29 and 1.30. What factors apart from the strong wind gusts might have contributed to the scale of the destruction caused by Super Typhoon Haiyan in Tacloban in the Philippines?
7. Using figure 1.31 below, consider the age distribution of people in Tacloban in the years prior to Super Typhoon Haiyan (a graph of Australia's age pyramid is provided for comparison; note these graphs use different parameters and are from different years, so you will need to consider them carefully). How might the age distribution of the population affect the way a city or region can recover from a natural disaster?

FIGURE 1.31 (a) Age and gender of Tacloban population (2010 census) and (b) age and gender of Australian population, 1995 and 2015

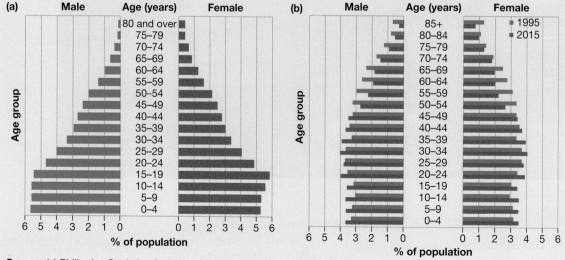

Source: (a) Philippine Statistics Authority, (b) Australian Bureau of Statistics

8. The Philippines is classified as a developing country by the United Nations. World Bank data from 2015 suggests that 21.6 per cent of Filipino people are living in poverty. What impact would the economic strength of a country have on the ability to recover from a significant natural disaster like Super Typhoon Haiyan?

Suggest recovery strategies

9. If the 2017 western Pacific typhoon season was below-average in terms of numbers and intensity than the two previous years. Predict what might happen in 2018 and 2019. Suggest and justify mitigation strategies that might need to be put in place by local authorities across the western Pacific to manage any potential hazards you predict.

FIGURE 1.32 Map the paths of Typhoon Hato and Super Typhoon Haiyan

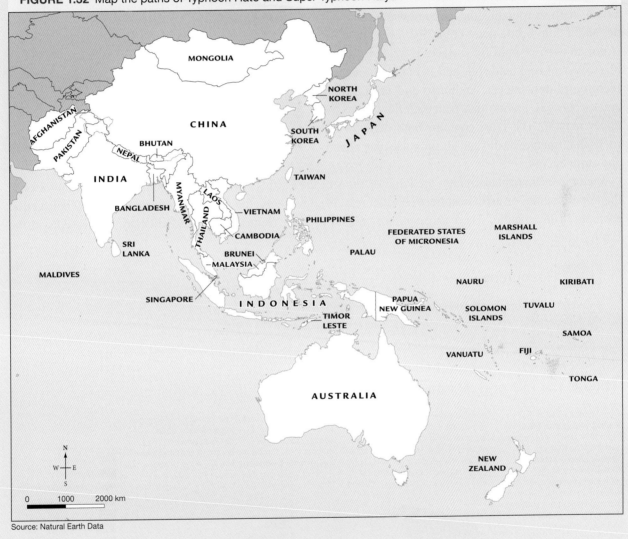

Source: Natural Earth Data

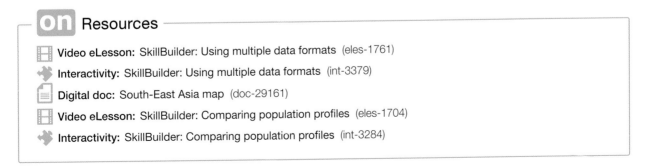

CHAPTER 1 Natural hazards 39

1.8 Geological hazards

1.8.1 Introduction

A geological hazard occurs or originates within the Earth, rather than on its surface or in its atmosphere. The most common geological hazards are earthquakes and volcanic eruptions, but tsunamis (large wave surges caused by earthquakes on the ocean floor) are also geological hazards.

FIGURE 1.33 The eruption of Klyuchevskaya Sopka in Russia, 2016

The Earth is divided into three parts: a dense core, a thick shell surrounding the core known as the mantle, and a thin, brittle outer crust. However, the crust is not an even or uniform surface. It is a very irregular cover of rock and soil, and is made up of large **tectonic plates**, uneven landforms and **faults** (huge cracks or weak points) that penetrate down into the upper mantle. The crust contains both hard, brittle rocks that fracture easily and elastic rocks that absorb and store energy.

Tectonic or lithospheric plates are large sections of the Earth's crust that float on the semi-molten rocks of the upper mantle, or **asthenosphere** (see figure 1.34). Over time the plates are moved around by convection currents in the mantle. These currents are created by heat from the radioactive decay of elements in the Earth's core.

Resources

Video eLesson: SkillBuilder: Interpreting a complex block diagram (eles-1746)
Interactivity: SkillBuilder: Interpreting a complex block diagram (int-3364)

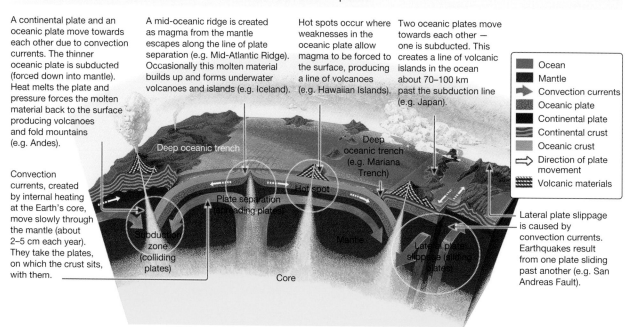

FIGURE 1.34 Fault types and the movements of the tectonic plates

1.8.2 Processes that create geological hazards

Earthquakes occur and volcanoes are formed when forces in the Earth move sections of tectonic plates, usually at faults. The Earth's lithosphere is divided into seven major tectonic plates and about 20 smaller ones. Most crustal movement happens when these plates push together, pull apart or pass each other in different directions.

Faults are named according to the way plates or blocks of crust move along the surface. There are generally three types.
- A diverging or normal fault creates a gap when two plates pull away from each other. These occur at diverging plate boundaries and a block either goes upward or downward, or **sea floor spreading** occurs on the ocean floor. An example of this on land is the East African Rift Zone.
- A converging or reverse fault, sometimes also called a thrust fault, occurs where two plates collide or are forced together at converging plate boundaries. On land, these collisions build fold mountains like the Himalayas whereas on the ocean floor the process is called **subduction**. Here, one plate may override another forming deep marine trenches, such as the Mindanao Trench near the Philippines.
- A transform or strike-slip fault occurs where two plates pass each other laterally and there is very little vertical movement. The San Andreas Fault is this type.

FIGURE 1.35 (a) Converging plate boundary (b) Diverging plate boundary (c) Transform fault boundary

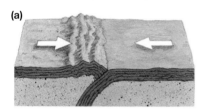
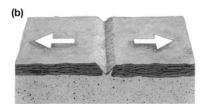

Converging plate boundary creates a reverse fault – two plates pushed together

Diverging plate boundaries creates a normal fault – two plates pulled part

Transform fault boundary creates a strike-slip fault – two plates slide past each other

This process of plate tectonic movement involves two different types of crust: oceanic and continental. Oceanic crust is heavier, thinner and younger than continental crust. Most commonly, two diverging plates are both composed of oceanic crust. As the oceanic plates drift apart, molten rock or **magma** rises from the mantle below, cooling and forming a new oceanic crust. Such magma is basaltic in composition and creates Icelandic-type volcanoes (see section 1.8.7) and mid-ocean ridges. Earthquakes in areas near diverging plates tend to have relatively small magnitudes.

Over time, this tectonic movement puts enormous strain on crustal rocks well beyond their level of strength. Eventually, these sections of plate slip or break, releasing an enormous amount of stored energy. This sudden and powerful slipping of sections of crust is called an earthquake. Because of the sheer size and force of the Earth's tectonic plates, surface movement is extremely powerful and there is nothing people can do other than take preventative action to survive. Even though most earthquakes happen in the crust, they can also occur in the upper mantle, mostly near **subduction zones** where plates collide. The site where rock begins to break is called the **focus**. Most earthquake tremors occur directly on the surface at the **epicentre** (the location on the ground immediately above the focus).

The strength of the earthquake depends on how much rock 'breaks' and how far it is moved. Following a major quake, several 'adjustments' called aftershocks or tremors occur, adding to the fear and uncertainty of people affected. Earthquakes below the ocean can displace large areas of sea bed and water, sometimes causing a tsunami — for example, the tsunami that hit the coast of Sulawesi, Indonesia, after a magnitude 7.5 earthquake in September 2018.

At the surface, seismic vibrations cause damage to infrastructure, particularly in populated areas. People can be killed or injured when buildings collapse or fires break out from the rupturing of fuel tanks and gas lines. They can also cause other types of hazards such as landslides and **liquefaction** (when the ground acts as though it is liquid, rather than solid).

FIGURE 1.36 World map of tectonic plates

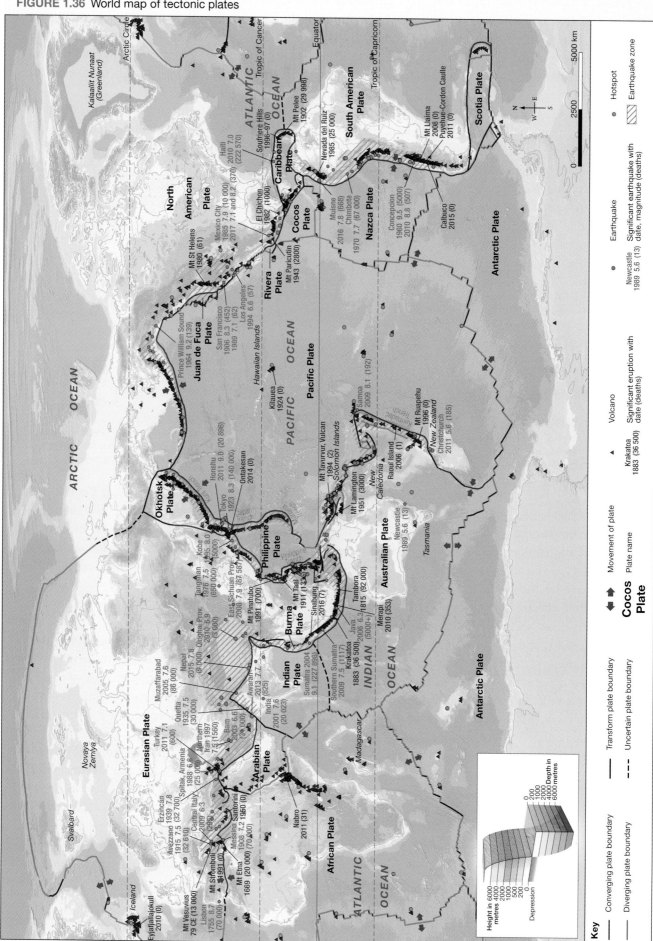

Source: Redrawn by Spatial Vision based on information from the Smithsonian National Museum of Natural History.

Many active volcanoes are also associated with pressures at convergent and divergent plate boundaries. Subduction occurs at convergent boundaries, when one plate is pushed below another. When this occurs, surface material is forced down into the mantle where it softens. At some other weak point in the crust, this molten material can be carried upwards to the surface through a fissure (crack) as a volcano. Earthquake clusters can also occur in association with magma movements. At divergent boundaries, where crustal plates pull apart, large underwater ridges (mid-ocean ridges) form as hot molten lava oozes out across the ocean floor and cools.

Activity 1.8a: Tectonic plate movement
Refer to the world map in figure 1.36.

Comprehend and explain tectonic plate movement and its impacts
1. Do the world's continents mainly sit on their own separate plates?
2. According to the map, which continents are largely unaffected by convergent faults and divergent faults?
3. On which plate is Australia located?
4. Which of the seven large plates covers the largest area?
5. Identify two continents that are severely affected by faults. From their position and the plate movement arrows on the map, suggest what type of faults affect these continents.
6. Identify which continent has a major subduction zone off-shore and is located close to the deepest ocean trenches.
7. If Australia continues to move north-east at about 6 cm per year, describe the approximate location of where Cape York be in 50 million years' time?
8. Scientists have established the Pacific Ocean is getting smaller, the Atlantic Ocean is getting larger and the Himalayas are getting taller. What could be causing these to happen?
9. With close reference to figure 1.36, explain the geological processes that result in earthquakes and volcanic activity in New Zealand.

1.8.3 Earthquakes

When an earthquake occurs, **seismic waves** (energy vibrating through the Earth) travel outwards from the focus (also called the hypocentre). The fastest waves (known as **P-waves**, or primary waves) travel at about 6 km/sec through the Earth. These P-waves pass seismic recording stations, a network of sensors located in over 150 different places around the globe – the Global Seismographic Network (GSN). The arrival time of the P-wave is detected and noted in real-time by the network's computers, which collate the information about the speed and direction of the P-waves' travel to calculate the location of the earthquake. The epicentre location is often available less than one minute after an earthquake occurs.

The magnitude of an earthquake is determined from the strength of the seismic waves detected at each station. There are several different formulas to calculate the magnitude. Most formulas depend on a measure of the

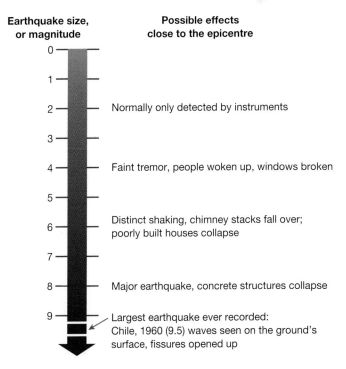

FIGURE 1.37 The Richter Scale

Earthquake size, or magnitude — Possible effects close to the epicentre

- 2 — Normally only detected by instruments
- 4 — Faint tremor, people woken up, windows broken
- 6 — Distinct shaking, chimney stacks fall over; poorly built houses collapse
- 8 — Major earthquake, concrete structures collapse
- 9 — Largest earthquake ever recorded: Chile, 1960 (9.5) waves seen on the ground's surface, fissures opened up

shear waves (S-waves, or secondary waves), which have the largest amplitude and carry the most energy. S-waves travel more slowly than P-waves so it may take a few minutes to calculate its magnitude. S-waves travel at between about 1 and 8 km/second, and P-waves between about 1 and 14 km/second. The speed of any seismic wave depends on variables such as the density and composition of the ground it is travelling through.

One common method of measuring the strength of an earthquake is the Richter Scale. Invented by Charles Richter in 1935, it compares the amount of energy released from an earthquake using numbers on a logarithmic scale – a magnitude 8.0 is ten times more powerful than a 7.0. The largest earthquake ever recorded was in May 1960, when a magnitude 9.5 was measured near Valdivia, in southern Chile.

Another method of measuring earthquake intensity is the Mercalli Scale. This is also a numerical scale shown in Roman numerals with I at the low-intensity end and XII at the high-intensity end. On the Mercalli Scale, a value is assigned to a specific location based on how people were affected or how much structural damage occurred. Lower numbers match how or what people felt, while higher numbers are based on observed structural damage to infrastructure such as buildings or roads. This means that an earthquake might be low on the Richter Scale, but if there is significant damage it may rate high on the Mercalli Scale. This might occur, for example, in a developing country where communities do not have the resources to construct larger buildings that are strong enough to withstand weaker earthquakes, or in mountainous areas prone to landslips.

TABLE 1.3 Modified Mercalli Scale, used to determine levels of earthquake damage (USGS)

Mercalli Scale	Shaking	Effects
I	Not felt	Not felt except by a very few persons under especially favorable conditions.
II	Weak	Felt only by a few persons at rest, especially on upper floors of buildings.
III	Weak	Felt quite noticeably by persons indoors, especially on upper floors of buildings. Many people do not recognise it as an earthquake. Standing motor cars may rock slightly. Vibrations similar to the passing of a truck. Duration estimated.
IV	Light	Felt indoors by many, outdoors by few during the day. At night, some awakened. Dishes, windows, doors disturbed; walls make cracking sound. Sensation like heavy truck striking building. Standing motor cars rocked noticeably.
V	Moderate	Felt by nearly everyone; many awakened during the night. Some dishes, windows broken. Unstable objects overturned. Pendulum clocks may stop.
VI	Strong	Felt by all, many frightened. Some heavy furniture moved; a few instances of fallen plaster. Damage slight.
VII	Very strong	Damage negligible in buildings of good design and construction; slight to moderate in well-built ordinary structures; considerable damage in poorly built or badly designed structures. Some chimneys broken.
VIII	Severe	Damage slight in specially designed structures; considerable damage in ordinary substantial buildings with partial collapse. Damage great in poorly built structures. Fall of chimneys, factory stacks, columns, monuments, walls. Heavy furniture overturned.
IX	Violent	Damage considerable in specially designed structures; well-designed frame structures thrown out of plumb. Damage great in substantial buildings, with partial collapse. Buildings shifted off foundations.
X	Extreme	Some well-built wooden structures destroyed; most masonry and frame structures destroyed with foundations. Rails bent.

Source: Abridged from The Severity of an Earthquake, USGS General Interest Publication 1989-288-913

> **Resources**
>
> 🔗 **Weblink:** Global Seismographic Network

1.8.4 Responding to earthquakes

Earthquakes can happen at any time. There is very little warning; many people are caught in the middle of the shaking. Most deaths and injuries occur from falling objects or buildings. People who live in earthquake zones practice safety drills, such as 'Drop (Duck), Cover, and Hold on!' Authorities suggest that when an earthquake strikes people should:

- keep calm
- if possible, turn off electricity, gas, and tap water
- protect yourself from falling objects such as signs, light fixtures and potted plants
- do not rush out of a building or use elevators.

In many communities at high risk of earthquakes, response and recovery efforts begin before an earthquake occurs, for example, by ensuring buildings are able to withstand predicted intensities of earthquakes in the area, ensuring people know what to do when an earthquake strikes and that they have the appropriate equipment ready.

After the earthquake, immediate concerns include finding and providing medical assistance to injured people, recovering the bodies of people who have been killed, ensuring collapsed or damaged buildings and infrastructure are made safe or cordoned off from the public, and managing safety issues such as fallen powerlines or broken gas, chemical spills and water mains. Displaced people then need to be given shelter and access to food, water and other necessities. As time passes, the recovery moves into a rebuilding phase where the debris is cleared and buildings and infrastructure can be repaired or replaced.

Part of the recovery process after any natural hazard is also repairing the social and economic impact of the event. Government agencies and community organisations might offer counselling and financial support, building assessment services or longer-term accommodation options. Systems and supports may also be put in place to help the community adapt in the longer term. For example, in New Zealand after the Christchurch earthquake, some people had to relocate permanently because their homes were no longer habitable.

The ability of a community to respond to an earthquake – at each stage: preparedness, mitigation, prevention and adaptation – often depends on their level of economic development and existing infrastructure. Countries with more money to spend on sturdier infrastructure, preparedness, education, communication and emergency response systems have fewer casualties and costs than countries that are unable to fund comprehensive programs to mitigate or prevent significant impact.

FIGURE 1.38 Earthquake damage to the road near Kaikoura, New Zealand, 2016

Recovery efforts and reconstruction of areas devastated by earthquakes, including supporting people who have been left homeless by a quake, are also expensive. Developing countries often do not have the financial means to do this as quickly and efficiently as developed countries, especially in remote or inaccessible areas. For example, a major city in a developed nation, like Christchurch, has highly-trained emergency response teams and sophisticated equipment accessible in the city that can be mobilised when an earthquake strikes.

Communications infrastructure also provided people in Christchurch with the means of seeking help; media reports and social media both featured footage and images of the 2011 earthquake within minutes.

In contrast, in the remote southern highlands of New Guinea in 2018 aid workers could not access many affected communities for weeks after the earthquake because roads were cut by landslides or dams bursting. The only way that aid agencies and the PNG government could assess the effect on some communities was through photographs taken from aircraft or satellite imagery. This did not just mean the the affected communities could not access much-needed aid, it also meant that the event was not reported on as widely by the media, meaning that public support and the resulting aid donations were slow to eventuate. Strong, shallow aftershocks measuring up to 6.5 on the Richter Scale also made it difficult for aid agencies to access affected areas. In some areas, the UN also suspended their operations for fears that localised violence made the region unsafe for their staff.

Access to information and levels of literacy and education can also be an important factor in recovery, as can restoring access to electricity and communications, such as internet access and phone towers. In the short-term, mobile communications allow people to access emergency help, but communications and access to information are important for long-term recovery too. For example, after the Christchurch earthquake in 2011, the Christchurch City Libraries posted pages of earthquake recovery information that collated links to all of the agencies and organisations that could provide people with information and assistance.

Resources

- **Weblink:** Recovery from the Great East Japan Earthquake
- **Weblink:** Papua New Guinea recovers from a remote major earthquake

---

Activity 1.8b: Earthquake impacts and responses

Using the information you have learned about earthquakes, answer the following questions.

Comprehend and explain the impacts

1. As with other natural hazards, the impact of an earthquake can be sorted into primary (immediate consequences), secondary (medium-term) and tertiary (long-term) impacts. Construct a table to divide the following effects into primary, secondary and tertiary impacts.

farms close
airports close
power lines fall
buildings collapse
sewage lines rupture
dams at risk of breaching
telecommunications out
schools close
outbreak of disease due to hygiene issues
water supply is closed
tourist industry downturn
towns not rebuilt due to cost
roads and rail lines are destroyed
looting and theft
liquefaction
people sleep outside for weeks
landslides block roads
tsunami
gas lines break causing fires

Suggest risk management strategies

2. If you were travelling in a car when an earthquake occurred, what action could you take to keep safe? Justify your answer with reference to the three variables that would affect your level of risk: the hazard, your vulnerability and level of exposure.
3. How might the geographical location and the level of development in an area affect earthquake responses? Write two paragraphs describing how the impacts of an earthquake might be managed differently in a remote location in a developing nation, compared with a remote location in a developed nation.
4. The San Andreas Fault is a continental transform fault running through California, the USA's most populous state (with over 37 million residents). The fault is about 1200 km long and forms part of the boundary between the Pacific Plate and the North American Plate. Both plates are moving to the north, but the Pacific Plate is moving slightly faster than the North American Plate. An enormous level of stress energy has been building up for over a hundred years, and seismic experts fear it has reached a critical state. With so many cities in the region, a large earthquake could be catastrophic. How might such a large population affect earthquake response? Propose two well-justified strategies to mitigate risk and to manage response if such an earthquake occurs.

FIGURE 1.39 The San Andreas fault in California, USA, is a continental transform fault.

1.8.5 Earthquakes in Christchurch, New Zealand

New Zealand has a high risk of earthquakes because of its position in relation to the Australia Plate and the Pacific Plate (see figure 1.41). New Zealand has had up to 15 000 tremors recorded each year, with more than 100 large enough to be felt. Many are low magnitude or happen in remote areas; however, when they happen close to populated and urban centres, the results can be disastrous.

FIGURE 1.40 Dust clouds rising from Christchurch during the 2011 earthquake

Source: © Gillian Needham

On 22 February 2011, much of inner city Christchurch was severely damaged and 185 people died when a magnitude 6.3 earthquake, with an epicentre at a depth of 5 km, struck 10 km from the city during the busy lunch period; many people were killed or injured by collapsing buildings and falling objects. More than 10 000 commercial buildings and 10 000 homes were destroyed, with an estimated 25 per cent of buildings in the centre of the city 'red tagged' (deemed unsafe to enter). This earthquake followed a 7.1 magnitude quake that had struck the previous September in the same Canterbury region – 40 km west of Christchurch at a depth of 11 km. The combined cost of these two earthquakes is estimated by the New Zealand Treasury to be NZ$15 billion.

As well as suffering extensive damage to road, rail and air infrastructure, utilities such as water supplies, electricity, gas and sewerage were made inoperable. Huge quantities of liquefaction created more than 400 000 tonnes of silt in the city's eastern suburbs, blocking roads and burying properties and vehicles. Many of the historic buildings which were the city's tourist attractions were destroyed and rendered unsafe to enter.

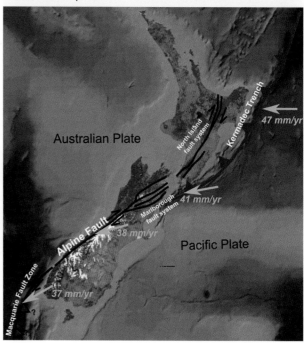

FIGURE 1.41 The boundary between the Australian and Pacific plates

FIGURE 1.42 Cars in liquefied soil on the road in Christchurch, 2011

Tremors continued for weeks, with many people afraid to enter buildings or travel. Residents relied on delivery of water by tanker and used chemical toilets for months until services were restored. As Christchurch is the main gateway for tourists visiting the South Island, the tourism industry and ski fields were shut down for the remainder of the year. For some, rebuilding was impossible and about 9000 people left the city to live elsewhere. Reports and research on the aftermath of the earthquake by the New Zealand government suggest that it can take five to ten years for people to socially and psychologically recover after a natural disaster of this size and significance.

A research paper commissioned by the New Zealand Parliament in October 2014 identified social impacts of the earthquakes that struck the Canterbury region in 2010 and 2011. Some of the findings are outlined below.

Social effects of Canterbury earthquakes

FIGURE 1.43 Earthquake damage to Christchurch Cathedral, 2011

Health
- 4 September 2010 earthquake: 377 people injured during and more than 1000 injured in the aftermath
- 22 February 2011 earthquake: 185 fatalities, 3129 injured and 1293 injured in the aftermath.
- The shock and effects of the disaster itself are followed by secondary, recovery-related issues: damaged homes, insurance claims and damaged roads and community facilities.
- Research in 2013 found that 80 per cent of respondents in the Canterbury area felt their lives had changed significantly, a third that it had caused financial problems, and 64 per cent felt guilty that others in their community had been affected more severely. By 2014, 66 per cent felt like their lives had been 'normal' over the last year compared to 60 per cent two years prior.
- Otago University research found that people from the area were 40 per cent more likely (than people from other areas) to experience major depression, post-traumatic stress disorder or an anxiety disorder.

Population
- In the two years ending June 2012, more children and their parents left the city than arrived and fewer young adults moved to Christchurch to study.
- In the year ending June 2013, the population aged 25 to 34 grew, partly due to the arrival of construction workers and engineers from overseas.

Crime
- Total recorded crime in the Canterbury District fell from 52 981 in 2009 to 40 393 in 2013.
- Reports of family violence offences increased (262 during February 2011 to 291 the following month before falling to 253 in May). Figures reported in 2013 suggest family violence reports continued to rise.
- There were changes in patterns of crime, including more crime being reported around shopping malls.

Source: Parliamentary Library of New Zealand, 'Social effects of the Canterbury earthquakes', 8 October 2014. Used under CC BY 3.0 licence

Some geologists predict that New Zealanders should be prepared for a very large earthquake (magnitude 9.0) to strike and warn of the possibility of a disaster like the 2011 earthquake in Japan, when 16 000 people died from the earthquake and the tsunami that followed.

> **on Resources**
>
> 🔗 **Weblink:** Earthquake recovery information
> 🔗 **Weblink:** Economic impacts of the Christchurch earthquake
> 🔗 **Weblink:** Impacts on tourism in New Zealand
> 🔗 **Weblink:** Earthquake building safety regulations in New Zealand

1.8.6 Earthquakes in Mexico

Mexico City, one of the world's most populated cities, is another at high risk for earthquake hazards. On 19 September 1985, a powerful earthquake, with a Mercalli Scale rating of IX and measuring 8.1 on the Richter Scale, struck close to the city causing many buildings to collapse and killing more than 10 000 people. Another 30 000 were injured and hundreds of thousands were left homeless. Because of this calamitous event, Mexico City improved its building and construction standards and established an emergency response team, *Brigada Internacional de Rescate Tlatelolco*, known as *Los Topos* ('the moles').

In 2017 on the same date, 19 September, a 7.1 tremor struck Puebla near Mexico City killing 370 people. This occurred only twelve days after an 8.2 magnitude earthquake struck off the coast of Mexico near Chiapas, generating a tsunami alert. As buildings shook and collapsed, thousands fled onto the streets in fear. Because this quake struck during office and school hours, many casualties were workers and children. Rescue teams of *Los Topos* and volunteers sifted through the debris and damaged buildings looking for survivors feared buried.

The following article, published in *The Conversation*, outlines some of the social and economic impacts of earthquakes for regions in Mexico.

Mexico's road to recovery after quakes is far longer than it looks

In the span of just 11 days, Mexico was devastated by two major earthquakes that destroyed buildings and claimed lives across southern and central Mexico. The official death count was higher than 400 as of Sept. 24, but it will continue to climb as relief efforts turn from rescue work to the recovery of bodies buried in the rubble.

In the days ahead, other measures of the disaster's extent will emerge, including the number of people who were physically injured and the estimated costs to the Mexican economy. No matter the measure, the disaster has clearly devastated many parts of Mexico. But, even then, those measures still obscure the true human cost of the disaster.

Long after the dust settles and new buildings are erected in the place of those that crumbled, tens of thousands of Mexicans will continue to feel the impact of the disaster. Many families, especially those living in poverty, will see their health, well-being and ability to escape poverty worsen for decades. Some will be affected for life.

I study how earthquakes and other natural disasters affect individuals, households and communities – and how to prevent natural hazards from becoming natural disasters in the first place. My research on past earthquakes and other natural disasters shows that these events exacerbate social disparities that are much more difficult to repair than the physical destruction.

The hidden consequences of disaster

Despite being the 15th-largest economy in the world Mexico's GDP per capita is only US$18,900, compared to $57,400 in the United States. To make matters worse, more than half of Mexico's population – 67 million people – live in outright poverty.

In southern Mexico, the region most affected by the twin earthquakes, the consequences are likely to be particularly severe: More than 70 per cent of people in Guerrero, Oaxaca and Chiapas states live in poverty. Many of those families live in extreme poverty, on less than $2 per person per day.

FIGURE 1.44 Damage in the streets of Juchitan, Oaxaca, Mexico, a month after the September 2017 earthquake

Losses caused by a natural disaster almost always affect the poor disproportionately and can even cause poverty. Beyond the devastating loss of a loved one, the loss of life is catastrophic for a household that struggles to put food on the table every day. For a poor family, the loss of a breadwinner threatens the future of everyone. For many families, even a modest loss of access to food can lead to malnutrition or affect the long-term health of family members.

And a minor loss in the ability to work or farm profoundly threatens the welfare in households that live close to the subsistence level.

What little savings poor households have are typically tied up in the value of their house, their livestock or some other physical asset. These life savings are often meant to support children through school or to invest agricultural equipment that could substantially increase yields. In developing communities where access to credit is limited, a household's ability to escape poverty depends almost exclusively on savings. In the blink of an eye, the life savings of thousands of Mexican families disappeared this month.

While shaking near the epicenters of the two earthquakes was 8.1 and 7.1 on the Richter scale, both of which can cause even modern buildings to crumble, shaking as low as 5.5 can cause noteworthy property damage. While fully collapsed buildings, fatalities and even injuries were fairly concentrated, at least nine states outside of Mexico City experienced widespread shaking high enough to ruin a poor household's assets.

The loss of property deteriorates a family's ability to sustain the agricultural output upon which their food security and other needs depend. The 2017 earthquakes came during the middle of a growing season for many households. It is too soon to know just how badly the agricultural capacity in southern Mexico has been affected. In other disasters, like the earthquakes in Nepal in 2015, there was a significant loss of crops.

Lower agricultural output will have widespread consequences across the region, inevitably affecting food prices. As the yield drops, or the price of sustaining the yield increases, food prices must rise. Poor families will in turn have a harder time sustaining a sufficient diet, or they will have to reallocate funds intended for long-term improvements to satisfy immediate needs. Many households that sustained no direct damage will be affected.

Beyond the misleading measurements

While the fatality count was higher for the second earthquake, which caused major structural collapses such as the collapse of a primary school with young children trapped inside, the first earthquake will probably have greater long-term consequences. It struck three southern states hardest, each near a 50 per cent poverty rate.

The United States Geological Survey predicts losses of between $100 million and $1 billion for the second of the earthquakes alone. However, these numbers almost certainly underestimate the long-run consequences that accrue, especially in the case of poor families.

As Mexico moves forward and the world responds, it will be important to remember that the total number of assets lost is not a meaningful indicator of how deeply lives are affected by the disaster. Losses of expensive luxury or vacation homes will quickly increase the total asset losses, while not affecting the food security of their owners. A $100 loss, while adding little to the total, can mean ruin for a subsistence-level household. Such a loss can cause not only short-term food insecurity but also an inability to escape poverty in the long run.

The emergency response will soon end and the world will turn its attention to the next disaster, but Mexican families will still feel the effects of the twin earthquakes for years to come.

Source: Morten Wendelbo, Lecturer, Bush School of Government and Public Service; Research Fellow, Scowcroft Institute of International Affairs; and, Policy Sciences Lecturer, Texas A&M University Libraries, Texas A&M University, published in *The Conversation*.

Activity 1.8c: Comparing the effects of earthquakes

Using the information you have learned about earthquakes in Mexico and New Zealand, answer the following questions.

Explain the impact of earthquakes in different communities

1. Use reports and available data in the text to list some of the main effects and responses to the earthquakes in Christchurch in 2011 and Puebla in 2017. Use a table like the example below to collate your findings.

	Christchurch, New Zealand 22 February 2011	Puebla, Mexico 19 September 2017
Population of country	4.6 million	112 million
Approx. GDP per capita	US$39 400	US$8201
Primary impacts (e.g. fatalities, injuries, impact on buildings and infrastructure)		
Secondary impacts (e.g. effects on water, food supply, sewerage, power supply, health)		
Tertiary impacts (e.g. effects on infrastructure, insurance, businesses, employment, residency)		

2. Write an extended paragraph describing the similarities between the earthquakes' impacts by providing examples from your table. Use the terms primary, secondary and tertiary in your response.
3. Were the primary, secondary or tertiary impacts of the earthquakes different for each place? Using data from your table, write a paragraph describing the different impacts and suggest what might have contributed to this difference.

Apply your understanding of earthquake responses

4. Suggest how communities in these places could manage the effects of future earthquakes.
5. Complete a table like the one below to compare primary, secondary and tertiary effects of two geological hazards that have occurred recently. (Refer to the **Smithsonian Institution Global Volcanism Program** and the **USGS Earthquake Hazards Program** weblinks in the Resources tab to find recent events.)

Effect	Earthquake	Volcano
Primary		
Secondary		
Tertiary		

on Resources

Weblink: 7 September 2017 earthquake
Weblink: Mexico's vulnerability to earthquakes

1.8.7 Volcanic eruptions

Volcanoes exist on every continent. At present, several thousand are extinct or dormant and pose no threat to life. However, there are about 600 volcanoes above sea level that are highly active. When they erupt, they may cause death and injury to people as well as destroy property and crops; however, volcanoes also produce mineral-rich soils, and so people settle around the base to make the most of the fertile farmland. In doing so, they put their lives and communities at risk.

Most volcanoes occur when magma can escape or be released at weak points of the Earth's crust. Over 60 per cent of all volcanoes occur along tectonic plate boundaries where either subduction or sea floor spreading is occurring. The most active volcanoes occur over subduction zones, many in the Pacific Ring of Fire (see figure 1.36). Volcanoes also form at isolated weak points in the crust known as 'hot spots', such as the Hawaiian Islands.

Volcanoes differ according to the type of lava that is discharged from them. There are two main types.
- Basic lava cones are low with gentle slope like those in Hawaii and Iceland. This lava is low in silica and runs out quickly, and may travel a considerable distance from the volcano. These cones are commonly called shield volcanoes.
- Acidic lava volcanoes have tall, steep cones. This lava is high in silica, and is viscous (thick); it moves slowly and may even clog some of the volcano's vents resulting in violent, explosive eruptions that can blast tonnes of rock and ash, and hot, poisonous gases into the air. Acidic volcanoes are an extremely dangerous hazard because of this explosive force. They are also referred to as **pyroclastic cones**.

Impacts of volcanic eruptions

The most obvious effect of volcanic eruptions is the threat to life, but eruptions can also destroy forests and wildlife, houses, farmlands and rivers, as well as contaminate the atmosphere. They create lava flows that travel long distances and burn, bury or harm anything in their path, and are often associated with earthquakes, mudflows lahars and avalanches. The gases they emit, such as carbon dioxide (CO_2) and sulfur dioxide (SO_2), are not only toxic and smelly, but also cause acid rain in regions some distance away. Large quantities of rock, ash and dust can bury things, cause roofs to collapse and make it difficult for living things to breathe. Volcanic ash also affects aircraft engines, sometimes even leading to the failure of key navigation equipment. If ash is sucked into aircraft engines and accumulates, it can potentially lead to engine failure. Because it is unsafe to fly through airborne volcanic ash, flight paths are often closed down near ash plumes, causing significant disruptions to travel and tourism. These hazards can cover significant distances, so volcanic exclusion zones (areas from which people are encouraged to evacuate in order to stay safe) can be quite large.

FIGURE 1.45 Pyroclastic cloud, Mount Sinabung, Indonesia

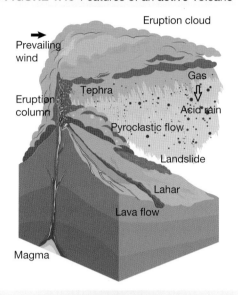

FIGURE 1.46 Features of an active volcano

TABLE 1.4 Impacts of ash falls

Ash depth	Impacts
<1 mm	Lung and eye irritation Airport closures due to potential damage to aircraft Vehicles, houses and equipment potentially damaged by fine abrasive ash Water supplies potentially contaminated, particularly roof-fed tank supplies Poor road visibility and traction due to dust or mud
1–5 mm	Amplification of impacts of less than 1 mm of ash Crops potentially damaged Feed for livestock reduced and possible contamination of water supplies Houses may sustain some damage from fine ash soiling interiors or blocking air-conditioning filters Electricity substations may short-out from wet ash (low-voltage systems more vulnerable than high-voltage) Water supplies may be cut or limited because of power cuts to pumps Water supplies may be contaminated by chemical leachates Water supplies may become restricted because of the increased water usage required for clean-up Roads may need to be cleared to reduce the dust nuisance and to prevent stormwater systems from becoming blocked Sewerage systems may be blocked by ash, or disrupted by failing or damaged electricity infrastructure Electrical equipment and machinery may suffer damage
5–100 mm	Amplification of impacts from less than 5 mm of ash Pasture and low plants may be buried, foliage may be stripped from some trees (most trees survive) Most pastures will be destroyed with falls of more than 50 mm of ash Urban areas will require major ash removal Weak roof structures could potentially collapse with 100 mm ash coverage, particularly if the ash is wet Road transport will cease due to the build-up of ash on roads, cars may stop working from clogged air filters Rail transport may cease due to signal failure brought on by short-circuiting if ash becomes wet

(continued)

TABLE 1.4 Impacts of ash falls (*continued*)

Ash depth	Impacts
100–300 mm	Amplification of impacts from less than 100 mm of ash Roofs may collapse if buildings are not cleared of ash, especially from large flat-roofed structures and if ash becomes wet Trees suffer severe damage, including being stripped of foliage and broken branches Electrical reticulation fails due to falling tree branches and shorting of power lines
>300 mm	Amplification of impacts from less than 300 mm Widespread death of vegetation Soil horizon is completely buried Significant numbers of livestock and other animals die or experience serious distress Aquatic life in lakes and rivers dies Many roofs collapse from ash loading Power and telephone lines break Roads are inaccessible

Source: GNS Science

Activity 1.8d: Interpreting isopach maps

An isopach map is one that shows lines connecting points of equal formation thickness. It could be used to show fallout from a volcano or thickness of mining seams.

Refer to data in table 1.4 and the isopach map in figure 1.47 to answer these questions. (Note that the map and table show measurements in different units – mm and cm – so you may need to convert some data to make accurate comparisons.)

Comprehend and explain the information

1. Name the volcanic peak in the centre of the isopach lines.
2. What is the interval used (in cm) to distinguish between the isopach lines?
3. Name two cities where ash could reach a depth of at least 25 cm.
4. What depth of ash could fall on Lake Taupo?
5. Why might more ash could fall to the east of the volcano?
6. Which urban centre had the most ash fall: Wellington or Wanganui?
7. Estimate the depth of ash at Te Araroa.
8. What effects would ash fall have on the aquatic life in Lake Taupo?
9. How would road and rail transport in and out of Te Araroa be affected?
10. What affects might Napier have suffered that Wanganui did not?

Apply your understanding and suggest risk management strategies

11. How might residents of Auckland, Wellington and Napier better prepare themselves for future eruptions of Mt Tongariro? Construct a risk mitigation plan that might be distributed in each community to help people prepare.

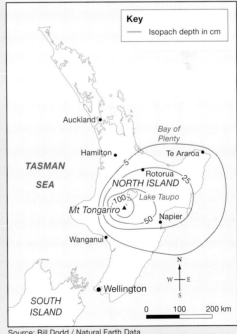

FIGURE 1.47 A hypothetical isopach map of the North Island of New Zealand showing potential depths of ash fallout

Source: Bill Dodd / Natural Earth Data

Resources

- **Video eLesson:** SkillBuilder: Constructing and describing isoline maps (eles-1737)
- **Interactivity:** SkillBuilder: Constructing and describing isoline maps (int-3355)
- **Weblink:** Unzen pyroclastic cloud
- **Weblink:** Smithsonian Institution Global Volcanism Program

1.8.8 Earthquakes and volcanoes in South America

The South American west coast has been one of the lithosphere's most active areas over the last century, particularly around Ecuador's mainland and the Galapagos Islands. The Galapagos Islands and Hawaii are well-known hot spots in the Pacific where magma has protruded through the crust to eventually form island chains. The ocean floor of the Nazca Plate between the Galapagos Islands and the coast (about 1000 km away) is being subducted below the South American Plate. However, a section of this sea floor about 450 km wide is an **aseismic ridge** (known as the Carnegie Ridge). An aseismic ridge is one that forms when a tectonic plate moves over a hot spot and undersea mountains and guyots (flat-topped undersea mountains) may form.

The Carnegie Ridge consists of rock that is less dense than the surrounding sea floor, and so is not subducted to the same depths as it passes into the Colombia–Ecuador trench and below South America. However, this uneven shallow subduction does cause several geological issues, such as:

- generating additional tectonic stress as old seamounts and guyots on the sea floor get dislodged into the plate
- tearing of the lithosphere, which increases the potential for earthquakes near the subduction zone (Ecuador has had six major earthquakes in the past 100 years)
- generating 'shallow' (close to the surface) earthquakes, which are more intense and dangerous
- developing **adakitic** volcanoes (where magma consists of partially melted sea floor and basalt, which can be highly explosive). There are 11 active volcanoes south of the Ecuadorian capital, Quito, and all are within 200 km of each other.

The tectonic threat to Ecuador is very high. Currently, Ecuador has more than 20 volcanoes that are either currently active or have been active recently. The best-known are Sangay, Cotopaxi, Reventador and Pichincha. Of most concern is the proximity of these dangerous volcanoes to highly populated, built-up areas such as Quito, which has more than 2.6 million inhabitants. At least 9 per cent of the total population lives around the base of volcanoes and about 66 per cent of the road network is in seismically active country. These volcanoes are not only close to major centres, they are also very active. Sangay has erupted almost daily since 1934 and there are frequent earth tremors because of the off-shore subduction.

Because they are part of the Andes Mountain chain, Ecuadorian volcanoes also have very high summits. Cotapaxi, for example, has an elevation of 5897 m above sea level and is covered with thick snow year-round. Cotopaxi also contains many glaciers, so the surrounding lowlands are at significant risk of lahars and flooding. In fact, lahars from Cotapaxi have been known to reach the headwaters of the Amazon in past eruptions. Water supply in Ecuador is dependent on alpine rivers and springs, so it is also quite vulnerable to pollution from the heavy metals found in volcanic ash. Because of the high levels of risk associated with Cotopaxi, the volcano is monitored for activity with seismographs, which monitor tremors, and

FIGURE 1.48 Cotopaxi, Ecuador, 2014

Source: Sandra Duncanson

for the increased pressure associated with the chamber filling with magma. The slopes are also continuously monitored, measuring any changes in nanometres (thousand-millionths of a metre).

Ecuador has a GDP per capita of less than US$6000, which affects the capital and resources available to reinforce and protect infrastructure, and to evacuate and protect citizens. The United Nations and humanitarian organisations such as the Red Cross and Oxfam play key roles in the planning for and response to volcanic eruptions in Ecuador. Emergency situations are coordinated through relevant government ministries that operate under a National Decentralized System, bringing together all government and non-government organisations (NGOs) for assistance and advice during times of heightened risk.

The 2015 eruption of Tungurahua created ash deposits of up to 6 mm in nearby towns and villages. The eruption affected an estimated 130 000 people, including 5000 who required direct assistance. Government agencies issued an orange alert, indicating residents should remain alert and prepare for the possible need to evacuate. Even though the threat from the volcano eased within a few days, the ash coverage still had a significant impact on the affected communities.

As soon as the government alert was issued, Ecuadorian Red Cross units were mobilised to coordinate efforts with community leaders, to set up emergency operations, assess and gather information about damage, and to ensure local communication channels were maintained and remained open in the event that the disaster worsened. This included the number of people who needed to be made aware of evacuation and emergency procedures, potential numbers of people who might have needed humanitarian assistance and thorough risk assessments based on data taken from previous eruptions and their impacts. Key objectives then took into account the vulnerability of groups in the area and potential scenarios from explosive pyroclastic flows and ongoing gas and ash emission to activity ceasing entirely.

NGOs such as the World Food Programme and UNICEF also operate in Ecuador outside times of emergency, developing contingency plans for significant volcanic eruptions or earthquakes.

Activity 1.8e: Volcano alert codes

In Ecuador, there are four levels of alert: white, yellow, orange, and red.
- White – almost no possibility of an eruption, although the volcano is not extinct
- Yellow – there has been minor seismic activity with release of some gases and ash. This is an early warning where the government may declare a state of emergency and warn people to prepare for possible evacuation.
- Orange – seismic activity has increased and eruption is imminent; residents near foothills are evacuated due to lahars or pyroclastic flows
- Red – a volcanic eruption has occurred.

Explain and analyse volcano alerts in Ecuador
1. Find out if any of these well-known volcanoes are currently on alert: Sangay, Cotopaxi, Reventador and Pichincha. Describe the features and extent of the volcanic activity that is occurring.
2. Why would it be important to be give a hazard zone map to tourists that shows where lahars may be possible?
3. Considering the history of seismic activity in Ecuador, what information should walkers climbing Cotopaxi be told before they begin their climb?
4. What challenges might communities in hazards zones around active volcanoes in Ecuador face that tourists would not need to consider? Make lists of primary, secondary and tertiary impacts for local people that would not affect visitors to the region.

1.8.9 Responding to volcanic eruptions

When there is a volcanic eruption, and if residents or tourists have time, it is safest to move beyond the exclusion zone. However, if this is not possible or people fail to leave, there are certain steps to take. People should protect themselves against ash and dust by wearing a mask or placing a wet cloth over their mouth and nose, wearing long-sleeved clothes and removing contact lenses. This helps to prevent acid-coated ash irritating their lungs and eyes. Keeping at least three days' supply of clean drinking water, a supply of food, a battery-operated radio, cash and a first aid kit is also important. People should also protect electronic equipment with plastic.

Responses to volcanic eruptions vary according to the severity and impact of the eruption, and the community's ability to respond. Consider the plan of action for the Tungurahua Volcano in Ecuador and the preparedness guide developed by the US Geological Survey found in the Resources tab.

Resources

- **Weblink:** Volcano Discovery: Live updates and webcam footage from Ecuador's volcanoes
- **Weblink:** Tungurahua Volcano emergency plan of action
- **Weblink:** US Geological Survey: Volcano preparedness

1.8.10 Mt Vesuvius, Italy

Italy is home to some of the world's most famous volcanoes, including Vulcano, the island which gives us the name used for all eruptive mountains, as well as for a type of volcanic eruption. Better known volcanoes, such as Vesuvius and Stromboli, have also provided names used to describe types of volcanic eruptions. Italy has about 30 volcanoes, seven of which are still considered active. Two Italian volcanoes, Etna and Stromboli, are almost continuously active, and Etna, at a height of 3323 m, is also Europe's largest continental volcano. Italy's volcanic activity is due to a collision between the African and Eurasian tectonic plates. As the African plate moves towards the Eurasian plate, subduction of the African plate occurs and magma rises to the surface through the weakened and fractured crust.

Even though Italy's volcanoes have provided its farmers with rich, fertile volcanic soils, especially around Mt Vesuvius and Mt Etna, they are also a grim reminder of Earth's awesome authority. In 79 CE, Vesuvius, a massive stratovolcano overlooking the Bay of Naples, erupted with overwhelming fury, destroying the Roman cities of Pompeii and Herculaneum, killing many of their occupants.

Vesuvius has erupted many times since 79 CE, most recently in 1944. It is still regarded as active and is monitored daily for gas and temperature changes. Today, Naples and the area adjacent to the base of Vesuvius is a heavily populated urban and tourist centre. More than 3 million people live within 12 km of its cone and are exposed to the wrath of Vesuvius, which continues to rumble and discharge smoke. Vesuvius's history of violent eruptions make it one of the world's most dangerous volcanoes.

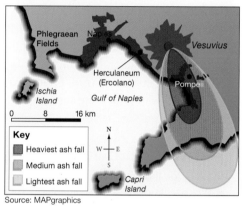

FIGURE 1.49 Mt Vesuvius and surrounding areas, 79 CE

Source: MAPgraphics

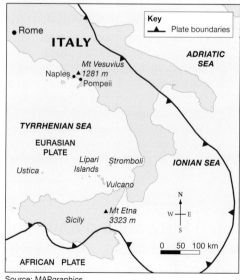

FIGURE 1.50 Plate boundary, southern Italy

Source: MAPgraphics

Vesuvius evacuation plans

Authorities in the areas surrounding Vesuvius have formulated plans to evacuate over 700 000 people if the volcano erupts. The Department of Civil Protection, which is charged with risk management and emergency response in Italy, has identified a 'red zone' for risk that includes 25 towns, all of which can be evacuated within 72 hours of a significant eruption. Another 63 towns, with a combined population of more than 1 000 000 people, lie in a 'yellow zone' that would be likely to suffer from falling ash and rock from the volcano.

The plan also includes four alert levels: basic, attention, pre-alert and alert. When the pre-alert level is activated, patients in care facilities and hospitals are relocated and heritage monuments are protected. Activation of the alert level means the plans to evacuate come into effect.

Risk management plans contain key strategies that need to be developed while the mountain is dormant. These include:
- delineating the three emergency zones based on proximity and potential fallout. The red zone is the area around the base of Vesuvius, where the risk of pyroclastic flow, heat and toxic gases is greatest. The yellow zone is further away and is the area most likely to be affected by fallout of ash and lapilli (rock fragments being launched from the volcano). Homes would be at risk of collapsing and residents would suffer from respiratory problems. The blue zone falls inside the yellow zone, but is considered at higher risk because of its topography. Here there is more chance of flooding or mudflows.
- educating and informing people for an orderly evacuation and removing complacency, without becoming alarmist. Money has been set aside for school education programs.
- having operational strategies to evacuate the entire red zone in 72 hours (12 hours for organisation, 48 hours for movement and a buffer of 12-hours if needed) using a fleet of 375 000 registered cars, 500 buses and 220 trains each day of the evacuation period.
- arranging evacuation centres. Each of the 25 local regions has been 'twinned' with another area of Italy where evacuated residents would be accommodated. Each township has been allocated a specific mode of transport depending on the destination of evacuees. For example, Pompeii residents would leave by boat to Sardinia, while Neapolitans would board trains that would take them to Lazio, the administrative region around Rome.

FIGURE 1.51 Looking across Naples towards Mt Vesuvius

FIGURE 1.52 Satellite image of the area around Mt Vesuvius

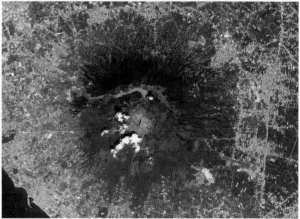

Source: © NASA Visible Earth

Emergency planning

Despite these emergency plans, and extra funding being assigned to improve infrastructure and help spread awareness of the emergency protocol, there are many unresolved issues and complexities when an active volcano is located so close to a major city and tourist area.

- Public officials and emergency services personnel might not be familiar enough with the plan to implement it immediately.
- Thousands of tourists visit the area daily and most would not be familiar with local conditions.
- Language barriers between locals and visitors might prevent people from understanding instructions.
- Much of the area in the hazard zone is still farmland accessed by smaller, rural roads.

A program has been set up offering money for families living near Vesuvius to relocate out of the danger zone, but this has not proved popular.

Activity 1.8f: Mt Vesuvius, Italy

Using the information you have learned about Mt Vesuvius, answer the following questions.

Analyse the data

1. Using the information available, complete a SWOT analysis of the National Emergency Plan for Naples. (A printer-friendly version of this analysis grid can be found in the Resources tab.)

FIGURE 1.53 SWOT analysis grid

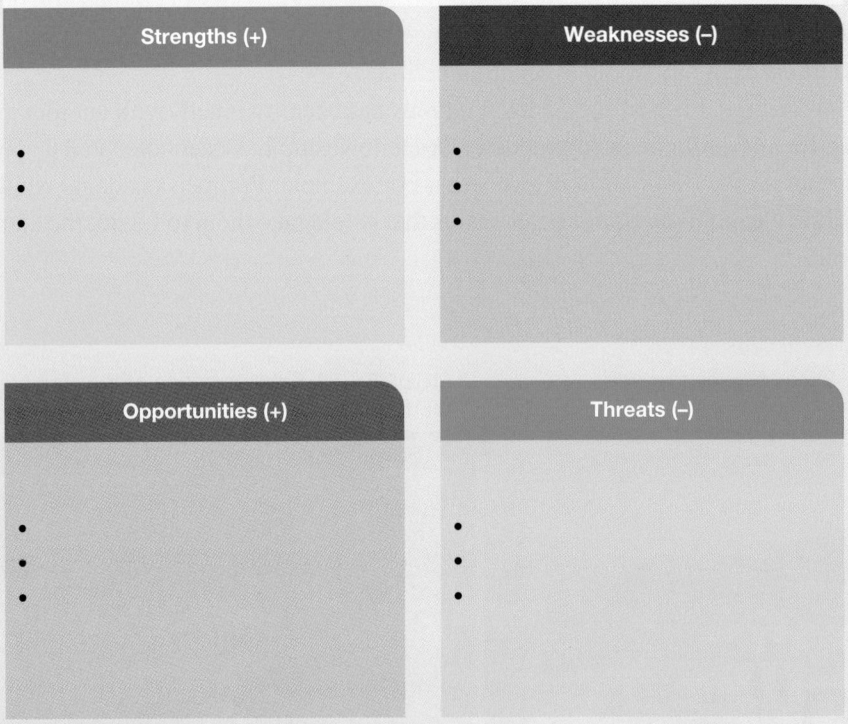

Strengths (+)	Weaknesses (−)
• • •	• • •

Opportunities (+)	Threats (−)
• • •	• • •

Comprehend primary sources

Read through the description of Vesuvius erupting in 79 CE, as written by Pliny the Younger, and answer the questions that follow.

Extract from Pliny the Younger's account of the eruption of Vesuvius (24 August 79 CE)

'One night the earth shocks became so violent that it seemed the world was turned upside down ... We decided to escape, and a panic-stricken crowd followed us ... When we were clear of the houses we stood still ... The sea seemed to roll back on itself ... the shore had widened, and many sea creatures were beached on the sand. In the other direction appeared a horrible black cloud ripped by bursts of fire ...

Soon the cloud began to descend upon the earth and cover the sea ... We were enveloped in night like the darkness of a sealed room without lights. We could hear the cries of women, the screams of children and the shouting of men. Some were calling to their parents, to their children, to their wives. Some lifted their hands to the gods, but a great number believed there were no gods now, and that this night was to be the world's last eternal one ...

We were immersed in darkness and ashes fell thickly upon us. From time to time we had to get up and shake them off for fear of being buried and crushed under the weight ... Finally, a real daylight came. Before our terror-stricken sight everything was covered by a thick layer of ashes like a heavy snowfall. We returned to Misenum and had no thought of leaving until we received news about my uncle.'

2. What was Pliny describing in his account?
3. Make a list of things that came out of the volcano.
4. What hazardous events could Pliny have been describing when he said, *'The sea seemed to roll back on itself ... the shore had widened, and many sea creatures were beached on the sand.'*
5. What was Pliny recalling when he said, *' ... horrible black cloud ripped by bursts of fire ... Soon the cloud began to descend upon the earth and cover the sea.'*
6. What evidence is there to suggest the ash fall was both substantial and light?
7. Using figure 1.52, identify where Pliny might have been standing to witness the events of the eruption as he describes them.
8. Use figure 1.49 to explain why Pompeii was completely buried by ash and its occupants incinerated by hot gas.

Resources

- **Weblink:** Mt Vesuvius Emergency Plan
- **Weblink:** What if Vesuvius erupted today?
- **Weblink:** Smithsonian Institution Global Volcanism Program
- **Digital doc:** Volcano warnings (doc-29160)
- **Digital doc:** SWOT analysis grid (doc-29162)

1.8.11 Mt Agung, Indonesia

Every year, more than 1 million Australians visit Bali for holidays or business. Visitors fly in to the main airport at Denpasar, the largest city, and stay at local resorts or at Kuta, a popular beach area. However, the tourist experience is sometimes overshadowed by the presence of a large stratovolcano called Mt Agung, which is situated about 50 km north-east of Denpasar. Rising more than 3000 m above sea level, Agung has provided Bali with rich lava soil, in which farmers cultivate rice, vegetables and tropical fruits.

The mountain holds special spiritual and cultural significance for locals, and several temples are built on its slopes. For the many Balinese people who hold Hindu beliefs, Mt Agung is the most sacred mountains on the island.

In late 2017, Agung erupted again, pouring ash and smoke from its crater and threatening people in the surrounding area. A safety exclusion zone of 12 km was declared and thousands of people were forced to evacuate the area, leaving their farms and homes.

FIGURE 1.54 Mt Agung erupts in Bali, Indonesia, 2017

The eruption also had a significant impact on tourism. Air travel in and out of Denpasar was severely disrupted by volcanic gas and ash clouds and many flights were cancelled for safety reasons. The airport became a temporary holding centre for hundreds of stranded passengers, many with little money. Hundreds of international holiday bookings were cancelled, so many resorts were left without customers, and local businesses were temporarily closed, leaving workers without income.

Resources

Weblink: The cultural significance of Mt Agung

Digital doc: Map of Bali (doc-30354)

Activity 1.8g: Risk assessment

Using the information you have learned about volcanic eruption risk assessment and figure 1.55, answer the following questions. (You will also need to complete some online research.) A printer-friendly version of this map can be found in the Resources tab.

Comprehend and explain the risks

1. Mark in the exclusion zone of 12 km around Mt Agung.
2. Bali is divided into eight administrative areas. Research these divisions and their population densities, and mark them on the map.
3. Research and mark in key infrastructure and facilities that might be affected or are vital to response efforts during or after a volcanic eruption, such as hospitals and airports.

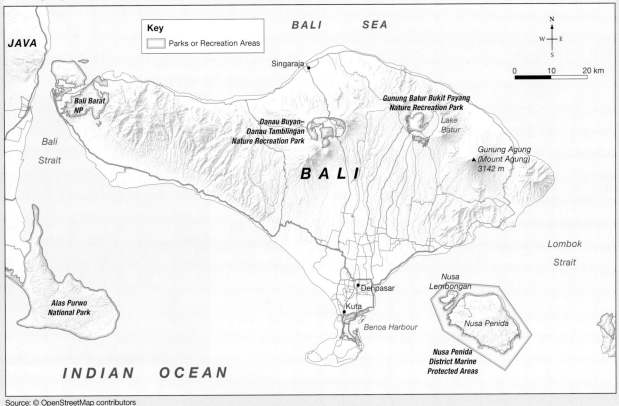

FIGURE 1.55 Bali is a popular tourist destination for many Australians but resorts' and flight paths' proximity to Mt Agung can pose a potential risk for tourists and locals

Source: © OpenStreetMap contributors

> ### Suggest and justify risk management strategies
> You have been asked by the Bali Tourist Bureau to undertake a risk assessment to determine if the National Parks and walk trails around the base of Mt Agung are safe for residents and tourists in the short- to medium-term. Create a proposal using the questions that follow to guide your assessment of the risks.
>
> 4. List the key criteria that you will need to consider to assess the risk. Specifically consider the vulnerability of different groups using the area, how risk might be mitigated, preparedness and the ability of tourists, locals and local authorities to respond quickly in the event of an eruption. Design a decision-making matrix to consider the various criteria upon which you will make your decision. Your criteria might include social, economic, geographical or other factors that affect people's level of risk, vulnerability or willingness to follow your proposed strategy.
> 5. Identify and prioritise which criteria are most important. Use an ordinal system.
> 6. Write your response to each of the criteria, listing positive and negative factors that might contribute to your decision and giving examples or evidence where possible. For example, considering the needs of tourism operators might be important because tourism now makes up 80 per cent of Bali's economy.
> 7. Once you have outlined your key points on the matrix, rank the key points listed under each criterion.
> 8. Write a paragraph about each criterion and its key points, providing evidence and elaborating on the positives and negatives of each of your points.
> 9. Using this evidence, propose a course of action for the management of the trails.

1.9 Geomorphic hazards

1.9.1 Introduction

Geomorphic hazards are the result of processes that occur on the crustal surface or lithosphere. Often called 'mass wasting', they refer to **slope failure** (which occurs when the pull of gravity causes a hill or mountain slope to collapse), **subsidence** (when part of the land sinks or collapses), or **flow movement** (when rock, soil or sand mix with water and air and move downhill in a flow). The most common geomorphic hazards are landslides, mudslides and avalanches (snow). These usually happen when the force of gravity cannot be withstood any longer because of one of two changes:
- geological movement, particularly in an already unstable area
- human activity that has caused changes to the topography, such as vegetation clearing, changes to drainage patterns and below-ground seepage or wild fires.

Even though mass wasting of rock, soil, mud or snow is less common than earthquakes and volcanic eruptions, it happens without warning and, in many cases, people are completely unprepared. The consequences can be catastrophic, causing widespread death and injury as well as burying towns and villages, destroying buildings and infrastructure (roads, tunnels, bridges and property), and even reshaping the surface topography. Across the world, landslides, mudflows and avalanches are some of the biggest killers of people in mountain regions, particularly where people live on or below hillsides.

The speed at which material travels downslope ranges from very slow (creep) to very fast (rockfall). Momentum and force of a flow is usually determined by the mass of material (size and weight), the mix of material (clay, rocks, soil, debris), the amount of water/fluid to lubricate and reduce friction, and the slope angle.

TABLE 1.5 Classification of flow movement

Average speed	Saturated flows	Non-saturated flows
Very slow (e.g. 1 cm/yr)	Solifluction	Soil creep
Slow (e.g. 1 m/day)		Earth flow
Moderate (e.g. 1–2 m/hr)	Debris flow	
Rapid (e.g. 50 km/hr)	Mudflow	
Very rapid (over 100 km/hr)		Debris avalanche

1.9.2 Types of mass wasting

A **landslide** is a geomorphic event where rock, soil, mud or artificial fill move down a slope under the force of gravity (slope failure). They may be caused by any number of natural processes, including long-term weathering of rocks and scree (producing regolith: loose weathered rock fragments), soil erosion, vegetation removal, earthquakes or volcanic eruptions. Human activities can also trigger landslides. The most common human causes are road building, construction on very steep slopes, poorly planned changes to drainage and disturbing old landslide sites.

The most lethal landslide in Australia occurred in July 1997, when a large section of steep mountainside collapsed in Thredbo, New South Wales, carrying the Carinya ski lodge at high speed onto the Bimbadeen ski lodge below. Both buildings were destroyed and 18 people died beneath the rubble. One injured man survived for three days within the landslide debris before being rescued. In comparison, a huge debris avalanche in Peru in 1970, which moved at speeds of up to 400 km/h, killed more than 20 000 people and destroyed the towns of Yungay and Ranrahirca.

A **mudflow** is a type of mass wasting where soil and debris become saturated (liquefied) and move rapidly down a slope (flow movement). Often, there is insufficient vegetation to hold the soil together due to land clearing or fires. In recent times, mudflows have happened due to very heavy rain falling in areas where slopes have been cleared for farming (the Philippines, Nepal) or razed by wildfires (California, USA). A mudflow is not the same as a mudslide.

Another common trigger for mass movements is heavy rainfall, such as that associated with tropical cyclones and severe thunderstorms. Large volumes of water add weight to soil and weathered rock particles, making slopes more unstable and susceptible to the influence of gravity. Water may also lubricate rock surfaces, thus reducing friction and allowing easier movement of overlying soil and rock. In countries that have cleared slopes from either farming or bushfire, heavy rain can trigger mudslides, often with tragic consequences.

Mass wasting is most common on uneven or hilly ground, particularly if there is heavy rainfall. In wet tropical areas such as South-East Asia and central Africa, heavy rainfall associated with thunderstorms or tropical cyclones adds weight to soil and rock particles, making slopes unstable and vulnerable to gravity. Water may also lubricate rock surfaces, thus reducing friction and allowing soil and rock to move easily.

FIGURE 1.56 (a) Types of slope failure (b) Types of flow movement

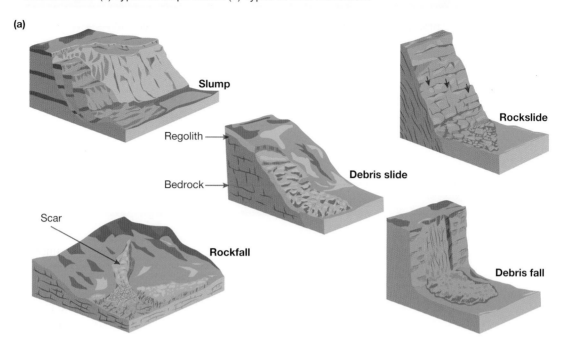

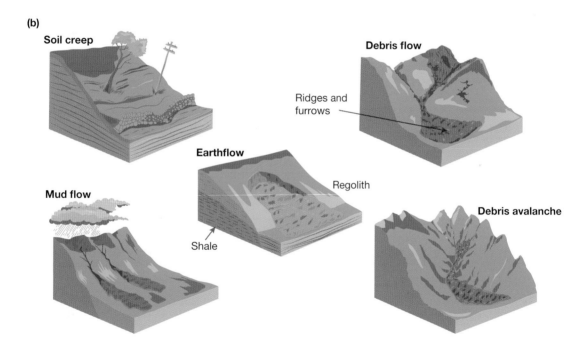

1.9.3 Responding to geomorphic hazards

As with all other hazards, the best response to geomorphic hazards is understanding how to reduce harm and preparedness for and after the event. Most people have limited choices about where they live, and humans have little or no control over natural events, so strong mitigation, prevention and adaptation strategies are required to minimise risk.

While the type, scale and scope of geomorphic hazards vary according to the location and proximity of people to the threat, they all have a common thread – they are potentially dangerous for everything in their path. The most significant variable in humans' ability to survive such a hazard is a country's level of development and the amount of money available for relief efforts when an event occurs.

Activity 1.9a: Landslides

Using the information you have learned about responding to geomorphic hazards, answer the following questions.

FIGURE 1.57 Information and advice on landslides for the Philippines

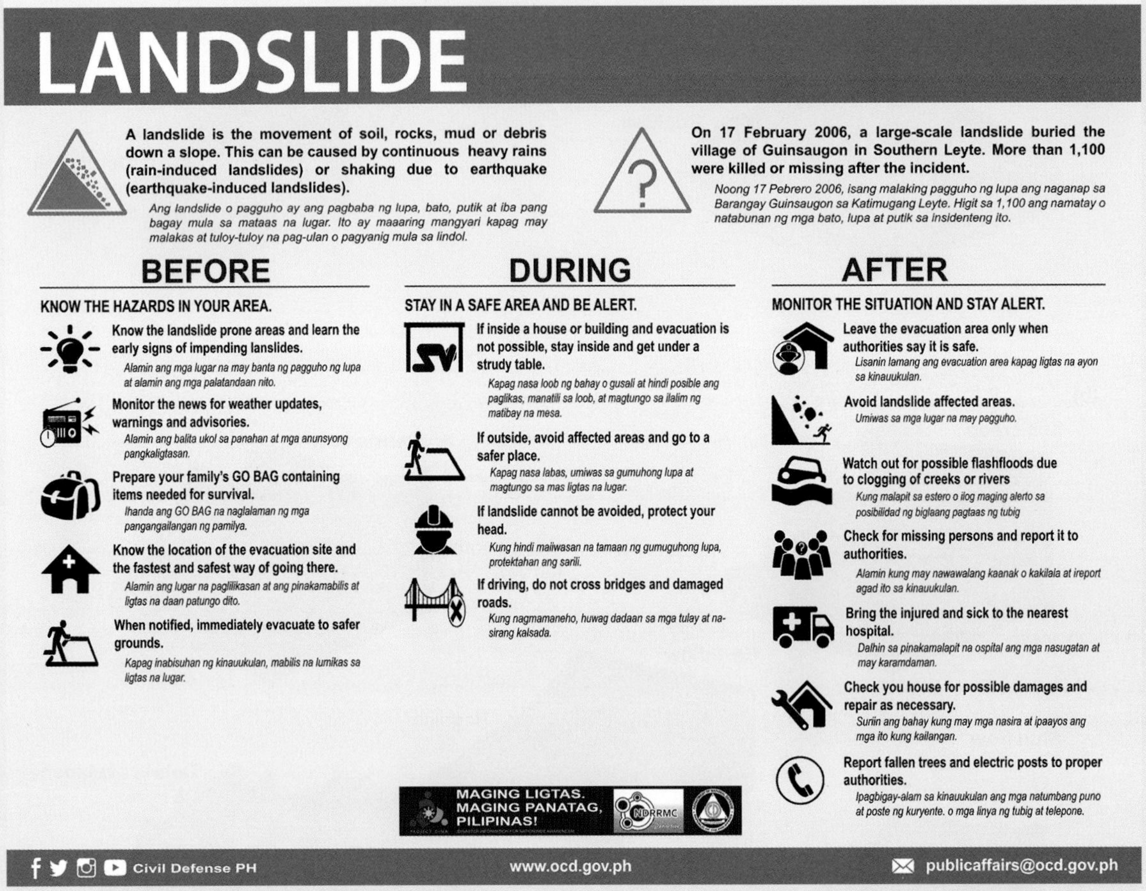

LANDSLIDE

A landslide is the movement of soil, rocks, mud or debris down a slope. This can be caused by continuous heavy rains (rain-induced landslides) or shaking due to earthquake (earthquake-induced landslides).

Ang landslide o pagguho ay ang pagbaba ng lupa, bato, putik at iba pang bagay mula sa mataas na lugar. Ito ay maaaring mangyari kapag may malakas at tuloy-tuloy na pag-ulan o pagyanig mula sa lindol.

On 17 February 2006, a large-scale landslide buried the village of Guinsaugon in Southern Leyte. More than 1,100 were killed or missing after the incident.

Noong 17 Pebrero 2006, isang malaking pagguho ng lupa ang naganap sa Barangay Guinsaugon sa Katimugang Leyte. Higit sa 1,100 ang namatay o natabunan ng mga bato, lupa at putik sa insidenteng ito.

BEFORE
KNOW THE HAZARDS IN YOUR AREA.

- Know the landslide prone areas and learn the early signs of impending lanslides.
  Alamin ang mga lugar na may banta ng pagguho ng lupa at alamin ang mga palatandaan nito.
- Monitor the news for weather updates, warnings and advisories.
  Alamin ang balita ukol sa panahan at mga anunsyong pangkaligtasan.
- Prepare your family's GO BAG containing items needed for survival.
  Ihanda ang GO BAG na naglalaman ng mga pangangailangan ng pamilya.
- Know the location of the evacuation site and the fastest and safest way of going there.
  Alamin ang lugar na paglilikasan at ang pinakamabilis at ligtas na daan patungo dito.
- When notified, immediately evacuate to safer grounds.
  Kapag inabisuhan ng kinauukulan, mabilis na lumikas sa ligtas na lugar.

DURING
STAY IN A SAFE AREA AND BE ALERT.

- If inside a house or building and evacuation is not possible, stay inside and get under a strudy table.
  Kapag nasa loob ng bahay o gusali at hindi posible ang paglikas, manatili sa loob, at magtungo sa ilalim ng matibay na mesa.
- If outside, avoid affected areas and go to a safer place.
  Kapag nasa labas, umiwas sa gumuhong lupa at magtungo sa mas ligtas na lugar.
- If landslide cannot be avoided, protect your head.
  Kung hindi maiiwasan na taaman ng gumuguhong lupa, protektahan ang sarili.
- If driving, do not cross bridges and damaged roads.
  Kung nagmamaneho, huwag dadaan sa mga tulay at na-sirang kalsada.

AFTER
MONITOR THE SITUATION AND STAY ALERT.

- Leave the evacuation area only when authorities say it is safe.
  Lisanin lamang ang evacuation area kapag ligtas na ayon sa kinauukulan.
- Avoid landslide affected areas.
  Umiwas sa mga lugar na may pagguho.
- Watch out for possible flashfloods due to clogging of creeks or rivers
  Kung malapit sa estero o ilog maging alerto sa posibilidad ng biglaang pagtaas ng tubig
- Check for missing persons and report it to authorities.
  Alamin kung may nawawalang kaanak o kakilala at ireport agad ito sa kinauukulan.
- Bring the injured and sick to the nearest hospital.
  Dalhin sa pinakamalapit na ospital ang mga nasugatan at may karamdaman.
- Check you house for possible damages and repair as necessary.
  Suriin ang bahay kung may mga nasira at ipaayos ang mga ito kung kailangan.
- Report fallen trees and electric posts to proper authorities.
  Ipagbigay-alam sa kinauukulan ang mga natumbang puno at poste ng kuryente. o mga linya ng tubig at telepono.

MAGING LIGTAS. MAGING PANATAG, PILIPINAS!

Civil Defense PH · www.ocd.gov.ph · publicaffairs@ocd.gov.ph

Source: NDRRMC Office of Civil Defense, Philippines

Comprehend and explain landslide response strategies

1. Read through the Landslide Information Sheet and make a summary of key points using a flow diagram.
2. Put the points in order of importance based on safety in each section.
3. What adjustments might be made to the plan if:
   (a) the landslide occurred to night instead of during the day?
   (b) visitors arrived the day before?
4. Consider figure 1.58, which shows a series of events that could contribute towards a landslide in a hot, wet region, such as South-East Asia.
   (a) What does the phrase 'potential for disaster' mean?
   (b) Identify a place in the flow chart where the first early warning signs might appear.
   (c) What interventions could be made to reduce the risk of a landslide?
   (d) Which of the physical variables might be magnified due to climate change?
   (e) Describe a situation where the two 'Social/Economic' changes might be interchanged with each other.

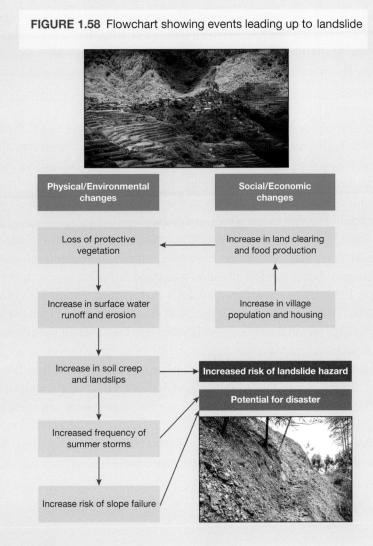

FIGURE 1.58 Flowchart showing events leading up to landslide

Assess and suggest risk management strategies

5. Which do you think has the greatest impact on the potential for landslides, social/economic changes or physical/environmental changes? Give reasons to support your answer.
6. What actions could you propose to mitigate the risks of landslide hazards in poor, mountainous regions like rural Nepal? Explain each mitigation strategy and justify your decision with clear reasons.

1.9.4 Landslides in Nepal

Nepal is a small country of around 141 000 km². It is roughly rectangular-shaped and lies north of India. The topography is dominated by the tallest mountains in the world, the Himalayas. These fold mountains formed because of a collision between two tectonic plates: the Indo-Australian to the south and the Eurasian to the north. They continue to grow today at a rate of two to ten millimetres a year. Because Nepal lies in such an active tectonic zone, where mountain building processes continue, the country is highly susceptible to hazards such as earthquakes, landslides, avalanches and glacial lake outburst floods. Avalanches and floods can be caused

FIGURE 1.59 Nepal lies to the north of India in the Himalayas

when unstable glacial lakes fracture and collapse, sending a torrent of water down into valleys. These floods have become more frequent due to increased melting of glaciers caused by climate change.

Nepal is a very mountainous country wedged between India and China. It comprises three regions: the Terai Lowlands (17 per cent), the Mid Hills/Mountains (68 per cent) and the High Himalayas (15 per cent). The population of Nepal was approximately 30 million in 2018 and was increasing at a rate of about 1.09 per cent per year. Nepal ranks among the poorest and least developed countries in the world with a GDP per capita of approximately US$835 reported by the World Bank in 2017.

Farming provides a livelihood for 70 per cent of the Nepalese people, but population growth is causing farmers to seek more land on higher and steeper slopes to terrace for farming. Forests are also being stripped to feed livestock, and for fuel for cooking and warmth. This land clearing, the naturally steep slopes and Nepal's heavy monsoonal rains make it vulnerable to landslides and flooding. Occasionally there are earthquakes, which also trigger landslides. It is estimated that every five years, between 10 and 25 per cent of Nepal's mountain roads are completely lost due to landslides or floods.

Low levels of education and generally poor construction standards make many Nepalese people more vulnerable to mass wasting hazards. Most homes are mud and timber and are built without the benefit of the modern engineering techniques that can help a structure withstand hazards.

In August 2017, Nepal received heavy monsoonal rains that created flooding and landslide hazards in half of the country's districts, most significantly in the Terai Lowlands region. The disaster affected around 1.7 million people. Despite widespread damage, only 70 lives were lost; a result largely attributed to the quick response of government search and rescue teams.

The cost of recovery from the disaster was estimated to be around US$705 million, but specific challenges hampered reconstruction efforts. According to the United Nations Office for Coordination of Humanitarian Affairs, by May 2018, most of the eligible families whose homes were completely destroyed (about 41 000 homes) or damaged (about 150 000 homes) were yet to receive their allocated rebuilding grant from the government. Food shortages were being experienced in the hardest hit areas of the Terai Lowlands and some education, health and sanitation facilities were still inoperable.

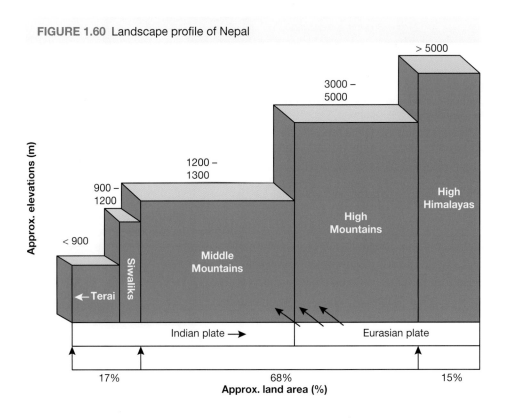

FIGURE 1.60 Landscape profile of Nepal

TABLE 1.6 Recent natural hazards in Nepal

Year	Landslides	Floods	Earthquakes
June 2013		39 killed 18 missing More than 1000 homeless	
August 2014		53 killed More than 29 000 displaced	
April 2015			9000 killed 22 000 injured
July 2016	73 killed	64 killed	
August 2017	70 killed		

Resources

Weblink: Nepal floods and landslides

Digital doc: Landslides in Nepal (doc-29159)

Reducing the risk of hazards in Nepal

Given Nepal's location, it is not possible to eliminate the risk of earthquakes, landslides, floods and other natural hazards. However, it may be possible to influence or change some people's actions, which have increased the risk of these hazards. Some examples include:
- better road construction would reduce exposure to landslides
- not using heavy machinery and blasting to cut through slopes
- using netting and grasses to stabilise slopes
- containing material from hillside cuts behind stone walls.

At present, the Annapurna Conservation Area Project is designed to seek broader goals in land rehabilitation and tree planting, but could assist with the reduction in landslides too.

Activity 1.9b: Landslides in Nepal

Analyse the information and apply your understanding

1. Streams flowing from the Middle/High Mountains and High Himalayas into the Terai are fed in part by glacial melt as well as rainfall and runoff. If warmer air temperatures and more intense monsoon rains are now occurring, how might this combination of factors affect denuded slopes and lowland rivers of the Terai?
2. What specific challenges might Nepalese authorities and non-government organisations working in the region face in mitigating the risks of landslides? Consider physical, economic and social factors.

1.10 Review

After reviewing the causes and effects of several natural disasters, it is apparent that many of the worst events occur in developing countries. Data supports the view that people living in countries with the least wealth are less able to prepare for or respond to hazards.

Some economists argue that natural disasters trigger economic growth and improve economic development. Evidence suggests that in a developed country, natural disasters may stimulate production and increase employment to meet construction activity and produce resources needed for rebuilding infrastructure. Retail sales also increase to replace goods lost in the misfortune. However, these arguments may not include lost opportunity costs. For example, could this money have been used for further improvement of infrastructure rather than rebuilding?

However, evidence shows that natural disasters in developing countries weaken already struggling economies. If people or property are lost, the natural disaster becomes a human disaster. As people are the driving force of production and growth, a human disaster is an economic disaster. No matter whether a country is rich or poor, a natural disaster does not deliver favourable economic outcomes for many.

Analysing spatial and statistical data

Data reveals that about 1.35 million people have died from natural hazards over the past 20 years. During the 7056 disasters recorded in that period, more than 50 per cent of fatalities occurred due to earthquakes (and tsunamis). By far, the greatest number of deaths were from low and middle-income nations (see table 1.7) According to the United Nations, the least developed countries had the most fatalities in terms of numbers killed per disaster and per 100 000 of population. However, several trends have emerged from figures gathered over the past two decades.

- The frequency of geological events such as earthquakes, tsunamis and volcanic eruptions remained constant and similar to previous patterns.
- There was a significant increase in climate/weather-related events such as storms, cyclones, blizzards and heatwaves. The number of these hazards was more than twice that of the previous twenty years.
- There was more extreme weather, and new temperature and rainfall records set, in many parts of the world.
- The number of fatalities doubled against the previous twenty years. It is uncertain whether this is due to more people living in disaster-prone areas or a spike from mega-disasters such as the tsunami in the Indian Ocean (2004) or the Haiti earthquake (2010).

Activity 1.10: Analysing data and constructing a scattergraph

A scattergraph is a diagram used to plot and compare two sets of data and investigate if there is a genuine link between them. The idea is to plot two sets of data – one using the x-axis and the other using the y-axis, and then determine if the distribution of points shows a pattern. You can draw in a 'line of best fit' in the most preferred position of the trend. The link between the two sets of data is referred to as a correlation. It may be positive or negative depending upon the slope of the pattern. Important things about scattergraphs are:

- if data on the y-axis increases as data on the x-axis increases, the correlation is positive
- if data on the y-axis decreases as data on the x-axis increases, the correlation is negative
- if points are scattered randomly with no clear trend, there is no correlation
- if a point is plotted and well away from the line of best fit, it is called an anomaly.

TABLE 1.7 Deaths caused by major natural disasters (Australia GDP per capita US$49 600 for comparison purposes)

Country	Deaths 1996–2015	Major natural disasters	GDP per capita ($US)
Haiti	229 699	earthquake, hurricane	740
Indonesia	182 136	tsunami, tropical storms	3570
Myanmar	139 515	tropical storms	1275
China	123 937	earthquake, floods	8123
India	97 691	earthquake, tsunami, floods	1709
Pakistan	85 400	earthquake, floods	1468
Russian Federation	58 545	extreme weather, earthquake	8748
Sri Lanka	36 433	floods, landslides	3835
Iran	32 181	earthquake	4957
Venezuela	30 319	floods	14 300

Source: EM-DAT: The Emergency Events Database — Université catholique de Louvain UCL — CRED, — www.emdat.be, Brussels, Belgium

Create a scattergraph

1. Use ordinal numbers to rank the ten countries in table 1.7 by
    (a) number of fatalities between 1996 and 2015
    (b) wealth of country as determined by GDP per person.
2. Rank the hazards in order of highest fatality first, followed by the others. You may decide that some terms are synonymous or go together.
3. What patterns or trends can you identify just by looking at the data?
4. What hazard stands out as being most lethal?
5. What atmospheric hazard is the biggest killer?
6. Is there an irregularity in the list of hazards you may not have expected? Explain.
7. Construct a scattergraph by plotting GDP per capita on the x-axis and fatalities on the y-axis.
8. Draw in a line of best fit and determine the correlation (positive or negative). Are there any anomalies? Suggest reasons to explain the correlation or anomalies.

Resources

Video eLesson: SkillBuilders: Constructing and interpreting a scattergraph (eles-1756)

Interactivity: SkillBuilders: Constructing and interpreting a scattergraph (int-3374)

UNIT 2
THE CHALLENGES OF CREATING SUSTAINABLE PLACES

A wide range of factors shape the identities of places over time: physical factors, such as the availability of clean water or resources for building; economic factors, such as trade and the value of local resources; and social factors, such as the accessibility of education or size of the settlement.

In this unit, you will study a range of different places, examining the factors that shape their identity and influence their sustainability. This includes looking closely at rural, remote and urban places in Australia – including your own local area through fieldwork – and studying the challenges and benefits of urbanisation, specifically in relation to megacities.

CHAPTER 3 Challenges for Australian places (Unit 2, Topic 1) 139

CHAPTER 4 Challenges for megacities (Unit 2, Topic 2) 215

CHAPTER 2
Ecological hazards

2.1 Overview

2.1.1 Introduction

The World Health Organization defines an **ecological hazard** as an interaction between living organisms or between living organisms and their environment that could have a negative effect. Ecological hazards are substances or activities that place people, habitats or an environment at risk of illness, injury, damage or disruption. This includes negative social and economic impact, as well as physical harm.

In this topic you will examine different types of ecological hazards and the zones in which they are most likely to occur. This includes examining the different factors that affect the onset and severity of a specific ecological hazard, and the potential ways of managing the different challenges they present.

You will also explore both biological and anthropogenic (human-generated) causes for hazards, and consider why some hazards pose more of a threat than others. This involves analysing vulnerability data and explaining strategies for hazard preparedness, mitigation and response.

As part of this unit of work, you will examine case studies of ecological hazards, make evidence-based observations about the impacts of specific hazards, and suggest effective strategies to reduce risk.

FIGURE 2.1 Oil spill clean-up operations in Thailand, 2013

2.1.2 Key questions

- What is an ecological hazard?
- Where do ecological hazards occur and why?
- What biological and anthropogenic factors influence a community's vulnerability to specific ecological hazards?
- What factors affect the severity of impacts of an ecological hazard?
- What actions can be taken to reduce risk?
- What factors affect a community's response to an ecological hazard?
- How are people in developed and developing communities affected differently by ecological hazards?
- Why might different communities seek different solutions to the challenges of ecological hazards?
- How does climate change affect the impact of some ecological hazards?

2.2 Ecological hazards

2.2.1 Types of ecological hazards

Ecology is a field of study that deals with the relationships of organisms to one another and to their non-living surroundings. Usually, different species and their communities are studied in their natural environments and habitats, rather than in laboratories.

In our everyday lives, there will be times when people are exposed to ecological hazards, either intentionally or by accident. For example, you may have a conversation with a sick person who coughs and sneezes frequently, or a veterinary surgeon may be attending to a sick fruit bat that has lyssavirus. These circumstances have an element of risk of the spread of pathogens, organisms like bacteria or viruses that can cause disease. Because these are living things that can be risky for people to have contact with, they are considered to be **hazards**.

This means that ecological hazards are sources of danger or harm to people and environments. Ecological hazards are potentially dangerous phenomena, substances, human activities or conditions that may cause loss of life, injury or other health impacts, property damage, loss of livelihoods and services, social and economic disruption, or environmental damage.

Ecological hazards are the result of **biological** and **anthropogenic processes**. Biological processes are processes that are vital for organisms to live, for example, the process of photosynthesis. Anthropogenic processes involve human activity, for example, the burning of fossil fuels to produce electricity. Ecological hazards involve human interactions with living organisms and the environment, or interactions between living organisms that have the potential to adversely affect the social and economic wellbeing of people. Ecological hazards might also include substances or activities that pose a threat to a habitat or an environment, such as acid rain caused by air pollution, toxic chemicals released by industry into waterways or high levels of radiation caused by nuclear accidents.

FIGURE 2.2 Biohazards are an ecological hazard.

FIGURE 2.3 A journalist checks radiation levels in the Fukushima region in Japan, where a tsunami led to a nuclear power plant meltdown in 2011.

The specific areas, spaces or places at risk of experiencing an ecological hazard are known as ecological hazard zones. At a local scale, it is possible that even schools might be at risk of becoming an ecological hazard zone. A school may be severely affected by influenza outbreaks during the 'flu season', and be forced to close when the risk of the spread of disease becomes too great.

Diseases, both **infectious diseases** such as influenza and **vector-borne diseases** such as malaria (see section 2.8), are one type of ecological hazard. In this case, the hazard is a result of human exposure to living organisms, such as bacteria, viruses and parasites, and has an impact on human health and wellbeing. While infectious diseases are an outcome of biological processes, human activities may contribute to their outbreak, spread and severity.

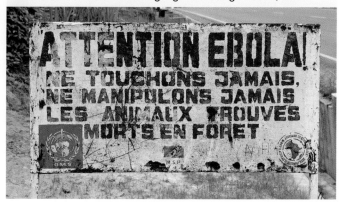

FIGURE 2.4 Ebola warning signs in Congo Africa, 2017

TABLE 2.1 Examples of hazardous diseases

Infectious diseases	Vector-borne diseases
influenza	malaria
measles	dengue fever
cholera	Ebola
hepatitis	yellow fever
HIV/AIDS	Lyme disease

TABLE 2.2 Examples of invasive plants and animals in Australia

Invasive plants	Invasive animals
lantana	cane toad
groundsel	fox
rubber vine	rabbit
prickly pear	carp
Paterson's curse	feral pig

Other ecological hazards may have an impact on the physical environment as well as on people. Such hazards include environmental **plant and animal invasions**, (see section 2.3), and **pollutants** of the lithosphere, atmosphere and hydrosphere (see section 2.4). Often, these ecological hazards are a result of people's activities, that is, of anthropogenic processes. People are responsible for many cases of plant and animal invasions, where the invasive species has been deliberately introduced, as were rabbits and the cane toad in Australia.

FIGURE 2.5 Invasive rubber vine

Pollution of the biophysical environment is frequently a result of human activities. For example, motor vehicles are a common source of air pollution in large cities, rising carbon dioxide levels in the atmosphere are contributing to climate change, and there is an increasing concern with the amount of plastic waste washing into our waterways and into the world's oceans. As the world's human population has grown and spread, and as economies have expanded, so too has the pollution associated with people's activities.

FIGURE 2.6 Pollution near the Panama Canal in South America

TABLE 2.3 Examples of environmental pollutants

Lithosphere	Atmosphere	Hydrosphere	Biosphere
salt	carbon dioxide	salt	herbicides
radiation spills	sulfur dioxide	plastics	pesticides
oil spills	nitrogen oxides	heavy metals	plastics
farm chemicals	particulates	fertilisers	oil spills

Activity 2.2a: Ecological hazards

How much do you know about ecological hazards? Reflect on your own experience and knowledge to answer these questions.

1. What ecological hazards commonly occur in Australia? Are they generally triggered by human or natural causes? Consider hazards in the physical environment (such a pollution) and hazards that are health-related (such as diseases).
2. Which ecological hazards present a risk where you live? Are there patterns that you can identify for when and where these hazards occur? (For example, is there typical flu season? Is there a time of year when a specific weed is more likely to affect crops?)
3. What kinds of ecological hazards present less of a risk in Australia, but are a significant risk in other parts of the world? Why might Australia be at less risk from these types of hazards?
4. What ecological hazards might affect people living in developing communities that might not be as significant in developed communities?
5. What factors might affect the way that different communities manage the following ecological hazards:
   (a) the introduction of a plant disease that damages important crops
   (b) the accumulation of plastics on beaches
   (c) the spread of a dangerous new strain of influenza (the flu).

2.2.2 Risk management

Risk is the chance that any hazard will cause harm to people or the environment. It is the probability of harmful consequences or expected losses (deaths, injuries, damage to property, damage to the environment, etc.), from ecological (and natural) hazards. Risk is closely associated with the **vulnerability** of particular people and places to hazards. Chapter 1 explained how the risk posed by a hazard might be mitigated by reducing the level of exposure and/or vulnerability.

Some places and people are more vulnerable to ecological hazards than others. For example, cholera, a water-borne infectious disease, rarely affects people in wealthier countries where treated water is the norm. People in poorer countries, where sanitation and water treatment may be inadequate, are much more vulnerable to outbreaks of cholera. Managing the impacts of ecological hazards involves actions that aim to prevent or reduce the damage or harm caused by the hazard.

These actions usually include:
- identifying the hazard
- if possible, putting in place measures to prevent the hazard
- where prevention is not possible, preparing for the hazard, including the use of hazard mitigation to control the risk.

Risk management of ecological hazards will depend on the nature of the hazard. Risk management for an invasive plant will be different from managing the risk of a disease outbreak. It may be possible to detect early outbreaks of potentially invasive weeds and to eradicate them before control mechanisms are needed. An outbreak of influenza may be impossible to prevent, although the use of vaccinations may mitigate the risk.

2.2.3 Factors affecting the severity of impact

Just as with natural hazards, the severity of an ecological hazard's impact will depend on a range of different factors.

Speed of onset

Hazards, both natural and ecological, can occur rapidly or develop over a longer period – that is, they may be rapid onset or slow onset hazards. The collapse of the Samarco iron ore mine tailings dam in Brazil in November 2015 is an example of a rapid onset hazard. Heavy-metal laden mud from the dam inundated the surrounding area, and contaminated community water supplies and the Doce River basin.

There is often a close link between the speed of onset of a hazard and its predictability. Rapid onset hazards may be difficult to predict. Communities are often more vulnerable to the impacts of hazards when there is little or no time to prepare. Because slow onset hazards tend to be much more predictable, there is generally more time to prepare for and manage the impact of the hazard, for example by evacuating people at risk.

Magnitude

Magnitude is a measure of the strength or extent of a hazard. A similar type of ecologically hazardous event may be more or less severe, depending on its magnitude. An oil spill from a recreational boat, for example, is of much smaller magnitude and therefore less severe than an oil spill from a large tanker.

Frequency

Frequency refers to how often a hazardous event occurs; it is the return interval of hazards of certain sizes. In most cases, smaller hazardous events tend to be more frequent than large-scale events and are usually less severe. For example, small outbreaks of influenza tend to occur frequently. However, an outbreak of a particularly virulent type of the flu is much rarer and much more severe in its impact. For example, the 1918–19 influenza pandemic killed more people than WWI because no one knew how to treat it.

Duration

Duration is the length of time that a hazard lasts. Usually, the longer the hazard lasts the greater the impact is likely to be, although this is not always the case. Some infectious diseases may have their most serious effects early in an outbreak when the most vulnerable people are at greatest risk and strategies to prevent spread have not been implemented yet. This was the case during an outbreak of cholera following the Haiti earthquake in 2010.

Temporal spacing

Temporal spacing refers to the sequencing and seasonality of events, i.e. whether the event is random or regular. Many diseases, such as influenza, tend to be seasonal, while others, such as measles, are much more irregular.

Mobility and location

The mobility and areal extent (the area affected) of a hazard can also have an impact on its severity. For example, a disease outbreak affecting just a single, isolated village is much less severe than an outbreak of disease in a large city or over several countries. Some invasive animal species are very mobile and adaptable, and can cover large distances in short periods of time, such as rabbits. Others may have limited ability to move to different locations because they require very specific conditions to survive.

Climate change

Climate change may affect the severity and incidence of some ecological hazards and increase risk. For example, climate change may increase the range of habitats available to disease-carrying insects such as mosquitoes, and therefore increase the areal extent and the risk of outbreaks of mosquito-borne diseases.

Climate change might also increase the range of invasive plants and animals and their associated risks. It might also be possible for the incidence of some types of pollutants to increase with a changing climate.

Activity 2.2b: Assessing the impact of hazards

Considering the different factors that affect the impact of an ecological hazard, answer the following questions.

Explain the features of hazards

1. Explain why influenza would be considered an ecological hazard. List three human processes or actions that might help an outbreak of influenza to spread.
2. Choose one hazard to research from each of tables 2.1, 2.2 and 2.3. Create a table to compare the ecological hazards based on their typical speed of onset, magnitude, frequency, duration, temporal spacing, and mobility and areal extent.
3. Vulnerability to an ecological hazard is affected by a range of physical, social, economic and environmental factors and processes. Choose three ecological hazards that might affect your community (such as measles, plastic pollution and feral pigs). Complete a table like the one below to demonstrate how people might be vulnerable to the risk of each hazardous event, based these factors.

Factor or process	Invasive species	Pollutant	Disease
Physical			
Social			
Economic			
Environmental			

Water extraction from the Murray–Darling Basin

The Murray–Darling Basin is Australia's largest river catchment. The basin extends into four states, as well as the ACT, and covers an area of more than one million square kilometres, equal to about 14 per cent of Australia's total land area.

The Basin contains some of Australia's most productive land. Ever since parts of the catchment were settled by graziers in the 1800s, it has consistently been a significant agricultural region. However, the system is also vulnerable to the risk of unsustainable or unregulated water extraction. Communities closer to the mouth of the Murray River in South Australia are particularly vulnerable to the impacts of this hazard.

Hydrologists have calculated that the annual inflow of water into the Murray–Darling is about 24 gigalitres (GL). Of this, an estimated 13 GL reaches the ocean, 11.6 GL is extracted for irrigation and water supplies for communities, 1.4 GL is lost through evaporation and 6.9 GL is lost through seepage and leakage. Most rain that feeds into the Murray comes in winter and spring, whereas the Darling receives most of its inflow from summer rains in the north. There is a seasonal imbalance between water inflow and water outflow. This is offset by the construction of several dams, weirs, barrages and diversion ponds from which water is taken from the river.

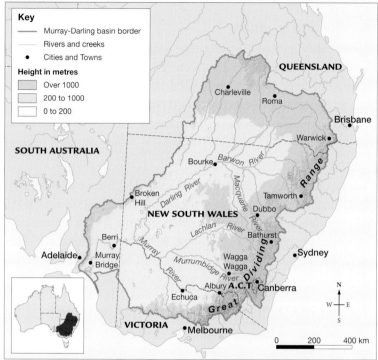

FIGURE 2.7 The Murray–Darling Basin is Australia's largest river catchment.

Source: © Commonwealth of Australia Geoscience Australia 2018. Redrawn by Spatial Vision.

Factors affecting the severity of impact

Water extraction is not just a hazard because it lowers the levels of water available for use in the Basin system. In conjunction with the increase in chemical-laden farming runoff, the potential for damaging algal growth in the system also increases with lower water levels. Blooms of blue-green algae, or cyanobacteria, not only produce toxins in the water, but also cause the river to become discoloured and smelly due to the unsightly scums that form on the surface. The algae can be harmful to humans and animals if it is ingested and can irritate the skin. This hazard poses specific challenges for communities who obtain their drinking water from rivers prone to algae blooms.

FIGURE 2.8 Irrigated canola crops near Lake Hume on the Murray River in rural NSW

Algae also consumes large amounts of oxygen from the water, which can damage already fragile environments and be harmful to river-dwelling animals. In 1991, the media attention received by a 1000-kilometre long algal bloom in the Darling River caused the New South Wales government to declare a state of emergency.

Reducing the risk from water extraction

The Murray–Darling Basin Plan (2014–2026) was passed by the federal government in 2012. This plan sets out how much water can be taken from the river system based on a long-term average of what is sustainable for key environmental sites and the river system. It also outlines strategies to ensure extractions levels are managed to protect water quality, reduce excessive water use, provide access to safe drinking water, and monitor water allocations and planning processes.

FIGURE 2.9 Weirs, such as the Mildura Weir, provide a stable pool for water diversion into nearby farming areas for extraction.

The key strategy of the plan is to mitigate the risk of extraction by limiting the amount of water that can be taken or diverted from the system. It outlines caps on water extraction levels for industry, agriculture and community needs. Diversion and extraction levels are allocated based on estimates of what is sustainable for the system in the long term: 10 873 gigalitres per year (GL/y). The plan also includes limits for groundwater extraction, offers a buy-back system for the voluntary reduction of water allocations, and other strategies to mitigate the range of hazards the Basin faces.

The implementation of the plan, however, has not been entirely smooth. In 2017, claims of significant illegal water extraction, especially in the upper reaches of the Basin, brought the Plan and its administration to the attention of the media.

The impact of climate change

The Basin Plan also considers the likely impact of climate change. While periods of drought and variable conditions do affect the flow of water, the plan allows for the conditions of water use from the Basin to be altered to ensure that water is available for 'critical human needs' such as drinking water. It also has a legally mandated 10-year cycle of review to ensure that current practices can manage the Basin in the event of greater climate change-induced hazards, and also places the responsibility on the states to put plans in place for managing extreme dry periods and other potential challenges.

While the Basin itself is very large and covers many climates and land uses, it is estimated that climate change will cause the Basin to be drier overall, but will increase the likelihood of extremes in flooding and drought conditions in certain areas. Current models suggest that a global temperature rise of only 1 per cent would reduce rainfall across the Basin by up to 9 per cent. This will result in an estimated reduction in the amount of surface water available in the system of about 10 per cent by the year 2030.

These reduced levels will also need to be taken into consideration in the management of extraction levels. Businesses and communities relying on water from the river will be more vulnerable to changes in the catchment system. These longer-term changes may reduce the viability of farming and other industries along the river. The types of farming activities and industry in areas relying on the Basin for water may change or existing land use may no longer be viable, and people may be forced to relocate.

Activity 2.2c: Water extraction from the Murray–Darling Basin

Considering the different factors that affect the impact of water extraction on the Murray–Darling Basin, answer the following questions.

Comprehend and explain the impacts of water extraction

1. Create a chart or mind map to demonstrate the primary, secondary and tertiary impacts of water extraction from the Murray–Darling basin.
2. Which states are at risk from the hazards of excess water extraction from the Murray–Darling Basin?
3. What strategies have Australian governments put in place to prevent unsustainable water extraction?

Analyse the data and apply your knowledge

4. What impacts might climate change have on communities that rely on the health of the Murray–Darling Basin? Write one paragraph about each of the following groups, explaining why they rely on the Basin and how they might be impacted by climate change.
   (a) Residents of Adelaide
   (b) Cotton farmers in the Bourke region
   (c) Fruit growers in the Berri region
   (d) Eco-tourism operators in the north of Victoria.
5. What might the potential impacts be for Australia if the water in the Murray–Darling Basin dries up or the water quality becomes unsuitable for its current uses? Suggest what impacts this might have on Australia as a country socially, economically, politically and environmentally. What solutions might you propose to prevent this from happening?

Resources

- **Weblink:** The Murray–Darling Basin Plan
- **Weblink:** The science behind the Murray–Darling Basin Plan
- **Weblink:** Pumped: Who's benefitting from the billions spent on the Murray–Darling? (2017)
- **Weblink:** Climate change impacts
- **Weblink:** Travel the length of the Murray–Darling system
- **Video eLesson:** SkillBuilder: Reading and describing basic choropleth maps (eles-1706)
- **Interactivity:** SkillBuilder: Reading and describing basic choropleth maps (int-3286)

2.3 Plant and animal invasions

2.3.1 Types of plant and animal hazards

An **invasive plant** or **animal** is defined by the Australian Department of Environment and Energy as 'a species occurring, as a result of human activities, beyond its accepted normal distribution and which threatens valued environmental, agricultural or other social resources by the damage it causes.'

Invasive plants and animals have three characteristics.

- They are not **indigenous** (native) to the area in which they are found (they are exotic or alien species).
- They are a consequence of anthropogenic activities (they have been introduced, either accidentally or deliberately, through the actions of people).
- They pose a threat to the environment and to people and their activities, i.e. they are hazards.

Since British settlement, many exotic plants and animals have been introduced to Australia, either deliberately or accidentally. Some were brought for economic reasons, for example as crops or farm and work animals. Others, such as cats, were brought as pets. Many garden plants and animals were introduced by acclimatisation societies in the 1800s to make Australia more like Europe. Domesticated species, such as lantana and donkeys, also became wild. These are known as feral plants or animals.

FIGURE 2.10 Rabbits were introduced to Australia by the British.

Unfortunately, many of Australia's introduced species have caused much environmental and economic harm. Foxes, feral cats, rabbits and feral pigs are among the worst of Australia's invasive animal species, while around half of our introduced plants have invaded native vegetation and around a quarter are regarded as, or have the potential to become, serious environmental weeds. These include the rubber vine, lantana, groundsel and water hyacinth. Scientists have identified 32 weeds of national significance based on an assessment of their invasiveness, potential for spread and environmental, social and economic impact.

TABLE 2.4 Some of Australia's most noxious weeds

alligator weed	European blackberry	fireweed
gamba grass	mimosa	athel pine
prickly pear	African boxthorn	prickly acacia
bitou bush	Chilean needle grass	mesquite

Invasive plants and animals will have many or all the following characteristics.

- They grow and mature rapidly, producing large numbers of seeds or offspring.
- They are highly successful at spreading to and colonising new areas.
- They can thrive in different types of habitats.
- They can outcompete native species and have few or no natural enemies.
- They are very costly or difficult to remove or control.

FIGURE 2.11 Invasive water hyacinth

2.3.2 The impact of invasive plants and animals

Invasive plants and animals have environmental, economic and social impacts. Table 2.5 provides an example of the possible costs of a several of Australia's invasive animal species, as estimated from analysis by the Cooperative Research Centre for Pest Animal Control. Even though impact of these species in some criteria could not be quantified, the costs that were measured indicate the huge impact that these species do have on Australia.

TABLE 2.5 Annual impact of pest species (in order of total cost)*

	Total	Economic impact (costs of control measures such as bait or fencing)		Environmental impact (cost to preserve biodiversity or repair damage)		Social	
	$m	Impact	$m	Impact	$m	Impact	$m
Foxes	227.5	♦	37.5	♦	190.0	♦	nq
Feral cats	146.0	♦	2.0	♦	144.0	♦	nq
Rabbits	113.1	♦	113.1	♦	nq	♦	nq
Feral pigs	106.5	♦	106.5	♦	nq	♦	nq
Dogs	66.3	♦	66.3	♦	nq	♦	nq
Mice	35.6	♦	35.6	♦	nq	♦	nq
Carp	15.8	♦	4.0	♦	11.8	♦	nq
Feral goats	7.7	♦	7.7	♦	nq	♦	nq
Cane toads	0.5	♦	0.5	♦	nq	♦	nq
Wild horses	0.5	♦	0.5	♦	nq	♦	nq
Camels	0.2	♦	0.2	♦	nq	♦	nq
Total	719.7		373.9		345.8		

Note: * Key: nq = not quantified   ♦ = bigger impact   ♦ = smaller impact
Source: McLeod, R. 2004 Counting the Cost: Impact of Invasive Animals in Australia 2004. Cooperative Research Centre for Pest Animal Control. Canberra.

Environmental impacts

Perhaps the most significant impact of invasive species is on Australia's biodiversity. Australia is one of the most biodiverse places in the world, one of 17 'megadiverse' countries. However, this biodiversity is under threat. Around 1700 species of plants and animals are listed by the Australian government as at risk of extinction. Already, 30 native mammals have become extinct since European settlement.

While land clearing and fire are in part responsible for the extinction threat, the most significant causes involve invasive species: predation by feral cats and foxes; habitat destruction by feral herbivores such as pigs, goats and rabbits; and the spread of invasive weeds. Invasive species threaten natural ecosystems and biodiversity. Frequently, they crowd out and replace natural species because they can outcompete, prey on or poison native species or may carry diseases. Invasive species also alter habitats and cause land and water degradation, making it difficult for native species to survive.

Economic impacts

Invasive plants and animals have both direct and indirect economic costs. There are the direct costs of monitoring and controlling pest species, including the physical removal of pests, baiting, poisoning, fencing and research. Indirect costs occur when agricultural production is reduced through damage done to fences and crops, degradation of soils and water, herbivore competition with farm animals and reduction in fish stock in marine environments.

Social impacts

The social impact of invasive species may include the loss of amenity in national parks, impact on Indigenous peoples and their way of life and traditional practices, health impacts and reduction in recreational fishing. Ecological hazards such as mouse plagues can also have a considerable social impact. For farmers who have to manage the impact on their crops, mice can take a significant toll on their wellbeing by infesting and reducing their yield.

2.3.3 Invasive plant and animal hazard zones

As figure 2.12 shows, every continent, other than Antarctica, is currently threatened by plant and animal invasions. Australia and New Zealand are especially vulnerable to the risk of invasion. As table 2.6 shows, island countries are over-represented among countries with the highest numbers of invasive species – six out of the top ten countries are islands. A recent study of the world's alien species hotspots identified the Hawaiian Islands, the north island of New Zealand and the Lesser Sunda Islands of Indonesia as the places with the highest numbers of established alien species. In fact, about half of New Zealand's plant life consists of invasive species.

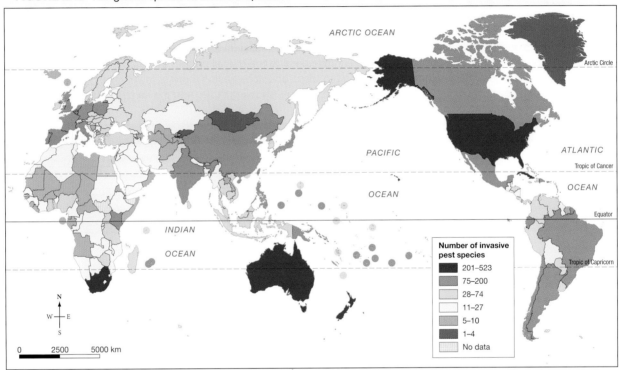

FIGURE 2.12 The global spread of invasive species

Source: 'Mapping the global state of invasive alien species: patterns of invasion and policy responses' by Anna J. Turbelin, Bruce D. Malamud and Robert A. Francis, Global Ecology and Biogeography, (Global Ecol. Biogeogr.) (2017) 26, 78–92, Fig 1.

In addition, all ten countries in table 2.6 have been European colonies. European settlers – in these cases the British, French, Spanish and Dutch – introduced alien invasive species to the countries they colonised, sometimes deliberately, such as the rabbit in Australia, and sometimes accidentally, such as the mouse.

TABLE 2.6 Countries with the highest number of invasive alien species

Country	Number of species	Density of species (species per 100 000 km²)
1. USA	523	5.7
2. New Zealand	329	124.9
3. Australia	322	4.2
4. Cuba	318	298.8
5. South Africa	208	17.1
6. French Polynesia	190	5191.3
7. New Caledonia	183	1001.1
8. Reunion	173	6889.7
9. Fiji	167	914.1
10. Canada	166	1.8

Source: Department of Environment, Land, Water and Planning

However, invasive plant and animal hazard zones are often not country wide. Like most species, invasive species are usually adapted to particular physical environments in their place of origin. They pose the greatest threat to similar environments in the invaded country. Many of Australia's invasive plant species are tropical, so mostly threaten the northern parts of Australia. Some invasive species are adapted to coastal environments, so predominantly threaten those environments. As the climate changes, the areas potentially susceptible to invasion also change.

Activity 2.3a: Invasive species hazard zones
Refer to figure 2.12 and table 2.6 to complete these questions.

Explain and analyse invasive species hazard zones
1. Describe the global pattern of alien invasive species numbers illustrated in figure 2.12. Refer to continents in your answer, and identify which continents have the most and least invasive species.
2. Identify the two countries that show the greatest difference to the overall pattern of their continent. Suggest features, trends or strategies that might have led to these countries having more or fewer invasive species.
3. Use the map key in figure 2.12 to complete column 1 of the following table. Add the names of three or four countries at each interval to column 2 of the table.

Number of alien invasive species (range)	Country examples

4. Create a new table, based on table 2.6, which re-ranks countries according to the density of alien species.
   (a) Describe how Australia's ranking changed compared with its overall numbers rank.
   (b) Make a list of factors that might influence the density of invasive species in these ten countries. Suggest which of these factors might apply specifically to Australia.
   (c) Digitally construct a multiple column or bar graph to illustrate the economic costs and total costs of the invasive species listed in table 2.6.
5. Identify the only invasive fish species in table 2.6. Suggest some of the possible environmental and economic risks that an invasive fish species might pose to Australia?
6. Suggest reasons why rabbits and feral pigs are responsible for the highest economic costs to Australia.
7. Suggest reasons for the high costs to Australia's natural environment of foxes and feral cats.
8. Which of the invasive species has the greatest social impact? Explain why this might this be the case.
9. Explain why the social impact of invasive species might be so difficult to quantify.

on Resources
Video eLesson: SkillBuilder: Creating and reading compound bar graphs (eles-1705)
Interactivity: SkillBuilder: Creating and reading compound bar graphs (int-3285)

2.3.4 Reducing the risk of plant and animal invasions
The best way to mitigate the risk of plant and animal invasions is prevention: keep the invaders out. In Australia, we now have strict biosecurity measures to reduce the risk of alien species being introduced. Anyone entering Australia from overseas is required to complete customs declaration forms. These declarations are part of our management of the risk of introduced species, and considerable penalties apply to people who illegally bring in banned products or do not declare them.

These rules are governed by the Biosecurity Act (2015), which requires people bringing goods into Australia, bringing vehicles (such as planes or boats) into Australia or Australian waters, or selling goods to people in Australia, to declare any prohibited or potentially prohibited goods, such as food, animal products or plant materials. It is also prohibited to take some kinds of food into some areas of Australia to prevent the spread of invasive species from one part of the country to another.

A second step in the management process involves the early control and eradication of any invasive species. It may be possible to completely eradicate invaders if they are detected early enough. If that proves impossible, early detection and control may be able to keep the numbers of the invaders at reasonably low levels. The ongoing attempt to keep invasive fire ants confined to small areas in south-east Queensland is an example of this, although success is proving difficult and expensive (see page 91). Once invasive species are established, there are three possible control methods: chemical, mechanical and biological.

FIGURE 2.13 Parts of Australia enforce quarantine regulations to prevent the spread of invasive species.

Resources

- **Weblink:** Australia's Biosecurity Act
- **Weblink:** Interstate quarantine in Australia

Chemical control

Chemical control involves the use of herbicides and pesticides to kill invaders. Chemicals can be an effective management technique, but there may also be negative consequences. In addition to killing the targeted invader, chemicals may also kill non-target species, including native species. They may also be expensive, and can lead to pollution of soil and water. Over time, some invasive species may also develop resistance to the chemicals used.

FIGURE 2.14 Chemical removal of invasive weeds

Mechanical or physical control

Mechanical control involves the use of machinery and people to remove the invaders. This is often effective in small areas, and can reduce local invasive populations. Physical control includes techniques such as trapping, shooting and fencing to control numbers of invasive animal species and weeding of invasive plant species.

FIGURE 2.15 Mechanical removal of aquatic weeds

Biological control

Biological control involves the use of an invasive species' natural enemy or a disease. Prickly pear cactus was brought under control in Australia with an introduced cactoblastis moth, whose caterpillar feeds on the cactus. Rabbits have been controlled through the use of myxomatosis and calici viruses. Other biological control methods involve genetic engineering and breeding intervention programs.

Figure 2.17 provides an overview of the costs associated with managing invasive species over time. The first step, prevention (stopping the species from invading), is the most cost-effective way of dealing with invasive plants and animals. The second step, eradication, is the complete removal or killing off of a species. Containment is the third step, and refers to measures taken to prevent the further spread of a pest species. The final step, asset-based protection, involves prioritising the control of a pest species based on the threat it poses to environmental assets in the area. As any invasion progresses, management techniques become more expensive and the benefits compared with the management costs decrease. It is important, then, that Australia continues with its program of prevention and early eradication of any new invasive species.

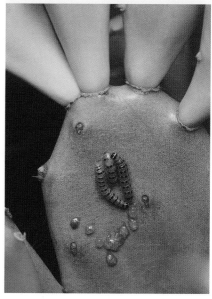

FIGURE 2.16 Caterpillar of the Argentinian moth *Cactoblastis cactorum* feeding on prickly pear

Source: Wayne Lawler/Science Photo Library

Activity 2.3b: Managing invasive species

Refer to figure 2.17 to answer the following questions.

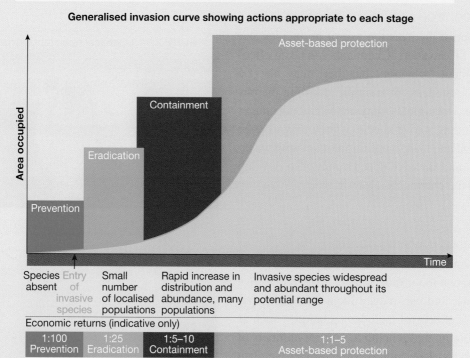

FIGURE 2.17 Cost curve for management of invasive species

Note: A ratio of 1 : x indicates that for every $1 spent, the economic return is $x. For example, a ratio of 1 : 100 means that for every $1 spent in prevention, the return to the economy is $100.

Source: Department of Environment, Land, Water and Planning

Investigate and suggest risk management strategies

1. Research what asset-based protective measures are currently being undertaken to manage the invasive species.
2. Choose one invasive plant and one invasive animal to suggest possible actions that could have been taken at the first three stages of the management curve for these species.

Lantana invasion

Lantana is one of Australia's most damaging invasive plant species, and among the top ten worst weeds worldwide. It was introduced to Australia in 1841 as an ornamental garden plant and quickly became an aggressive invader of natural ecosystems and of grazing and farming land. Today, lantana covers over 5 million hectares in Queensland, New South Wales, Northern Territory and Western Australia. Nearly 80 per cent of the area affected by lantana is located east of the Great Dividing Range in Queensland.

FIGURE 2.18 Lantana (*Lantana camara*) is one of Australia's most damaging invasive plant species.

Lantana invasions have a serious environmental impact. Lantana has toxic effects on other plants. Mature lantana also forms into dense thickets 2–4 metres in height, and is able to smother and replace native plant species. It is also capable of climbing trees and can smother their canopies. Lantana is therefore responsible for a reduction in Australia's biodiversity. At least 1400 native plant and animal species are at risk because of lantana invasions, many of which are already identified as threatened.

Lantana also has a significant economic and social impact. Invasions of grazing and farming land lead to the costs involved in controlling infestations, the loss of land for grazing and farming, stock poisoning and reduction in land values. Social costs include the reduction in recreational activities such as camping and bushwalking, and in the aesthetic appeal of natural landscapes.

Lantana was first declared a noxious weed as early as 1920. In order to prevent further introductions in Queensland, lantana has been classified as a restricted invasive plant species, so cannot be given away, sold or released into the environment without a permit. Strategies involved in mitigating the impact of lantana invasions include manual and mechanical removal of the plants, use of herbicides and fire, trampling and grazing by livestock and biological controls. In most cases, no single method of lantana control is successful by itself, so an integrated approach using two or more techniques is most effective in managing the invasion.

Activity 2.3c: Lantana invasions

Comprehend and explain distribution patterns

1. Go to the **Lantana distribution and control in Queensland** weblink in the Resources tab. Describe the spatial pattern of potential lantana distribution in Queensland. Why might some locations (in Queensland/Australia) be more susceptible to lantana invasions than others?

Apply your understanding

2. Explain why the use of one method for controlling the spread of lantana might not be effective.

3. Identify some of the possible positives and negatives of each of the methods used to control lantana invasions. Draw up a table like the one below to summarise.

Control method	Positives	Negatives
Manual removal of plants		
Mechanical removal of plants		
Use of herbicides		
Use of fire		
Trampling and grazing		
Biological controls		

Resources

Weblink: Lantana distribution and control in Queensland

Mouse plagues

Common house mice are not native to Australia; they arrived by ship during the early days of European settlement. Australia is one of the few places in the world where mice populations sometimes reach plague proportions. In addition to presenting a risk to human crop production, mice spread diseases such as salmonella, hantavirus, leptospirosis and even bubonic plague. These diseases can be spread to people or other animals directly through bites and by the insects that mice carry, such as fleas, lice, mites and ticks. Some diseases are also spread when people handle or eat food contaminated by mice faeces or urine. Businesses, especially in food production, are also vulnerable to damage from mice. An estimated 14 per cent of food stores around the world are destroyed by mice each year. Mice also chew through cables and wires, placing businesses at risk of fires and power outages.

Mouse plagues, however, are temporally and spatially constrained by environmental conditions, and eventually the risk passes when food sources or favourable breeding conditions reduce.

Factors affecting severity of impact

Populations of mice become a hazard because of a combination of biophysical and anthropogenic factors. Mice breed quickly, especially in a year following a strong cropping season when there is ample food left in paddocks after harvest. One female mouse can produce a litter of pups every 20 days, with up to 10 pups, on average, per litter in optimal breeding seasons. Human farming practices, food production and homes provide mice with access to the food and shelter that they need to thrive.

Mouse plagues have occurred in urban centres, but the hazard zones for mice plagues in Australia are predominantly grain-growing areas where excess production provides sufficient food for a population explosion. These areas include the Darling Downs in Queensland, the central west and Riverina areas of New South Wales, the Mallee in Victoria and south-eastern South Australia.

FIGURE 2.19 Mouse plagues erupt in the grain-growing regions of Australia, like the Darling Downs, causing massive disruption to communities and losses to farmers.

Source: Grant Singleton

Fluctuations in crop yields have a significant impact on the risks posed by mouse populations. In years following low crop yields, mouse populations can increase rapidly. In 2016, Australian wheat farms produced a record wheat harvest of just over 30 million tonnes. In 2017, crop production fell to about 20 million tonnes, one of the lowest in recent decades because of drought conditions in some grain-growing areas. In 2018, many farmers in the Victorian Mallee and the Adelaide plains region of South Australia experienced a mouse plague (see figure 2.20).

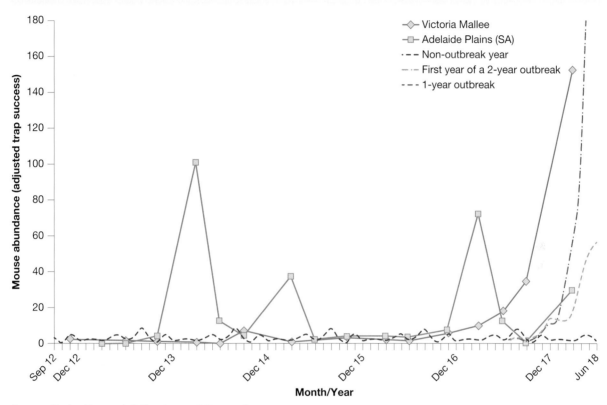

FIGURE 2.20 Mouse population in the Mallee (Vic.) and Adelaide Plains (SA), 2012–18

Source: Grains Research & Development Corporation

Outbreaks also occurred in 2017 and 2018 on the Eyre and Yorke peninsulas, near Ravensthorpe and Esperance in Western Australia, in parts of central New South Wales and on the southern Darling Downs in Queensland.

Mouse plagues can have a significant impact on the physical environment, especially in farming and rural communities, and can also affect human health and wellbeing. While the risk of mouse plagues can be mitigated with pest control measures and are generally short in duration, creating a relatively short-term primary impact, the magnitude of a plague can increase the impact of the secondary and tertiary effects. Financial losses from destroyed crops, the cost of pest control and the emotional impacts of these costs can be significant, even though the hazard itself may only last for a short period of time (see table 2.7).

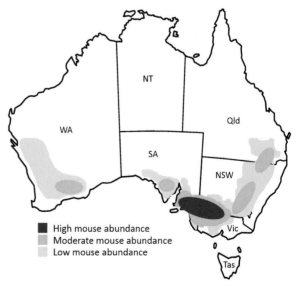

FIGURE 2.21 Mouse abundance in Australia, March 2018

Source: Dr Peter R. Brown and CSIRO

Hazard prevention and mitigation strategies

Mouse population reduction and breeding prevention strategies are best implemented at the first signs of larger than usual mouse populations. Strategies to reduce the risk of mouse populations becoming a hazard generally rely on early detection and prevention, such as inspecting paddocks regularly for mouse activity; minimising levels of grass, weeds and ratoon stubble after harvesting seasons; and managing population levels during sowing with baits and traps. These strategies prevent breeding by reducing available food sources, levels of protection from predators and the numbers of nesting sites available to mice. In addition to reducing food and ground cover, traps and baits are an important strategy during planting because they can prevent mice eating seeds before they can germinate.

TABLE 2.7 Economic, environmental and social impacts of mouse plagues on communities

Economic	Environmental	Social
• producers lose crops, production time and equipment • grain customers may have to find supply elsewhere at increased costs • producers need to purchase poisons and baits • additional funds needed for scientific research to try to mitigate problem	• soil erosion due to removal of vegetation and soil disturbance • increased mice numbers attract predatory birds (owls, eagles, falcons, kites, kookaburras, magpies) • poisons and residual toxic chemicals in soil affect insects, worms and micro-organisms, and can be ingested by other animals	• financial difficulty creates emotional stress • fighting the plague creates emotional stress and fatigue • mice reduce wellbeing and quality of life if they infest homes • mice spread diseases to humans and other animals

Resources

- **Weblink:** CSIRO 'Mouse tracker'
- **Weblink:** Infested
- **Weblink:** The bizarre mystery of Australian mouse plagues

Activity 2.3d: Mouse plagues

Using the information you have learned about mouse plagues, answer the following questions.

Explain the factors affecting plague patterns

1. Describe the biophysical and anthropogenic processes that contribute to mouse populations becoming a hazard.
2. What potential implications might a high yield cropping year have for mouse numbers?
3. Categorise the impacts of mouse plagues into primary, secondary and tertiary impacts.
4. Describe the spatial pattern of mouse abundance shown in figure 2.21.
5. What biophysical factors might have contributed to the peaks in mice populations shown in figure 2.20?

Suggest and justify ways to manage the risk

6. MouseAlert data from early 2018 showed increasing numbers of mice being reported in both the Mallee and Adelaide Plains regions. What rodent risk management steps could farmers implement to control the rising levels?
7. Propose and justify three strategies that farmers in the Victorian Mallee could implement in 2018 to reduce the risk of mouse plagues in 2019.

Fire ants in Australia

Fire ants (*Solenopsis invicta*) are native to South America and are considered one of the most invasive and damaging species of insect in the world. These insects are about 2–6 mm in length and form nests that look like small mounds of disturbed earth, often in open places such as lawns or paddocks.

In the USA, the red and black variety of fire ant has taken over about one third of the country and is one of the most dangerous species. Despite considerable efforts to control them in the USA, fire ants have invaded more than 1.2 million hectares of land, particularly in Florida, Texas and other southern states. For some years, they have been the dominant pest, not only destroying crops, but also killing livestock and wildlife.

Fire ants were first noticed in Australia at the Brisbane shipping port in 2001, but authorities are uncertain exactly how and when they arrived. It is thought they may have come in on a shipping container from the USA. Now, fire ants are the target in one of Australia's biggest biosecurity eradication programs in its history. In fact, the government has agreed to spend more than $400 million over ten years to ensure this potentially destructive insect is eliminated.

FIGURE 2.22 The fire ant (*Solenopsis invicta*)

Source: James H. Robinson/Science Photo Library

These small, copper-coloured ants have spread quickly to a small number of coastal locations in Queensland and New South Wales. Following their initial sighting in Brisbane, further incursions have been identified at Yarwun, near Gladstone, at Brisbane Airport in 2015, and again at the Port of Brisbane in 2016.

Prompt action and a widespread publicity campaign led to the eradication of colonies at the Port of Brisbane, Yarwun and Port Botany in New South Wales. However, the original infestation is thought to have spread to several areas in southern Queensland and possibly into northern New South Wales. Their nests have been found in suburbs of Brisbane, Ipswich, Redland, the Lockyer Valley, the Scenic Rim and parts of the Gold Coast.

The impacts of fire ants

Fire ants have the capacity to spread to most parts of Australia (except the driest deserts, high altitudes and very cold areas). In humans, their venom causes severe irritation and painful, burning stings that may cause an adverse reaction in some people. Typically, their bites are not deadly to humans (unless a person has an insect allergy), but fire ants are hazardous to the natural and human environments in many ways.

Primary impacts of a fire any hazard include reducing plant populations and competing with native fauna and insects for food – ultimately controlling the balance of species within some insect communities. They are resourceful feeders that consume almost anything. These omnivorous insects scavenge seeds, and prey on plants, small animals (such as frogs and lizards) and other insects, potentially displacing or eliminating some native species over time. Colonies of fire ants also breed very rapidly and can create 'super colonies' with numerous queens, a feature of a colony that enables them to accelerate their breeding cycle.

Their mounds and nests damage outdoor lifestyle facilities such as ovals, sports fields and golf courses. They are attracted to electrical fields or currents, and may damage electrical equipment and infrastructure, particularly below-ground cables. Consequently, they can also damage power supply and utilities.

Fire ants also have secondary economic impacts in certain agricultural sectors by attacking young animals and stinging around the eyes, mouth and nose, leading to blindness and suffocation. They can also damage and kill some lawn and turf species, nursery plants, and fruit and vegetable plants by tunnelling through roots and stems.

Fire ants also have a social impact, potentially limiting outdoor activities, such as picnics and sporting events if ovals, parks and playgrounds become infested.

Responding to the hazard of fire ants

Because it is considered technically feasible to eradicate fire ants from south-east Queensland, several strategies have been put in place.

First, the Biosecurity Act 2014 declared the species noxious and invasive as well as limiting or restricting the movement of nursery items such as plants, garden bark, potting mixes and agricultural machinery that could carry fire ants. The initial response was to minimise the spread of ants as best as possible.

Figure 2.23 shows the three fire ant eradication biosecurity areas in which the movement of various organic materials is restricted to minimise the spread of the existing fire ant hazard zone. For example, material such as soil, manure, mulch, hay or potted plants cannot be moved between zones unless it is being taken to an approved waste facility, moved within 24 hours of arriving, moved following specific biosecurity regulations, or is moved with the permission of a biosecurity inspector.

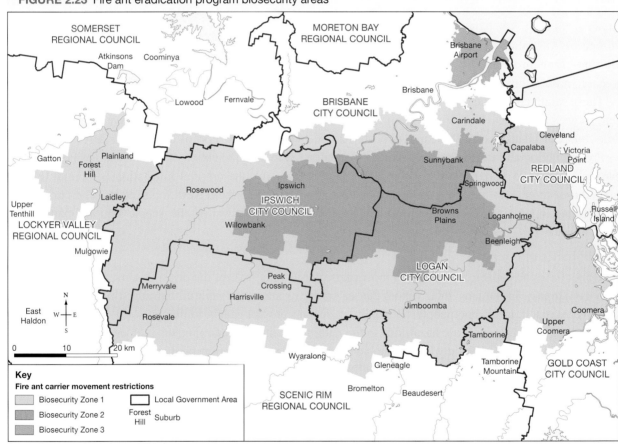

FIGURE 2.23 Fire ant eradication program biosecurity areas

Source: Department of Agriculture and Fisheries

Second, a publicity campaign was developed to inform the public about the dangers of fire ants and step up domestic vigilance throughout the south-east. A task force of inspectors was established to search suburban areas for nests where ants may be present.

Third, a program was developed to bait nest areas, with follow-up treatment using liquid insecticides and mound drenches.

The government predicted that spending $38 million a year for ten years would yield a 95 per cent chance of eradicating the ants from Australia. Of this amount, about $24 million was allocated per year on baiting using aircraft and field visits. Pesticide treatments are very expensive. When residual baits are used, there is the risk of killing other species that compete with fire ants. Precautions are also important to ensure children and domestic pets are not at risk of inadvertently eating or coming into contact with baits. The remaining $14 million was allocated for other eradication activities, such as community education, research and training odour-detection dogs.

Resources

- **Weblink:** Biosecurity Queensland Fire Ant Identification
- **Weblink:** Queensland Government fire ant information
- **Weblink:** Fire ants' aggression
- **Digital doc:** Fire ant hazard risk assessment (doc-29164)

Activity 2.3e: Fire ants in Australia

Using the information you have learned about fire ants and by conducting some research, answer the following questions.

Explain how fire ants present a hazard

1. Explain the variables that affect the severity of impacts that fire ants have on the environment or people by completing the following table. (You can find a Word version of this table in the Resources tab.) Give reasons for your suggestions.

Variable	Explanation
Causes	
How did fire ants enter Australia?	
Frequency and duration	
How quickly do infestations develop and spread?	
Does it appear as if fire ants will be here for some time or will they be eradicated?	
How do fire ants spread to new areas?	
Predictability	
Did people predict that it was possible for the species to arrive in Australia?	
Is it possible to predict how fire ants will spread within Australia?	
Controllability and potential for impacts	
What control strategies are in place to stop the spread of fire ants?	
What factors will influence whether the strategies will be successful?	
What are the likely impacts of not implementing control strategies?	
Response	
Is it too late to eradicate fire ants?	

Analyse the extent of the risk

2. In three paragraphs, explain the economic, environmental and social impact of fire ants and give examples.
3. (a) Choose another introduced animal species and compare its impact with that of the fire ant. Create a table for your findings that includes the following information about each species:
    - Type of animal (insect/mammal/amphibian/reptile)
    - Average size
    - Place of origin
    - Arrival location
    - Arrival date
    - Reason introduced
    - Environmental impacts
    - Economic impacts
    - Social impacts
    - Means of control
   (b) Based on the information in your table, suggest which species has the greater potential for causing long-term damage. Write a paragraph to justify your answer.

Suggest ways to manage the risk

4. In 2017, Australian governments pledged $411.4 million over 10 years to fight fire ants in south-east Queensland. Do you think the cost is justified? Give reasons to support your decision.
5. Plan, write and edit a proposal for one fire ant eradication strategy. Your proposal should include the following four sections:
    - Introduction (explain the type and extent of the hazard)
    - Evidence (include supporting data about impact and vulnerability)
    - Elaboration (explain your proposed strategy)
    - Conclusion (explain why your strategy will be successful)

2.4 Pollutants

2.4.1 Types of pollutants

Pollutants are substances introduced into the environment that are potentially harmful to human health and to the natural environment. While pollution may occur naturally, for example from the ash and gases from volcanic eruptions, most pollutants are a result of human activity. Virtually every human activity generates waste and therefore is a potential source of solid, liquid or gaseous pollution.

Anthropogenic pollution can come from a variety of sources, including:
- industrial sources, such as pollutants released from factories into the air or water, or leached into the soil
- transport, such as the exhaust emissions from various types of motor vehicles or spills from transport accidents
- agricultural sources, such as farm chemicals and animal wastes
- mines and quarries, such as dust and mining wastes
- domestic sources, such as smoke from cooking fires or household waste.

Pollution sources are usually divided into two categories: **point sources** and **non-point sources**. Point sources are particular locations and include an industrial site, a mine or quarry, a sewerage treatment plant or an oil storage tank. Non-point sources involve broad areas and include runoff from agricultural or urban areas. Because motor vehicles release pollutants into the air, which can then spread over wide areas, transport is an example of a non-point source.

There are three broad types of pollution:
- air pollution: the release of chemicals and particulates into the atmosphere
- water pollution: the release of wastes, chemicals and other contaminants into the hydrosphere (surface and groundwater)
- soil pollution: the release of wastes of various types on or into the lithosphere.

2.4.2 Reducing the risk of pollutants

There are two broad approaches to reducing the risk of pollution: prevention and control. Prevention and control of pollution will vary, depending on the type of pollutant and the source. Point sources of pollution are much easier to manage because they can usually be identified and monitored. Non-point source pollution often comes from a large number of small sources. These small sources can build up pollution over a large area, sometimes to unmanageable levels. Urban air pollution from motor vehicles is such a case.

The extent to which anthropogenic wastes constitute an ecological hazard depends on a range of factors. The amount of waste produced is important. Very small amounts of even potentially harmful chemical wastes may not be hazardous, while larger concentrations are hazardous. Some wastes **biodegrade** (break down through natural processes), so may not pose a risk to people or the natural environment, but some pollutants are very persistent and remain hazardous for long periods of time.

A further issue for human health is that some pollutants, especially water pollutants, can enter the food chain in a process known as **bioaccumulation**. The pollutant may start by being consumed by plankton, which are in turn eaten by fish, which are then consumed by humans. If there is a process of biomagnification, pollutants increase in concentration as they move along the food chain, and this may then increase the risk for humans at the end of the chain.

Some pollutants also create secondary pollutants. Acid rain, for example, is a secondary pollutant. It is produced when sulfur dioxide and nitrogen oxides (the primary pollutants) react in the atmosphere to produce sulfuric acid and nitric acid, which are absorbed into rain water. Sulfur dioxide and nitrogen oxides are produced when fossil fuels, such as coal and oil, are burnt in factories, power stations or motor vehicles.

FIGURE 2.24 (a) Los Angeles blanketed in pollution (b) Water pollution from copper mining (c) Oil pollution from pumps near Baku, Azerbaijan

Activity 2.4: Pollution hazard zones
Analyse figure 2.25 to answer the following questions.

Explain and analyse the impact of pollution hazard zones
1. Where do the worst incidences of acid rain generally occur?

FIGURE 2.25 World pollution map

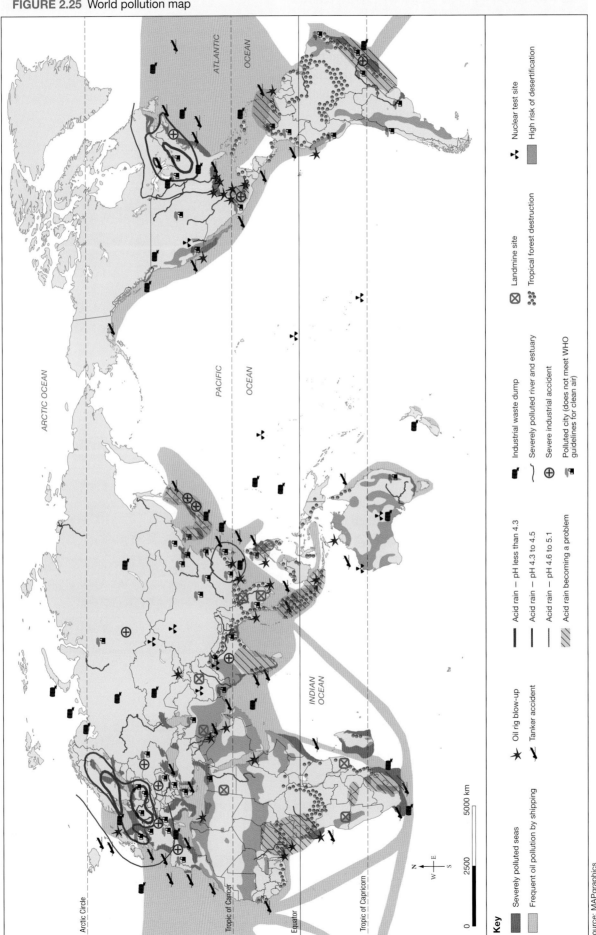

Source: MAPgraphics

2. Describe and account for the locations where acid rain is becoming a problem. Suggest what factors might be causing acid rain in these areas.
3. Where are the areas of severely polluted sea located? Suggest why these areas suffer greater pollution.
4. What patterns are evident in areas frequently polluted by oil from shipping?
5. What types of pollution have occurred in Australia? Describe the pattern of sea pollution around Australia. Suggest reasons for this pattern.
6. Based on your analysis of the figure 2.25, choose one of the areas you have identified as being highly polluted. Make a list of the social, economic and environmental challenges that might be faced by people living in that hazard zone because of the pollution levels.

2.5 Marine hazard zones

2.5.1 Types of marine ecological hazards

Marine environments are susceptible to a range of ecological hazards. Large-scale anthropogenic hazards, such as spills from tankers and oil rigs, cause significant damage and present complex challenges for clean-up crews. Some of the worst examples of anthropomorphic hazards exist in the remote parts of the ocean: floating garbage patches, largely made of plastic rubbish. The biggest, known as the Great Pacific Garbage Patch (or Pacific trash vortex), is a conglomeration of waste in the ocean. Estimates of the size of the patch vary from about 700 000 km^2, almost the size of New South Wales, up to more than 1.6 million km^2, just smaller than Queensland. First identified around 1985, the main body of rubbish (the Western Garbage Patch) is located between 35 and 42 degrees north of Japan. There is also a connected, smaller accumulation of rubbish (the Eastern Garbage Patch) between Hawaii and California.

This huge collection of rubbish has amassed because most items are non-biodegradable. About 80 per cent of the rubbish is thought to be plastic bags and bottles, as well as other domestic products. Most originated as land-based litter from streets and local suburbs, that was washed down drains and creeks after heavy rain. The remaining 20 per cent is thought to be items dumped at sea or lost from boats, much of which is discarded or broken fishing nets and equipment.

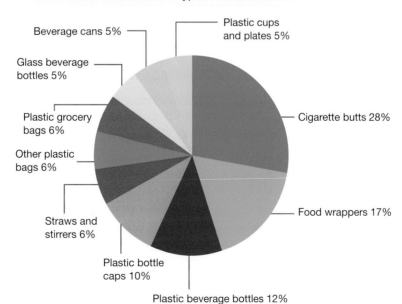

FIGURE 2.26 Breakdown of types of marine debris

- Beverage cans 5%
- Plastic cups and plates 5%
- Glass beverage bottles 5%
- Plastic grocery bags 6%
- Other plastic bags 6%
- Straws and stirrers 6%
- Plastic bottle caps 10%
- Plastic beverage bottles 12%
- Food wrappers 17%
- Cigarette butts 28%

Note: Data is the result of 25 years of surveying debris collected by volunteers in annual debris clean-ups in over 100 countries.

2.5.2 Factors affecting the severity of impact

When litter is washed into the sea, it is picked up by ocean currents and becomes part of the ocean circulation system caused by the Coriolis effect. When two ocean currents meet, they tend to slow and form **gyres** (circular currents). The circular motion of the gyre captures the debris, which is trapped towards the middle. Debris in the ocean is also moved on the Global Ocean Conveyor Belt by **thermohaline circulation** (deep currents in the ocean caused by changes in water density created by changes in water salinity and temperatures).

As well as creating a form of aesthetic pollution much of the plastic breaks increasingly smaller pieces, known as **microplastics** (measuring smaller than 5 mm in diameter) or even smaller **nanoplastics**. These nanoplastics and microplastics pose a hazard to wildlife, which can mistake the small plastic particles for food.

This process of breaking down plastic objects into small pieces occurs through exposure to the movement of the ocean and exposure to the sun and light, which breaks down the plastic in a process called **photodegradation**. Samples taken from several sites around the world show that each square kilometre of ocean now contains more than 5 kg of rubbish.

Images from space do not show a giant floating rubbish tip as much of the microplastic is not easily seen, although there are still large items that have not broken down, such as containers, plastic chairs, Styrofoam cups or thongs. At times, it gives the surface a dull appearance.

At the surface, and particularly in the gyres, these plastics are a hazard for marine creatures and birds. Sea turtles can mistake plastic bags for jellyfish, a major food source, dolphins and rays become entangled in waste fish

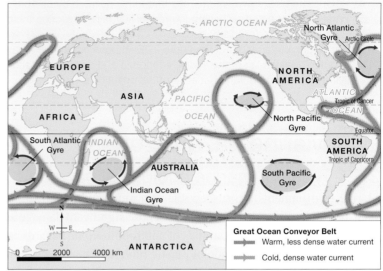

FIGURE 2.27 The five main ocean gyres and the Global Ocean Conveyer Belt

Source: Natural Earth Data map by Spatial Vision.

FIGURE 2.28 Turtles can mistake discarded plastic bags for jellyfish, one of their sources of food.

FIGURE 2.29 At high tide, rubbish from the ocean settles at the high-water mark.

net, and large birds such as albatrosses eat small pieces of plastic after mistaking them for fish eggs, then feed them to their chicks, which die of starvation.

If too much microplastic floats at the surface, it blocks off sunlight from algae and zooplankton, which are important food sources for many fish, turtles and baleen whales. Large whale sharks and blue whales also ingest small pieces of plastic in their food. As well as contaminating food sources, plastics also leach colourants and other chemicals into the water. Eventually, much of the debris sinks to the ocean depths and even to the seafloor. Scientists have recently found nanoplastics in some of the deepest parts of the ocean.

Large fishing nets or parts of damaged nets are also lost or jettisoned from boats or from jetties during stormy weather. Scientists estimate that more than 700 000 tonnes of lost fishing net are floating in the ocean in various states of decay. These nets are swept to sea in the current, catching fish and other sea life as they drift, as phenomenon termed 'ghost fishing'.

Table 2.8 shows the sources of marine debris by region. Land-based debris includes general litter such as wrappings and food containers discarded on land and washed into the ocean. Ocean-based litter is rubbish discarded from boats and oil rigs, such as nets, lines from fishing boats or litter from rigs. Smoking-related litter includes cigarette butts, lighters and packaging. The illegal and legal dumping of waste accounts for rubbish dumped deliberately into the ocean directly. Medical and personal hygiene waste includes syringes, disposable nappies and tampons.

TABLE 2.8 Sources of marine debris by region

Source	Africa (%)	North America (%)	Central America (%)	South America (%)	Caribbean (%)	South-East Asia (%)	Western Asia (%)	Europe (%)	Oceania (%)
Land-based litter	76.1	55.2	84.9	69.7	82.6	72.4	60.5	60.4	75.2
Ocean-based litter	12.7	5.0	4.8	12.0	6.6	12.7	9.7	24.9	5.2
Smoking-related litter	8.4	37.2	8.0	15.4	7.7	11.2	27.7	11.1	19.7
Legal and illegal dumping of garbage and waste	1.8	1.9	1.1	2.0	1.8	1.6	1.5	2.8	1.8
Medical and personal hygiene litter	1.0	0.7	1.2	1.0	1.3	2.0	0.6	0.8	1.1

Activity 2.5a: The Great Pacific Garbage Patch hazard zone

Using the information you have learned about the Great Pacific Garbage Patch, answer the following questions.

Explain and present the data
1. Explain why there is so much plastic in the ocean when just as much paper, cardboard and cans are also washed into the ocean.
2. One of the most common items found in the oceans is plastic water bottles. What might be the main point source for this?

3. On 11 March 2011, a massive earthquake hit the northern coast of Japan, followed by a tsunami that carried millions of tonnes of household debris out into the Pacific Ocean. Try to calculate how long it may have taken some of these items to reach the west coast of the USA if the average speed of the North Pacific Current is about 0.05 m/s (approximately 1.8 km/h).
4. Create a ternary graph using data from table 2.8 to represent three types of marine debris in the ocean for three regions. To complete this task you will need to transform the data, creating a new table displaying the total (as a per cent) for land-based, ocean-based and other sources of litter for your three chosen regions. Refer to **SkillBuilder: Constructing ternary graphs** in the Resources tab to complete this task.

Resources

Video eLesson: SkillBuilder: Constructing ternary graphs (eles-1728)
Interactivity: SkillBuilder: Constructing ternary graphs (int-3346)

The impacts of marine pollution in Bali

The impact of marine pollution is not contained to the Great Pacific Garbage Patch. Bali is popular for its sandy beaches, coral reefs and volcanic walk trails; however, sections of coastal water around the island, including near Nusa Penida, an important feeding ground for manta rays, are contaminated by tonnes of plastic: plastic bags, drink bottles and straws.

These levels of waste present in the ocean appear to fluctuate with the seasons, with most litter washing into the sea after rain.

Filter-feeders such as manta rays, whale sharks and whales are particularly at risk from plastic hazards. They feed by swallowing large volumes of sea water to capture plankton and other tiny organisms. With this intake of sea water, filter-feeders also ingest any plastic or rubbish present in the water, especially microplastics. Ingesting these can be fatal.

FIGURE 2.30 The habitats of manta rays are often contaminated by marine pollution.

Source: Elitza Germanov, Marine Megafauna Foundation

The impacts of marine pollution in Senegal

The coastal waters off Senegal in tropical western Africa are also experiencing significant impacts from marine pollution and waste hazards. Located between the Atlantic Ocean and the western fringe of the Sahara Desert, Senegal has an area about the same as Victoria and a population of 15.4 million people. Senegal's warm temperatures, pristine waters and sandy beaches of the Bay of Hann once made it a safe and popular tourist destination.

FIGURE 2.31 Senegal is located on the west coast of Africa, on the Atlantic Ocean.

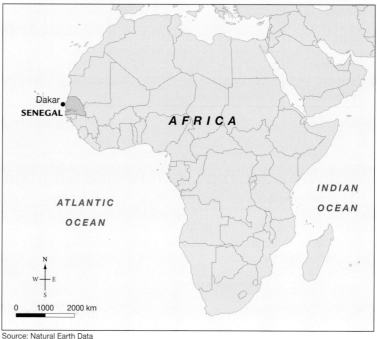

Source: Natural Earth Data

Senegal's coastal areas and larger cities, especially around Dakar, are threatened by the impacts of industrial pollution and poor waste disposal, with more than 70 factories discharging effluent into the Bay of Hann and raw sewage going directly into the ocean. Until 2011, the city was serviced by one single sewage treatment plant, and local services were sporadic, with some garbage collections occurring weeks apart. Water pollution is at such a high level, it is considered dangerous to swim in any of the coastal waters near Dakar; consequently, the tourism and fishing industries in the area have suffered significant losses.

To reduce the existing pollution hazard and to prevent it increasing, the government has banned the use of plastic bags. The construction of a new US$9 million treatment plant to convert sewage into potable (safe to drink) water, partially funded by the Bill & Melinda Gates Foundation, also helped to reduce the risk of the pollution hazard. Local environmentalists have also called for clean-up days to remove litter and fishing nets from the coast, and education programs to raise awareness of issues.

Activity 2.5b: Patterns in pollution

FIGURE 2.32 Pollution in Saint-Louis, Senegal, on the Senegal River, where coastal tourism makes up a significant part of the local economy.

Using the information you have learned about the marine pollution and research, answer the following questions.

Explain and analyse the challenges posed by marine pollution
1. List the items can you recognise on the riverbank in figure 2.32.
2. Explain why the amount of rubbish on this riverbank is an ecological hazard.
3. Complete the following table to compare marine pollution in Bali and Senegal.

	Plastic pollution off Nusa Penida, near Bali	Beach and coastal areas near Dakar, Senegal
Type of ecological hazard		
Likely causes		
Potential impacts on the marine environment		
Potential impacts on humans		
Possible solutions		

4. The tourist industries of Senegal and Bali are important to their economies. How might the pollution hazards in the oceans and rivers pose a risk to tourists?
5. If beaches and waterways in these countries continue to be polluted with rubbish, what might be some of the impacts of this ongoing ecological hazard?

Resources

- **Weblink:** Bali waters polluted with plastic
- **Weblink:** Pollution levels in Senegal
- **Weblink:** Microplastics a threat to wildlife
- **Weblink:** Bacteria evolve to eat plastic
- **Digital doc:** Marine pollution in Bali and Senegal (doc-29165)
- **Weblink:** The Great Green Wall

2.5.3 Monitoring marine hazards

Microplastics in the Pacific

A detailed collection and analysis of plastic concentrations in the eastern section of the garbage patch between California and Hawaii was undertaken in 2015 by a team of international marine scientists. The team not only used satellite imagery and aerial surveys, but also completed a detailed series of trawl runs to collect samples. A manta trawl (which collects debris from near the surface) was used to measure microplastics, and mega trawls (which measure a wider range of debris) were also conducted over the area.

Using a carefully designed mathematical formula, they were able to determine an accurate calculation of the volume of plastic in the selected areas, as well as identify where the highest concentrations of plastic had accumulated. The survey covered over 3.5 million square kilometres of ocean, and found that the patch contains around 80 million kilograms of plastic. Figure 2.33 shows the field monitoring area for plastics samples in the Pacific Ocean — grey and dark blue lines, and light blue lines, track ships and aircraft, and circles show locations where data was collected. Figure 2.34 shows the edges and concentrations of rubbish in the patch.

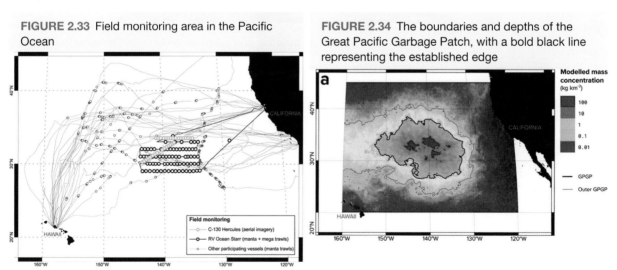

FIGURE 2.33 Field monitoring area in the Pacific Ocean

FIGURE 2.34 The boundaries and depths of the Great Pacific Garbage Patch, with a bold black line representing the established edge

Source: L. Lebreton, B. Slat, F. Ferrari, B. Sainte-Rose, J. Aitken, R. Marthouse, S. Hajbane, S. Cunsolo, A. Schwarz, A. Levivier, K. Noble, P. Debeljak, H. Maral, R. Schoeneich-Argent, R. Brambini & J. Reisser, Evidence that the Great Pacific Garbage Patch is rapidly accumulating plastic https://doi.org/10.1038/s41598-018-22939-wUsed under CC BY 4.0 licence https://creativecommons.org/licenses/by/4.0/

Microfibres and medication

Microfibres and residue from medications that wash into waterways are hazardous to marine environments. Many everyday clothes are made from synthetic materials such as acrylic, nylon and polyester fabrics that release thousands of tiny fibres where they are washed. These fibres are washed into the wastewater system and to sewerage treatment plants, but microfibres pass through the filtering and wastewater processing systems and are released with the treated water into the marine environment. As yet, it is uncertain which synthetic fibres shed the most microfibres making it difficult for manufacturers and the fashion industry to make changes to minimise their hazardous impact on the environment.

Some experts claim that this anthropogenic hazard is a bigger problem that microplastics in the ocean because the fibres' tiny size allows them to enter and stay in the marine food chain through **bioaccumulation** (when an organism absorbs matter at a faster rate than it is able to break it down and excrete it). Microfibres have been found in fish samples from the Great Lakes in the USA – initially consumed by smaller aquatic organisms, then by larger fish and eventually by humans.

Commonplace bathroom chemicals used in shampoos, conditioners, toothpastes, antibacterial soaps and cleansers, deodorants and sunscreens also wash into the marine environment from treatment plants, or directly from our bodies. In 2018, Hawaii banned the use of sunscreens containing oxybenzone and octinoxate; the two chemicals act as UVA filters but are believed to be a hazard for coral reefs, worsening the process of coral bleaching.

Reefs are made up of the skeletons of organisms called polyps held together by a limestone-like calcareous substance, which is produced by single-celled algae. This algae needs light for growth, salinity of about 2.7 per cent and water temperature about 25°C. The algae produce the scaffolding of the reef through a process of calcification. This involves using the calcium and carbonates dissolved in seawater to produce solid calcium carbonate. Coral bleaching occurs when coral expels the algae, *zooxanthellae*, in reponse to abnormal environmental conditions, such as rising water temperatures. In conjunction with this bleaching, rising concentrations of sunscreen chemicals cause the coral to release algae at lower temperatures. The chemicals also reduce the coral's ability to reproduce and impede the healthy growth of young coral.

Widely prescribed medications such as birth control pills, cholesterol tablets (statins), painkillers, antidepressants and antibiotics may also pose a risk. They are excreted into the sewerage network and pumped to treatment plants for the removal of contaminants. However, most treatment plants are not designed to extract 'specialty chemicals', so a great many pass through the normal filter processes and into the creeks, rivers and oceans.

Even in tiny concentrations, these chemicals are capable of being toxic to other organisms or causing hormonal effects on aquatic creatures. They are capable of leaching into groundwater from local septic systems and have been known to kill the bacteria required in some septic units.

It is not only legal drugs that find their way into our wastewater to pose a hazard to marine environments. Recently, the National Wastewater Analysis Drug Monitoring Program released its test results from monitoring fifty urban sewerage treatment sites across Australia. The wastewater analysis revealed some interesting data about drug use in Australia.
- Pharmaceuticals such as oxycodone and fentanyl were found in high levels in some regional centres in New South Wales, Queensland and South Australia.
- The Northern Territory recorded the highest alcohol consumption per person.

- Melbourne and Darwin had the highest levels of cocaine detection per capita.
- Nationally, more methamphetamine was being consumed than ecstasy.

Activity 2.5c: Monitoring and reducing ocean waste

Using the information you have learned about monitoring ocean waste, answer the following questions.

Explain the data and identify the patterns

Refer to figure 2.33.
1. Briefly explain what this map illustrates.
2. Estimate the approximate area from where samples of plastic were collected and determine which grid squares were sampled the most.

Refer to figure 2.34.
3. Give a precise location of sections of ocean with the most plastic using latitude and longitude to identify the location of the hazard zone.
4. Explain what the different shades on the map show. What do the different shades show about how plastics coalesce within the hazard zone?
5. Determine where the highest concentration of plastics (in kg/km^2) is located. Identify the latitude and longitude of the highest concentrations of plastics. Do these places match with a particular gyre? Refer to figure 2.27.

Evaluate and propose strategies to reduce the hazard

6. Rank these strategies for reducing the hazard posed by plastic in the ocean from the most to the least cost-effective, and list the potential benefits and problems of each.
   - Use large nets to physically collect ocean waste.
   - Pressure manufacturers into using biodegradable plastics as containers.
   - Limit the use of plastics and ensure they are disposed of carefully.
   - Fund more research into enzymes that can break down plastic.
7. Review and research some of the variables that affect the severity and impact of ecological hazards in the marine environment. Create a table or diagram to summarise your understanding of marine debris hazard zones. Include the following information:
   - causes (e.g. land litter, fishing vessels, chemical waste)
   - frequency (do the seasons influence the amount of litter?)
   - duration (how long does litter take to degrade?)
   - distance (how far and fast does the rubbish travel?)
   - predictability (did people predict this would happen – who and when?)
   - controllability (how can we control the amount of rubbish in the ocean?)
   - damage (what is the extent of the damage and how long will it last?)
   - response and management (what strategies might effectively fix, slow or prevent the damage?)

on Resources

- **Weblink:** Microfibres in the ocean
- **Weblink:** Drug monitoring in wastewater
- **Weblink:** Sunscreens and coral health

Human impact on the Great Barrier Reef

There are several hazards that pose a very significant threat to the future of coral reefs worldwide including the increased levels of runoff from farmland and the introduction of invasive marine species that damage the habitat or are predators of native species. The UN Intergovernmental Panel on Climate Change identified several hazards that threaten Australia's Great Barrier Reef specifically, including more intense tropical cyclone and rainfall events, and the combined effect of warming and acidification of the Pacific Ocean, which is likely to increase coral bleaching, disease and significant change in the composition and structure of the reef.

A survey conducted by the Great Barrier Reef Park Authority in 2014 found that 20 per cent of the study area showed evidence of coral bleaching. Since 1985, 10 per cent of the decline in total coral cover has been attributed to bleaching. In 2016, during a significant coral bleaching event caused by an extended marine heatwave, surveys of 83 reefs in the region showed that 29 per cent of the 3863 reefs in the Great Barrier Reef system were adversely affected.

The decline of the inshore reefs of the Great Barrier Reef, however, has been attributed to a different cause: deteriorating water quality. Farming is a major source of harmful pollution that causes damage to reefs through the increase of sediment and nutrient enriched runoff from agricultural land. Phosphate fertilisers are particularly harmful to the reproduction, skeletal calcification and framework development of coral.

It is estimated that about half of freshwater wetlands in the Tully and adjacent Murray River catchments have also been drained for agricultural land use and urban development. Loss of wetlands makes the problem worse because they provide a natural filter for rivers discharging sediment into the sea. During the summer months, heavy rainfall flows into the rivers along the northern coast of Queensland, heavily laden with top soil washed from farmland. In 2013, researchers at the Australian National University found a clear link between the health of porites coral and phosphorous levels from the Tully River. Porites coral is a stony coral with a characteristic finger structure. It is found extensively throughout the Great Barrier Reef.

FIGURE 2.35 Porites coral

Source: Department of Environment, Land, Water and Planning

The Tully catchment, 100 km south of Cairns, receives more than 4000 mm of rainfall each year. In some years, it can exceed 5000 mm. There are distinct wet and dry seasons, with rainfall more intense during the summer, particularly when tropical cyclones occur.

Farming is the mainstay of the Tully economy. Sugarcane production is the most important activity by value followed by livestock farming. The forest clearance required for commercial farming has substantially increased the sediment load of the Tully River. It is estimated that the annual sediment discharge of the river is more than 90 kilotonnes. This represents a fourfold increase since forest clearance in the 1860s.

The sheer intensity of the rainfall during the wet season is a major factor responsible for the surge in sediment load. Intensification of agriculture has also increased the amount of fertiliser used on farms and, as a result, the particulate phosphorous load of the Tully River. The seasonal pulses of high discharge generate plumes of nutrient-rich water discharged into the Coral Sea. These migrate north from the Tully estuary towards Dunk Island, more than 13 km away. During the wet summer season suspended sediment concentrations around Dunk Island of more than 300 mg/L are not uncommon.

FIGURE 2.36 Storm water sediment detention traps

Source: Barron Catchment Care

One strategy for preventing hazardous levels of runoff and sediment entering the reef is the implementation of sediment detention traps. These have been used in the Barron River catchment, to the north and inland from Cairns, to reduce the sediment and nutrient runoff into rivers by slowing down the flow of water. Landcare groups in the area initiated a number of other management strategies in the Barron catchment including restoring natural habitat, clearing marine debris from beaches, and monitoring and removing invasive weed species.

Activity 2.5d: Managing water quality

Examine the data from the Wet Tropics Report Card (2017) in tables 2.9, 2.10, 2.11 and 2.12 to answer the questions below.

The overall score for each location shows the total score out of 100 for that indicator. Standardised scores are translated to the following grades: <Very Poor = 0 to <21, <Poor = 21 to <41, <Moderate = 41 to <61, <Good = 61 to <81, <Very Good = 81 to 100.

TABLE 2.9 Index scores and grades, and overall basin scores and grades, 2014–15*

Basin	2014–15 Water quality	Habitat and hydrology	Basin grade	Basin score
Barron	71.22	38.93	Moderate	55.08
Tully	69.45	49.11	Moderate	59.28

TABLE 2.10 Index scores and grades, and overall basin scores and grades, 2015–16*

Basin	2015–16 Water quality	Habitat and hydrology	Basin grade	Basin score
Barron	82.83	43.33	Good	63.08
Tully	64.97	57.08	Good	61.03

Note: *The overall water quality indicator measures the levels of hazards such as the presence of pesticides, sediment and nutrients in the water. Habitat and hydrology measures the extent of the wetland and river extent, the flow of water, habitat modification and extent of invasive weeds. Basin scores for each location are the mean of individual indicators.
Source: Wet Tropics Report Card 2017 Results — Reporting on data July 2015 to June 2016. Wet Tropics Healthy Waterways Partnership, Cairns.

TABLE 2.11 Results for indices, and overall scores and grades for inshore marine zones (the area located between the coast line and the reef edge), 2014–15[#]

Inshore zone	Water quality score	Coral score	Seagrass score	Inshore zone	Inshore grade
North	57.20	44.30	21.29	40.92	Poor
Central	67.60	59.20	No data	63.39	Good
South	56.10	48.20	18.27	40.87	Poor
Palm Island	74.90	43.20	No data	59.05	Moderate

TABLE 2.12 Results for indices, and overall scores and grades for inshore (the area located between the coast line and the reef edge) marine zones, 2015–16[#]

Inshore zone	Water quality score	Coral score	Seagrass score	Inshore zone	Inshore grade
North	79.86	60.84	30.91	57.20	Moderate
Central	64.48	46.41	No data	55.45	Moderate
South	60.32	49.70	18.44	42.82	Moderate
Palm Island	69.15	55.78	No data	62.47	Good

[#]The overall water quality indicator measures the levels of hazards such as the presence of pesticides, sediment and nutrients. Coral indicator scores reflect factors such as coral cover and change. Seagrass indicators take into account factors such as area, species composition, reproductive effort, abundance and biomass (quantity in a given area). Zone scores for each location are the mean of individual indicators.
Source: Wet Tropics Report Card 2017 Results — Reporting on data July 2015 to June 2016. Wet Tropics Healthy Waterways Partnership, Cairns

Explain and analyse factors affecting water quality

1. Explain how poor landcare practices present a hazard to coral reefs.
2. Write a paragraph summarising how the overall water quality changed in the Barron and Tully basins between 2014 and 2016, and the impact these changes may have had on the reef.
3. The rivers of the Tully Basin run into the South inshore marine zone, and the rivers of the Barron basin run into the North inshore marine zone. Write a paragraph summarising how the overall health of each of these inshore zones changed between 2014 and 2016.
4. List the factors that might have contributed to the changes in the health of the inshore zone. Consider factors that are not measured in these tables that may have contributed, such as rainfall levels.
5. Research the strategies that have been put in place to reduce the impact of runoff in the Great Barrier Reef. Choose one strategy that you think has been effective in reducing runoff. Explain the process of how this action reduced the levels of runoff, and justify why you think this strategy is effective.

on Resources

- **Weblink:** The Wet Tropics Healthy Waterways report card
- **Weblink:** Tully Catchment Story
- **Weblink:** The Barron Catchment
- **Weblink:** Great Barrier Reef coral bleaching map

2.6 Atmospheric pollutants

2.6.1 Types air pollutants

Primary air pollutants come directly from a polluting source, such as motor vehicles, factories or power stations. Secondary pollutants are formed from reactions between or combinations of primary pollutants. Ozone is an example of a secondary pollutant. It forms in sunlight through chemical reactions between nitrous oxides and hydrocarbons, and is a component of photochemical smog. Nitrous oxides and sulfur dioxide can combine with rainwater to form nitric or sulfuric acid. This results in acid rain, another example of a secondary pollutant. Solid particles or particulate matter can also pollute the air, so air pollution can be a gas, a liquid or a solid.

Particulate matter is usually measured according to the size of the particles. PM2.5 is 2.5 micrometres or fewer in diameter; PM10 ranges between 2.5 and 10 micrometres, (in comparison, a human hair is around 100 micrometres). Because very fine particles can be inhaled and travel deep into the lungs, they are a significant cause of lung and heart diseases.

Air quality indexes differ around the world and measure a wide range of air pollution hazards, including specific pollutants such as carbon dioxide levels and general levels of particulate matter in the air. One commonly used measure of air quality is the US-EPA 2016 standard, shown in table 2.13, which measures fine particulate matter (PM2.5). These particles are typically heavy metals and compounds released into the air from car exhaust, burning landfill and industrial processes. Table 2.14 provides a summary of the main types of air pollutants and their sources and effects.

TABLE 2.13 Air Quality Index (AQI) scale using PM2.5 levels (US-EPA 2016 standard)*

AQI	Air Pollution Level	Health Implications	Cautionary Statement
0–50	Good	Air quality considered satisfactory; air pollution poses little or no risk	None
51–100	Moderate	Air quality acceptable; some pollutants may present moderate health concern for very small number of people (e.g. those with asthma or respiratory disease).	Active children and adults, and people with respiratory disease, should limit prolonged outdoor exertion
101–150	Unhealthy for sensitive groups	Members of sensitive groups may experience health effects; general public unlikely to be affected	Active children and adults, and people with respiratory disease, should limit prolonged outdoor exertion
151–200	Unhealthy	Everyone may begin to experience health effects; members of sensitive groups may experience more serious health effects	Active children and adults, and people with respiratory disease, should avoid prolonged outdoor exertion; everyone, especially children, should limit prolonged outdoor exertion
201–300	Very unhealthy	Health warnings of emergency conditions; entire population is more likely to be affected	Active children and adults, and people with respiratory disease, should avoid all outdoor exertion; everyone, especially children, should limit outdoor exertion
300+	Hazardous	Health alert: everyone may experience more serious health effects	Everyone should avoid all outdoor exertion

Note: * Air Quality Index shows levels of PM2.5 as μg/m3 (micrograms per square metre)
Source: © 2008–2016 World Air Quality / United States Environmental Protection Agency

TABLE 2.14 Sources and effects of major air pollutants

Air Pollutant	Main sources	Main effects
Nitrogen oxides (NO_x)	• motor vehicles burning diesel or petrol • power stations burning fossil fuels • industry	• respiratory problems • throat and lung infections • a major contributor to photochemical smog and acid rain • can affect growth/damage plants

(continued)

TABLE 2.14 Sources and effects of major air pollutants (*continued*)

Air Pollutant	Main sources	Main effects
Sulfur dioxide (SO_2)	• power stations burning fossil fuels • industry	• respiratory problems and severe coughing • eye irritation • circulatory and heart problems • major contributor to acid rain
Carbon monoxide (CO)	• motor vehicles burning petrol	• reduces oxygen in the blood • causes headaches and vomiting • large amounts are lethal • forms carbon dioxide (greenhouse gas)
Hydrocarbons	• incomplete burning of petrol in motor vehicles • industry • petrol stations and oil refineries	• contribute to photochemical smog (haze)
Ozone (O_3)	• motor vehicle exhaust fumes • other pollutants in the presence of sunlight	• component of photochemical smog (haze) • causes eye, throat and lung irritation • breathing difficulties • large amounts can be fatal • affects growth of plants
Particulate matter (PM2.5, PM10)	• combustion of fossil fuels in industry and motor vehicles • building and road construction	• eye irritation, breathing difficulties and lung damage • discolours paint and fabrics

2.6.2 The impact of air pollution

Air pollution can be hazardous to human health. According to the World Health Organization, air pollution is the cause of over one-third of deaths from stroke, lung cancer, and chronic respiratory disease, and one-quarter of deaths from ischaemic heart disease worldwide. The WHO estimated in 2018 that 91 per cent of the world's population lives in places where air quality does not meet recommended levels. People living in low and middle income countries are most vulnerable to the hazard of air pollution. China and India, with the world's largest populations, also have the largest number of deaths associated with air pollution.

FIGURE 2.37 The impact of acid rain on coniferous forests

Air pollution is most hazardous in large cities. Air polluting industries are frequently located in cities. More importantly, they have the greatest concentration of motor vehicles, which are the leading source of many air pollutants. Many cities around the world monitor their air quality and, if necessary, provide health warnings based on air quality index levels.

Air pollution also has a negative impact on the natural environment. Animals can be affected in similar ways to people and plants are vulnerable to changes in ozone, nitrogen oxides and sulfur dioxide, which can produce acid rain. Acid rain has a negative impact on soils and water courses, as shown in figures 2.37 and 2.38.

FIGURE 2.38 Deaths from air pollution, 2013

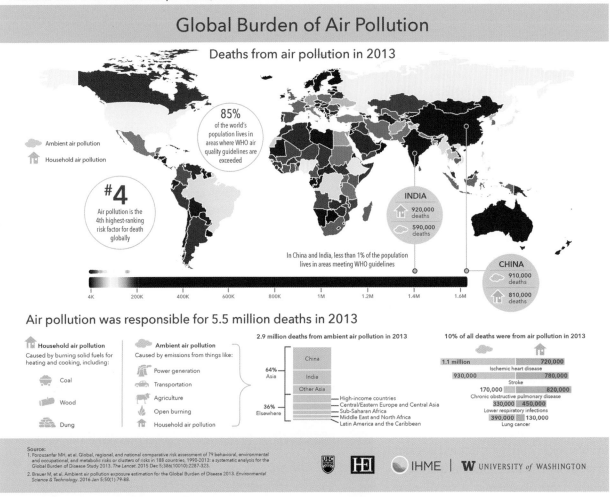

Source: Institute for Health Metrics and Evaluation

2.6.3 Factors affecting air pollution

The most hazardous levels of air pollution occur in the megacities of middle- and low-income countries. Delhi, Beijing, Cairo and Mexico City are among the most polluted of these cities. This is a consequence of the number of people and vehicles, as well as of the difficulty of controlling emissions in poorer countries, especially with large populations. Often, these very large cities are also the location of the country's major industries and commerce, and these also contribute to air pollution.

Air pollution in Mexico City and Brisbane

Mexico City is the capital and major political, economic, financial and cultural centre of Mexico. Because of this, it has attracted, and continues to attract, large numbers of people (see table 2.15) seeking employment, and educational and social opportunities. Mexico City is also highly car dependent. There are around 6 million motor vehicles in the city and more than 30 million vehicle trips are made each day. As figure 2.40 shows, transportation is the major source of air pollution in Mexico City.

FIGURE 2.39 The effects of acid rain on the natural environment

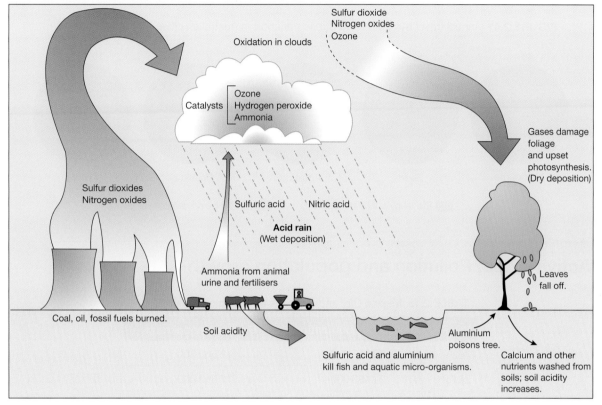

FIGURE 2.40 Sources of air pollution in Mexico City

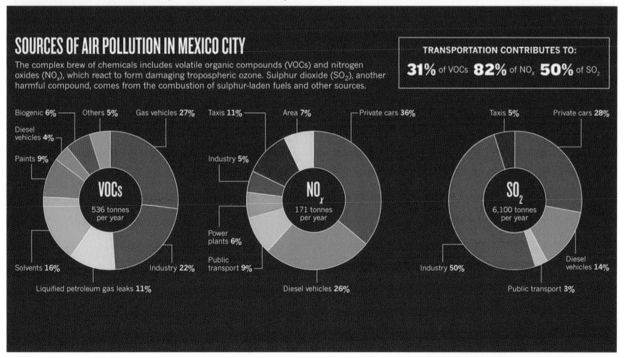

Source: Environment: Mexicos scientist in chief, Jeff Tollefson, Published online 20 October 2010, Springer Nature.

Motor vehicles are also the main source of air pollution in wealthy countries such as Australia. Figure 2.41 shows that all of our capital cities are car dependent. Brisbane had around 2.2 million motor vehicles in 2017, and 70 per cent of people's journeys to work were by car. Approximately 70 per cent of Brisbane's air pollutants are produced by motor vehicles.

FIGURE 2.41 Annual travel by private and public transport in Australian cities

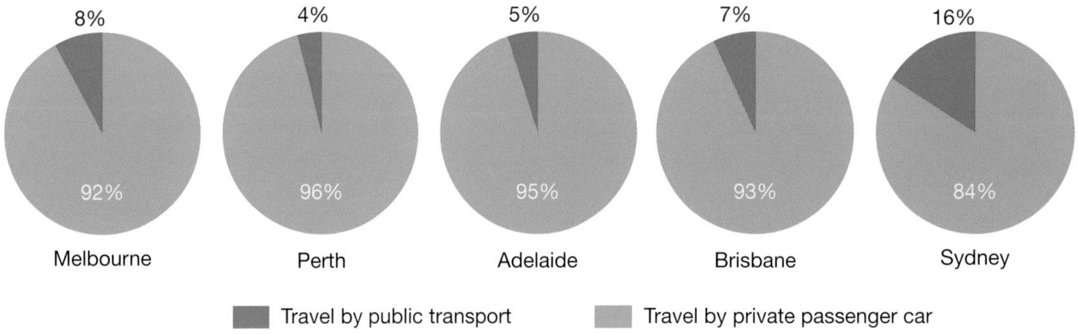

- Melbourne: 8% public, 92% private
- Perth: 4% public, 96% private
- Adelaide: 5% public, 95% private
- Brisbane: 7% public, 93% private
- Sydney: 16% public, 84% private

Travel by public transport | Travel by private passenger car

Activity 2.6a: Pollution and population growth

TABLE 2.15 Mexico City and Brisbane's population growth

Year	Mexico City population	Brisbane population
2025*	22 916 000	2 560 000
2018*	21 493 000	2 313 000
2015	20 999 000	2 202 000
2005	19 276 000	1 866 210
1995	17 017 000	1 527 888
1985	14 278 000	1 245 075
1975	10 734 000	1 042 100
1965	6 969 000	759 085
1955	4 294 000	527 500

Note: * Projected
Source: Australian Bureau of Statistics

Refer to **SkillBuilder: Constructing multiple line and cumulative line graphs** in the Resources tab to complete this activity.

Comprehend and explain the impacts of population growth

1. Use the data in table 2.15 to construct a graph to illustrate population change over time in Mexico City and Brisbane.
2. Compare the rate of change in the two cities' populations over time.
3. Suggest some reasons for the differences in growth rates.
4. What impact on air pollution might the cities' population growths have? Explain why.

Resources

Video eLesson: SkillBuilder: Constructing multiple line and cumulative line graphs (eles-1740)
Interactivity: SkillBuilder: Constructing multiple line and cumulative line graphs (int-3358)

The role of topography

Figure 2.42 illustrates another factor responsible for hazardous air pollution in Mexico City: the city is surrounded by mountain ranges. This creates a basin in which pollutants can be trapped by temperature inversions. The impact of this physical setting on air pollution can be seen in figure 2.42.

Brisbane's topography has some similarities to that of Mexico City. There are mountain ranges to the north-west, west and south of Brisbane, which form a basin-like structure, so Brisbane also experiences temperature inversions, especially in winter. In summer, air pollution is affected more by the recirculation of polluted air. Evening sea breezes move polluted air inland along the Brisbane Valley towards Ipswich, where it stagnates. Early the next morning, breezes move the polluted air back towards the south-western parts of Brisbane.

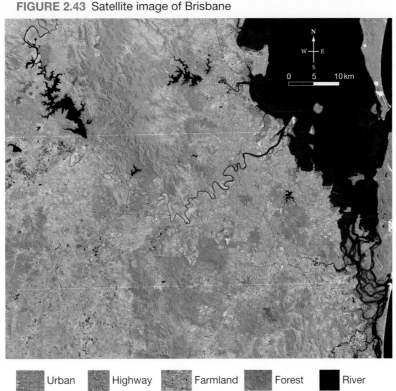

FIGURE 2.42 Mexico City air pollution

FIGURE 2.43 Satellite image of Brisbane

Urban   Highway   Farmland   Forest   River

Source: ACRE/000003/001W, "LANDSDAT imagery produced by Australian Centre for Remote Sensing ACRES AUSLIG, www.auslig.gov.au

CHAPTER 2 Ecological hazards

FIGURE 2.44 Brisbane metropolitan area

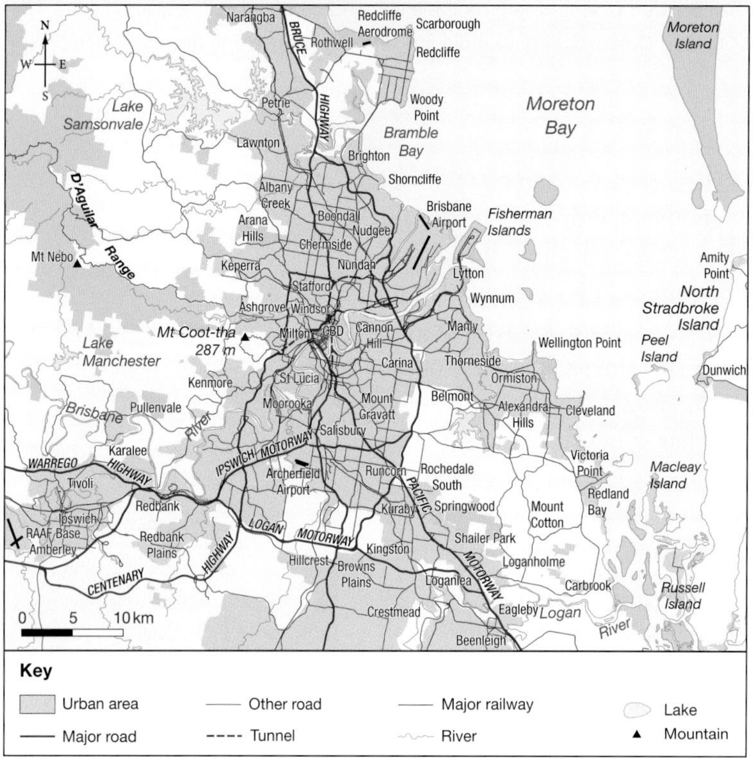

Source: MAPgraphics

Activity 2.6b: The impact of topography on air pollution

Refer to **SkillBuilder: Reading topographic maps at an advanced level** and **Creating a transect of a topographical map** in the Resources tab. (A printer-friendly version of figure 2.45 can also be found in the Resources tab.)

Explain and analyse the role of typography in pollution

1. Locate Cerro Madin (2470 m above sea level) and Xaltepec Volcano (2690 m above sea level) in figure 2.45. Draw a cross section between these two locations. What is the compass direction from left to right on the cross section?
2. Using your cross section, calculate the approximate altitude of the centre of Mexico City. How far below the summits of the two mountains does the city lie? (The interval between contour lines on the map is 100 m.)
3. Identify five more volcanoes on the map. Determine their altitudes.
4. Use your cross section and the contour patterns on the map to describe the topography of Mexico City and its surroundings.
5. Using a digital topographical map of Brisbane (for example, from the **QTopo** weblink in the Resources tab), draw a 100 km north–south and east–west cross section of Brisbane. Create a transect of each cross section. (Use the SkillBuilder in the Video eLessons of your online Resources to revise this skill.)
6. Using your transects and the topographical map, describe the topography of Brisbane.

Identify the impact of topography

7. Write a paragraph describing how Brisbane's topography might have an impact on the city's air pollution.

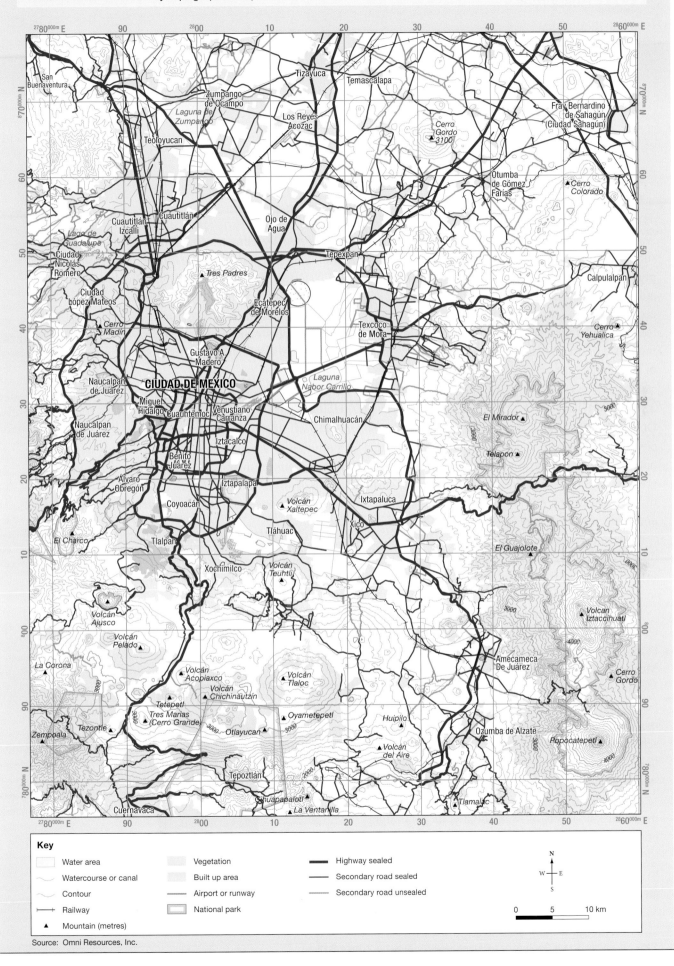

FIGURE 2.45 Mexico City topographic map

Resources

- **Video eLesson:** SkillBuilder: Reading topographical maps at an advanced level (eles-1749)
- **Interactivity:** SkillBuilder: Reading topographical maps at an advanced level (int-3367)
- **Video eLesson:** SkillBuilder: Constructing and describing a transect on a topographical map (eles-1727)
- **Interactivity:** SkillBuilder: Constructing and describing a transect on a topographical map (int-3345)
- **Video eLesson:** SkillBuilder: Understanding satellite images (eles-1643)
- **Interactivity:** SkillBuilder: Understanding satellite images (int-3139)
- **Weblink:** QTopo

The role of climate

Air pollution in both Mexico City and Brisbane is also affected by climatic conditions. Both experience sub-tropical climates with intense sunshine all year round. This provides the necessary conditions for the production of photochemical smog. In summer, heavy rain and storms do wash many pollutants from the atmosphere. In winter, however, there are lower rainfall totals. High pressure systems dominate both cities, producing clear, sunny and relatively calm weather conditions. The colder weather also contributes to the formation of temperature inversions. This means that air pollutants can build up over many days, so the incidence of ozone (a component of photochemical smog) and other pollutants tends to peak in winter months in both cities.

2.6.4 Mitigating the hazard of air pollution

During the 1980s and 1990s, Mexico City was considered the world's most polluted city. Since that time, the city has been able to significantly reduce hazardous air pollution to a point where high alerts now occur on only three or four days a year, rather than almost every day. Because transportation, especially cars, was the greatest source of air pollution, much of the hazard mitigation involved motor vehicles. This included vehicle restrictions, stricter emissions standards, the use of fewer polluting fuels and catalytic convertors, and regular emissions testing of cars.

Mexico City's vehicle restriction program is known as *Hoy No Circula* (Cars Don't Circulate), or 'no drive days'. This program, introduced in 1989, restricts a fifth of vehicles on rotating days between Monday and Friday, based on the last number of the car's registration plate. In 2008, the restrictions were extended to include Saturdays (figure 2.46).

In conjunction with vehicle restrictions, Mexico City has also invested heavily in public transport to provide cleaner alternatives to motor cars. Its Metro system has been greatly expanded and a new Metrobus network, with rapid bus transit lines, has been established. Taxis and minibuses were renovated and the city also has a large bike sharing scheme. Pollution from factories and power stations has also been reduced through the use of natural gas rather than more polluting fuel oil.

Although Brisbane's air pollution has not reached the hazardous levels of Mexico City, its continuing population growth, car dependence, and climatic and topographic conditions mean that the city will need to continue to take action to maintain and improve air quality.

FIGURE 2.46 Mexico City's *Hoy No Circula* system

Day	Last number of license plate			
Monday	5		6	1st Saturday of current month
Tuesday	7		8	2nd Saturday of current month
Wednesday	3		4	3rd Saturday of current month
Thursday	1		2	4th Saturday of current month
Friday	9		0	5th Saturday of current month

Australia already has high vehicle emission standards and high-quality fuels, and hybrid and electric vehicles are growing in number. The investment in road infrastructure, such as Brisbane's network of tunnels, does assist in reducing congestion and air pollution, but does not reduce car dependence.

Public transport networks play an important role in reducing car dependence and air pollution in all cities. Since the 1990s, Brisbane has developed of a busway network with separate rapid transit bus-only corridors, adding to Brisbane's already existing rail and bus networks. A Brisbane Metro subway system and a Cross River Rail project have also been planned to ease congestion in the inner parts of the city. The City Council has introduced a CityCycle scheme and invested in bikeways to encourage more people to cycle to work, rather than use a private motor vehicle. Brisbane City planning now incorporates transit-oriented developments, mostly high-density residences close to train stations, in order to encourage people to use public transport rather than cars.

As table 2.15 shows, the populations of both Mexico City and Brisbane will continue to grow. If people also rely on motor vehicles for transport, it is likely the risk of air pollution will increase in both cities. This means that effective methods of mitigation will be essential to the cities' future sustainability and liveability.

Air quality monitoring

Air quality monitoring involves the measurement of the quantity and types of pollutants in the surrounding air. This means that the extent of air pollution in a particular place can be assessed. Monitoring of air quality in cities is especially important if levels of pollution become hazardous, and warnings need to be provided to people at risk (see table 2.13). Monitoring also provides data to evaluate the effectiveness of any mitigation techniques that have been put in place to reduce levels of air pollution.

Table 2.16 shows data from one of 34 automatic monitoring stations located throughout Mexico City in a network known as *Red Automática de Monitoreo Atmosférico*, or RAMA. Data is collected for a number of different air pollutants and an overall air quality index is provided for each location. Weather data also forms part of the dataset for each monitoring station.

TABLE 2.16 Mexico Real-time Air Quality Index (AQI) 28 March 2018, 18:00 hours (Hospital General de México, México AQI: Hospital General de México)*

Overall AQI: 83 (Moderate)		Past 48 hours	
	Current	Min.	Max.
PM2.5 AQI	97	9	122
PM10 AQI	51	3	62
O_3 AQI	106	5	106
NO_2 AQI	21	10	65
SO_2 AQI	-	2	12
CO AQI	7	3	15
Temp. (°C)	22	12	25
Pressure (hPa)	1025	1025	1033
Humidity (%)	32	23	72
Wind (m/s)	6	3	17

Note: *AQI figures are calculated on a scale of 1–500 using the US-EPA 2016 standard (see table 2.13)
Source: © 2008–2016 World Air Quality

Brisbane has eight air monitoring stations located in the CBD, South Brisbane, Woolloongabba, Cannon Hill, Lytton, Rocklea and Wynnum, and Wynnum West. Measurements from these stations are then used by various local councils to provide a clean air index, similar to the air quality index used by Mexico City. Table 2.17 provides an example of the data from these monitoring stations.

TABLE 2.17 Brisbane air quality data for 27 March 2018, 9:00am–10.00am*

Station	CO (8hr av. ppm)	NO_2 (1hr av. ppm)	O_3 (1hr av. ppm)	SO_2 (1hr av. ppm)	PM10 (24hr av. ug/m^3)	PM2.5 (24hr av. ug/m^3)	Temp. (°C)	Humidity (%)	Wind (m/s)
Brisbane CBD	NA	NA	NA	NA	19.7	NA	25.5	78.4	0.7
Cannon Hill	NA	NA	0.009	NA	18.9	10.5	25.6	78.9	1.3
Lytton	NA	NA	NA	0.001	22.8	9.7	24	87	3.5
Rocklea	NA	0.008	0.001	NA	16.2	8.8	26.7	69.9	3.3
South Brisbane	0.2	0.014	NA	NA	16.4	7.2	26.3	77.6	1.5
Woolloongabba	0.3	0.024	NA	NA	22	13.1	26.2	75.2	1.9
Wynnum	NA	0.005	NA	0.004	25.9	7.9	23.2	96.1	2.1
Wynnum West	NA	NA	NA	NA	23.7	7.9	NA	NA	NA

Note: *not all air quality monitoring stations measure all pollutants
Source: © The State of Queensland Department of Environment and Heritage Protection 2012–2018

Activity 2.6c: Analysing air quality data
Refer to tables 2.16 and 2.17 to answer the following questions.

Explain and analyse patterns in pollution
1. Compare the types of air pollutants monitored by Mexico City and Brisbane. Suggest reasons for the similarities or differences.
2. Compare the level of each pollutant in Mexico City and Brisbane. Suggest reasons for any similarities or differences in the pollution levels.
3. Use the weblinks in the Resource box to obtain the current air quality data from five monitoring stations in both Mexico City and Brisbane to complete the following tasks. (You could also complete this task by comparing data from the monitoring station closest to where you live with Mexico City's data.)
   (a) Draw up tables to show the data for each city.
   (b) Draw multiple column or bar graphs to illustrate the data shown in the table.
   (c) Calculate the mean level for each pollutant in each city.
   (d) What similarities and differences do you notice in the cities' data sets? Suggest reasons for these.

Resources
- **Weblink:** Impacts of air pollution
- **Weblink:** Mexico City air quality
- **Weblink:** Queensland air quality

2.7 Lithospheric pollutants

2.7.1 Types of lithospheric pollutants

Pollution and contaminants are not always as visible as smog or floating islands of garbage in the ocean. Chemicals and other contaminants such as metals can also seep into the ground, contaminating the soil, and making their way into water-courses and eventually into the sea. On a small scale, this can occur from domestic chemicals leaching into the soil, or from lead-based paints flaking from homes into the soil. On a larger scale, long-term land and soil pollution can be caused by industrial spills and accidents, meltdowns at nuclear power plants or the burning or dumping of toxic waste. In some urban areas, homes have been built on the sites of former factories, mining leases or older buildings that contained lead paint or asbestos sheeting, which has degraded or been left mixed into the soil.

Hazards in backyard gardens

Not all hazards occur on a large scale that puts whole communities at risk. Studies of Melbourne backyard soil contaminants released in 2018 found that 21 per cent of samples collected from 136 backyard gardens contained levels of lead above the recommended levels of 300mg/kg for home gardens. These levels were found to be highest in the gardens of homes built before 1970 that were painted several times outside with lead-based paints, which used to contain up to 50 per cent lead. Many of these houses were in the inner city, near major roads and congested areas so the emissions from years of lead-based petrol were also thought to play a role.

While these lead levels were not considered high enough to prevent basic recreational activities, lead can cause serious health problems if it is breathed in or ingested in large doses, and can build up in the body over time. Many of the tests conducted were done on soil taken from backyard vegetable gardens, so people eating the produce from those vegetable patches could potentially have been ingesting higher levels of lead than was considered safe. Ingesting lead has been shown to affect children's development and acts as a neurotoxin. Studies have shown a link between exposure to low levels of lead and lower average IQ in children.

In addition to older paint and chemical hazards found in some backyards, some homes are also build close to former mining and industrial sites, which pose a potential risk to homeowners many decades after the site ceased to operate.

In Indooroopilly, a former lead mine site owned by the University of Queensland, which is surrounded by homes, was found to contain more than ten times the accepted standard of lead contamination in 2013. Testing on six different sites in the old mining area showed lead levels of up to 17 300 mg/kg. The acceptable level for safe recreational use is 600 mg/kg.

Similar studies have found high levels of lead contamination in Sydney backyard vegetable gardens, with some estimates suggesting that up to 40 per cent of backyard gardens have higher than the 300 mg/kg of lead. In some inner urban areas, such as Leichhardt, the mean soil lead contamination was 960 mg/kg.

Because lead poisoning is often caused by the breakdown of old lead-based paints into the soil, natural hazards can increase risk of exposure. The clean-up after the 2011 Queensland floods, for example, saw many older weatherboard homes requiring renovation as old paint flaked from walls after the floodwater subsided. However, homeowners were not always aware that the old paint contained lead, which meant that removing the old paint exposed them and their families to high levels of risk.

Agbogbloshie Dump, Accra, Ghana

The largest e-waste dump in the world, Agbogbloshie, is in Accra, Ghana. The site receives discarded electronic goods and household appliances – such as computers, phones, microwaves, monitors and other electronic goods – that cannot be repaired. Roughly 70 per cent of the electronic products imported into Ghana comprise second-hand equipment from other parts of Africa. An estimated 52 per cent of the e-waste imported to Ghana from around Africa is refurbished and resold, but in 2009, a UN study found that in that one year alone 22 575 tons of e-waste imported to Ghana was bound for the Agbogbloshie Dump.

There is a significant industry in repairing and selling these products, but about 15 per cent of the electronic equipment is unusable or unrepairable, and is recycled for component parts. This work is done by the estimated 40 000 individuals who work the dump site, including young children and teenagers from poor families who come to the capital seeking work.

FIGURE 2.47 Ghana is a country on Africa's west coast.

Source: Natural Earth Data

The process of recycling e-waste involves workers dismantling some parts of the products, but to extract the most precious metals from many of the components requires burning off the casing and insulation – largely plastics – in fires set around the dump. The remaining melted metals are collected from the ashes and sold on to traders. The dump also 'disposes' of tyres, which are burnt for the metal reinforcements embedded in the rubber. While there are other more environmentally friendly ways of extracting the metals from these products, burning the waste is the quickest and most cost-efficient method. Many of the workers in the dump site suffer from serious illnesses, such as cancers, from exposure to the components of the electronics.

The burnt e-waste releases a range of toxic gasses and substances, contaminating the air and soil. The dump is located on a former wetland, in the banks of the Odaw River, so some of the e-waste components also end up in the river system.

Increasingly, developed nations are passing laws that mandate minimum requirements for the recycling of e-waste and prevent the export of e-waste to developing nations. For example, since 2014, the EU has required member countries to recycle 45 per cent or more of their e-waste.

Shipping out hazardous e-waste is illegal in the EU, as it is in Australia, but there is an increasing temptation for companies to illegally export their

FIGURE 2.48 Tyres burnt to retrieve the metal components, Agbogbloshie Dump, Accra, Ghana

e-waste to developing nations. Safely disposing of e-waste is expensive, but traders in Africa and South-East Asia are offering to buy e-waste for repair, resale or recycling.

> **Resources**
>
> - **Weblink:** Lead in Australia
> - **Weblink:** The disposal of e-waste in Ghana
> - **Weblink:** Australian e-waste found in Agbogbloshie

Activity 2.7a: Comparing hazard mitigation strategies

Using the information you have learned about lead contamination in Australia and the Agbogbloshie Dump in Ghana, answer the following questions.

Compare and propose hazard mitigation strategies

1. What do the two examples have in common? Make a dot point list of the common risks and challenges that each hazard presents to people living or working in the vicinity.
2. Make a list of the factors that contribute to the existence of the Agbogbloshie Dump. Propose and justify two actions that the international community and/or local authorities could take to mitigate or remove the risk to people working and living at the dump.
3. Investigate Australia's laws regarding the disposal of e-waste and the disposal of lead-based products, and the testing and collection services that exist in your area. Do you think these laws and processes could be successfully implemented to mitigate the risks of hazards in less-developed nations? Might they help to manage the challenges workers face in Agbogbloshie? Justify your response with examples.

Contaminant hazards in the Aral Sea

The Aral Sea, located in Central Asia, lies within the countries of Kazakhstan and Uzbekistan. Almost all of its water comes from two rivers, the Amu Darya and the Syr Darya, whose catchment areas extend across parts of seven countries, as shown in figure 2.49. As a case study, the sea provides an important insight into how anthropogenic processes can cause irreparable damage to a physical environment. In 1960, the Aral Sea was the world's fourth largest lake, with an area of 66 100 square kilometres (an area similar to the size of Tasmania) but by 2007, the sea had shrunk to just 10 per cent of its original size. All that now remains of the Aral Sea are three residual lakes.

The upper reaches of the Syr River lie in the mountainous regions of Kyrgyzstan. From here, the Syr briefly flows through Tajikistan and Uzbekistan, before crossing the arid and semi-arid steppe regions of southern Kazakhstan and entering the Aral Sea. The Amu River has its headwaters in Tajikistan, and flows across the deserts of Turkmenistan and Uzbekistan to the south Aral Sea. Before 1991, these five Aral Basin countries were republics within the Union of Soviet Socialist Republics (the USSR, or Soviet Union).

The Aral Sea is the lowest point of a closed **inland drainage basin** measuring around 2.5 million km^2 in area. The two main rivers of the basin drain water from their headwaters in the Tien Shan and Pamirs mountains. The Aral Basin has a continental climate, with low and unreliable precipitation, large diurnal and seasonal temperature ranges, and low humidity, similar to the climate of central Australia. Lowland parts of the basin are arid, receiving only 80 to 200 mm of precipitation annually, while mountain areas receive up to 800 mm. Despite low precipitation levels even in the mountain areas, the amount of rainfall and snowmelt feeding the Amu and Syr rivers did provide enough water to keep the Aral Sea relatively stable until the 1960s.

FIGURE 2.49 The Aral Sea is located on the border of Kazakhstan and Uzbekistan.

FIGURE 2.50 The Aral Sea in (a) 1973 and (b) 2007

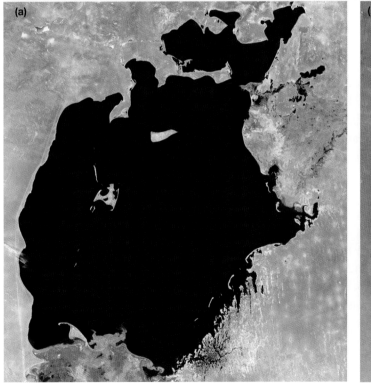
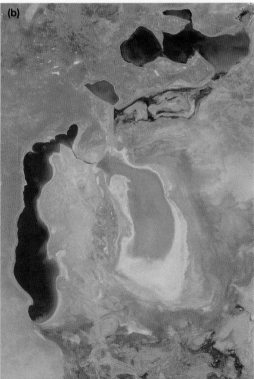

Source: australscope

Anthropogenic activities affecting the Aral Sea

The countries in the dry lowland regions of the Aral Basin have irrigated their food crops since ancient times, and during the period of Russian and Soviet control, irrigation areas continued to be expanded for growing food crops. Nevertheless, despite water extractions from the Amu and Syr rivers for irrigation, the Aral Sea remained stable, both in volume and area. The sea itself was also the centre of a thriving commercial fishing industry: catching over 40 000 tonnes of fish annually and supporting thousands of families.

During the 1960s, however, the central government of the Soviet Union embarked on a program of agricultural development in its central Asian republics, focusing on water-intensive cash-crops of cotton and rice. They built 94 water reservoirs and 24 000 km of canals.

As the area under irrigation expanded, the demand for water increased, and the Aral Sea began its dramatic decline (see figure 2.53). By the 1980s, seven million hectares of agricultural land were being irrigated, and almost 90 per cent of the water flowing to the Aral Sea was being diverted.

To make the issue worse, it is predicted that climate change will also worsen the effects: the Amu and Syr rivers feed from melting mountain glaciers that are expected to be diminished by rising global temperatures.

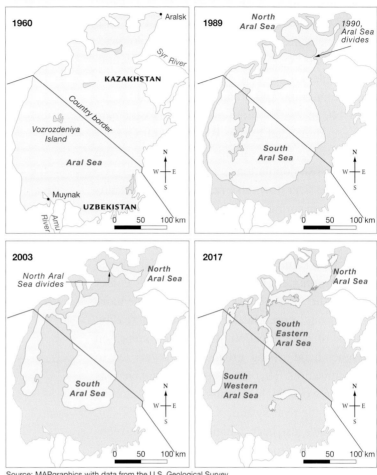

FIGURE 2.51 Changes to the Aral Sea, 1960–2017

Source: MAPgraphics with data from the U.S. Geological Survey

FIGURE 2.52 Ships on the dry bed of the Aral Sea

Environmental, social and economic impacts

There have been many severe environmental impacts of the extraction and diversion of the Aral Basin's water resources. As water levels dropped, salinity levels began to rise. In the southern section, for example, salinity levels rose from 14 grams per litre to more than 100 grams per litre in 2007, and up to 140 in 2017 – more than six times the safe level recommended by the World Health Organization. Rising water salinity also led to a severe decline in fish populations.

Reduced river flows also meant that the wetlands that relied on spring floods were lost, and the birds, mammals and other animals dependent on the wetlands also disappeared. The Aral Sea's decline has been so significant that even the local climate changed; summers became hotter, winters colder and humidity levels lower. Droughts became more common, and groundwater levels fell, increasing the risk of desertification. In irrigated areas, on the other hand, rising groundwater levels led to soil salinisation. This environmental damage was also made worse by local mining industry leaching waste into the river system.

There have also been a number of social and economic impacts from the over-extraction of water from the Aral Sea. Initially, some fishing villages attempted to dig trenches to get their boats to the water but their attempts were futile. Entire towns and villages that were previously reliant on the sea

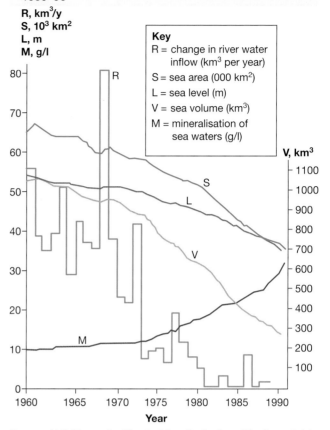

FIGURE 2.53 Changes in water inflow to the Aral Sea, 1960–90

Key
R = change in river water inflow (km³ per year)
S = sea area (000 km²)
L = sea level (m)
V = sea volume (km³)
M = mineralisation of sea waters (g/l)

Source: N.F. Glazovsky, The Aral Sea Basin, from "Regions at risk: comparisons of threatened environments", 1995 Jeanne X. Kasperson, Roger E. Kasperson, and B.L. Turner eds. Reproduced by permission of the United Nations University Press

for their fishing industries were stranded, ending the traditional livelihoods of more than 60 000 people. Drought conditions in 2000–01 also worsened the impact on the economy and increased the levels of migration from the area, as people left to find work and better living conditions.

While these effects of excessive water extraction might have easily been predicted, the impact on the population's health may not. People's health was also affected. Life expectancy of people living in communities in the area declined and infant mortality rates rose. One of the factors influencing levels of health was exposure to lithospheric pollution. As the sea receded, it exposed salty seabed that was contaminated with toxic chemicals washed down from agricultural areas. Strong winds picked up the contaminated dust, and it was carried for hundreds of kilometres. Local food and water exhibit very high levels of these contaminants. As a result, higher levels of birth defects, stunted growth, and heart and kidney disease have been found in children in the basin region. The lack of freely available fresh drinking water from former river sources also means that levels of parasite-borne illnesses, such as hepatitis, are higher than in other parts of the region.

Reducing the level of risk

Several strategies have been tried to reduce the level of risk faced by communities in the region. These include the construction of an earth dam to raise the level of the Northern Aral Sea by stopping the water flowing southwards and evaporating. While this strategy failed when the dam collapsed in 1999, the World Bank provided financial assistance that allowed Kazakhstan to subsequently replace the dam and modernise parts of the irrigation system to increase efficiency.

Once construction of the dam and a series of dykes and canals was finished in November 2005, water flow into the Northern Aral Sea increased by around 1.3 billion m³ a year and water levels rose by two metres. In addition, the surface area of the sea has increased from 2300 km² to 3250 km², salinity has dropped to its original level and local climate has improved. Encouragingly, fish numbers have also increased, so fishermen are again catching fish in large quantities. Given the success of this project, the Kazakhstan government is now considering an ambitious plan to connect the large, stranded port city of Aralsk with the sea. At the sea's lowest level, the city was 100 km from the shore, but it is now only 25–30 km away.

Activity 2.7b: Ecological hazards in the Aral Sea region
Using the information you have learned about the Aral Sea, answer the following questions.

Comprehend and explain the hazard
1. List and write a short explanation of three processes (either natural or anthropogenic) that contributed to the ecological hazards created by the decline of the Aral Sea.
2. (a) Analyse figure 2.51. Describe how the size and location of the Aral Sea changed between 1960 and 2017.
   (b) Identify and describe factors that may have contributed to these changes, using data to support your explanation.
   (c) Explain how these changes may have contributed to ecological hazards in the region.
3. Create a table to summarise the economic, social and environmental impacts of the damage to the Aral Sea on the local community. Give examples of primary, secondary and tertiary impacts for each.
4. Explain one action that has been taken in the Aral Sea region that helped to manage the needs of both the farming and fishing communities and also reduced the impact of ecological hazards. Explain the specific needs of each community and how the action helped to ease the challenges faced by each group.

2.8 Infectious and vector-borne diseases

2.8.1 Types of infectious and vector-borne diseases

Disease is a condition that causes harm to, or interferes with, the normal functioning of a living thing. Many diseases are caused by pathogenic (disease-causing) micro-organisms such as bacteria and viruses and by parasites. Infectious diseases, (also known as contagious or communicable diseases), can be passed from one person to another. Vector-borne diseases are carried by organisms capable of transmitting pathogenic bacteria, viruses and parasites from one person to another, for example, a house fly, mosquito or flea.

FIGURE 2.54 Polluted drinking water is a common source of infectious disease for poor communities.

FIGURE 2.55 Mosquitoes can act as vectors for many diseases.

Other types of diseases can be caused by genetics, nutrition and lifestyle, or by environmental factors. Infectious diseases are usually transmitted via water, air or food, or by a vector that contains the disease-causing organism. Water-borne infectious diseases are contracted when contaminated water is swallowed, or when food washed in contaminated water is eaten. Examples include cholera, typhoid, botulism, polio and giardia.

Air-borne infectious diseases occur when people breathe in bacteria or viruses attached to dust particles, smoke or water vapour, causing them to become ill. These diseases include influenza, tuberculosis, smallpox and chickenpox. Vector-borne diseases are transmitted mechanically or biologically. Mechanical transmission occurs where an insect picks up an infectious agent outside its body and passes it on, such as when a fly lands on rubbish or manure and then contaminates human food. Biological transmission happens when an insect harbours pathogens inside its body and passes them on to another animal or a person, such as when a mosquito passes on malaria through saliva and blood. Other vector-borne diseases include dengue fever, yellow fever, Chagas and sleeping sickness. More than 90 per cent of all deaths caused by infectious diseases are due to only a small number of them, including pneumonia, diarrhoeal diseases and tuberculosis (see table 2.18).

TABLE 2.18 Global Health Observatory data (WHO), 2015

Disease	Deaths	DALYs (disability-adjusted life-year) *
Lower respiratory infections	3 190 300	142 384 000
Diarrheal diseases	1 388 600	84 928 000
Tuberculosis	1 373 200	56 037 000
HIV/AIDS	1 000 000	62 759 000
Malaria	429 000	38 520 000
Measles	89 780	12 278 708

Note: *DALY is the disability-adjusted life year, a measure of overall disease burden.
Source: ©WHO 2018 http://www.who.int/sustainable-development/news-events/breath-life/air-pollution-by-numbers.jpg

2.8.2 Disease hazard zones

Infectious disease hazard zones can range in scale, from a single village to world-wide. When a disease is prevalent across a wide geographic area it is known as a **pandemic**. The Black Death, which swept across Asia and Europe in the fourteenth century, is one of the worst pandemics in human history, killing an estimated one-third of Europe's total population. A global pandemic of influenza (the Spanish flu), which began in 1918, infected around 500 million people. Estimates of the number of deaths in this pandemic range from 20 to 100 million people, compared with between 18 and 20 million deaths during WWI, making it one of the world's worst ecological disasters.

Another tragic outbreak of an infectious disease occurred in West Africa between 2013 and 2016. This smaller scale disease outbreak was an **epidemic** of the highly infectious Ebola virus. Over its course, the disease was responsible for more than 28 000 deaths in Guinea, Sierra Leone and Liberia. Ebola is both highly infectious and very dangerous – often it kills around 70 per cent of those infected. The WHO calculated that the West African epidemic killed around 40 per cent of those who contracted the disease, although also admitted that this was probably a significant underestimation of the numbers killed.

FIGURE 2.56 Location of recent serious disease outbreaks

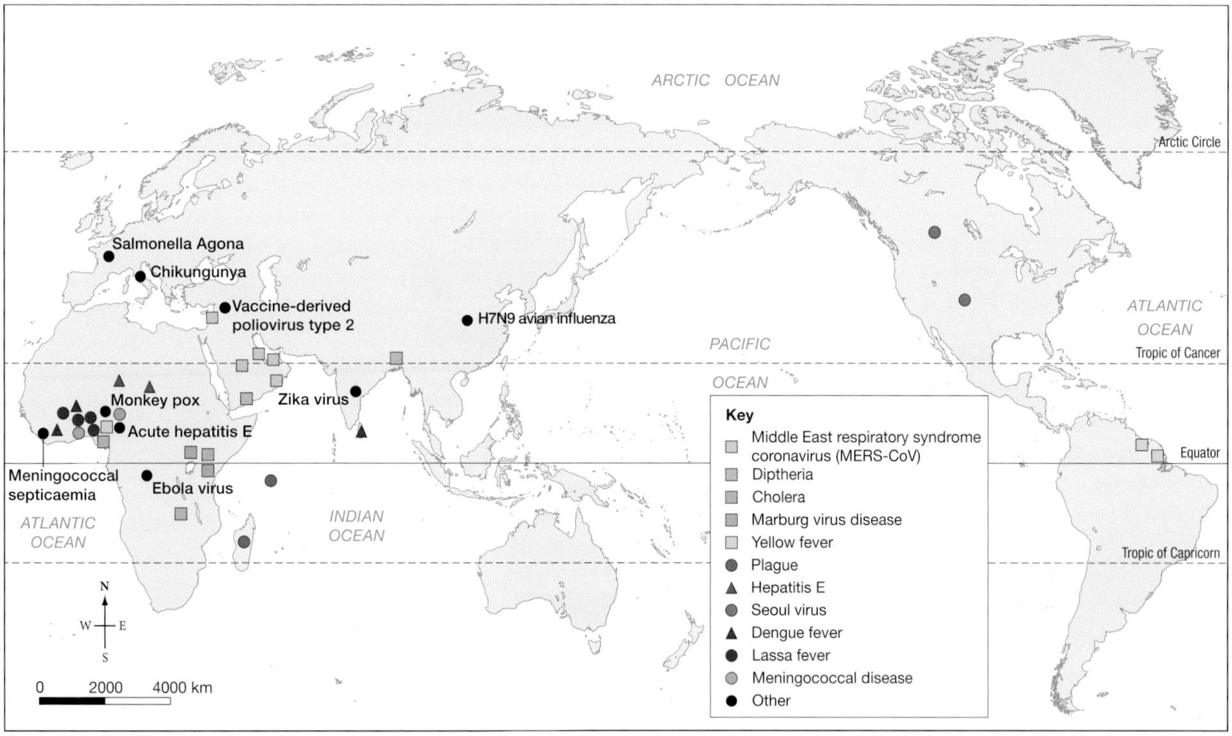

Source: © WHO 2018

Poverty and infectious disease

It is not uncommon for poorer countries such as Guinea, Sierra Leone and Liberia to be much more vulnerable to infectious diseases. In fact, around 63 per cent of deaths in the world's poorer countries in 2015 were a result of infectious diseases, compared with only 12 per cent in high-income countries. Of the top ten causes of death in wealthy countries, only one, lower respiratory infections, was an infectious disease; five of the top ten causes in the world's low-income countries were infectious diseases. In wealthier countries, the main causes of disease are often lifestyle-related, such as lack of exercise, obesity, smoking and excessive alcohol consumption, rather than infections (see figure 2.57).

The main factors that make people in developing countries more vulnerable to health problems and disease include:

1. Physical factors, for example extremes of climate, natural disasters, limited access to safe water and lack of fertile soils
2. Socioeconomic factors, for example lack of adequate sanitation, lack of adequate health infrastructure, lack of education, skills and technology, and urban overcrowding
3. Political factors, for example poor governance, corruption and limited human rights (especially for women and children) and civil unrest.

A combination of these factors means that more than half of all deaths in low-income countries are caused by infectious diseases, nutritional deficiencies and maternal causes (conditions arising during pregnancy and childbirth). In wealthier countries, less than 7 per cent of deaths are a result of these causes.

Cholera

Cholera is a severe diarrhoeal disease caused by the bacterium *Vibrio cholerae,* which enters a person's small intestine through the ingestion of contaminated water or food. Starting with nausea, fever and vomiting, cholera victims gradually develop severe diarrhoea, which leads to losing large amounts of fluid and results in failure of the circulatory system. Cholera can kill a person within a few days if it is left untreated.

FIGURE 2.57 Top ten causes of death in (a) low- and (b) high-income countries, 2015

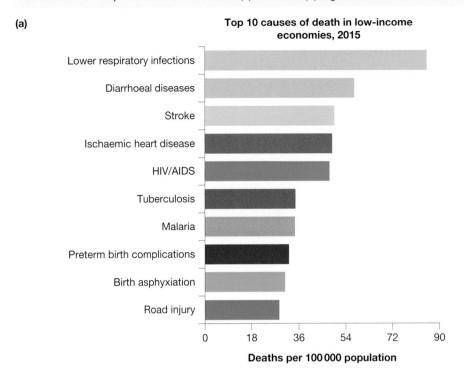

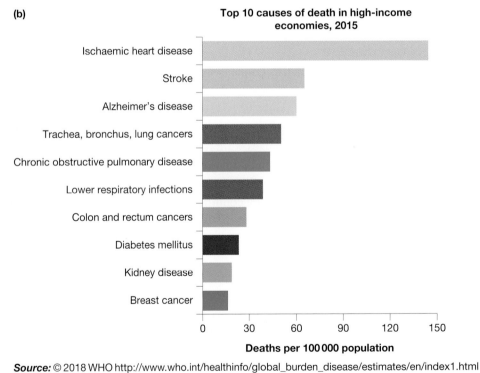

Source: © 2018 WHO http://www.who.int/healthinfo/global_burden_disease/estimates/en/index1.html

Most outbreaks occur in countries where there is substandard sanitation and low-quality drinking water. While cholera is rarely contagious from person to person, it spreads quickly in areas where people live in overcrowded conditions and have poor hygiene. Generally, the most common reason for an outbreaks is faecal contamination of water supplies from untreated sewage entering waterways. Cholera is also common in developing countries after earthquakes and cyclones because clean water is in short supply.

FIGURE 2.58 Deaths from communicable maternal, neonatal and nutritional diseases

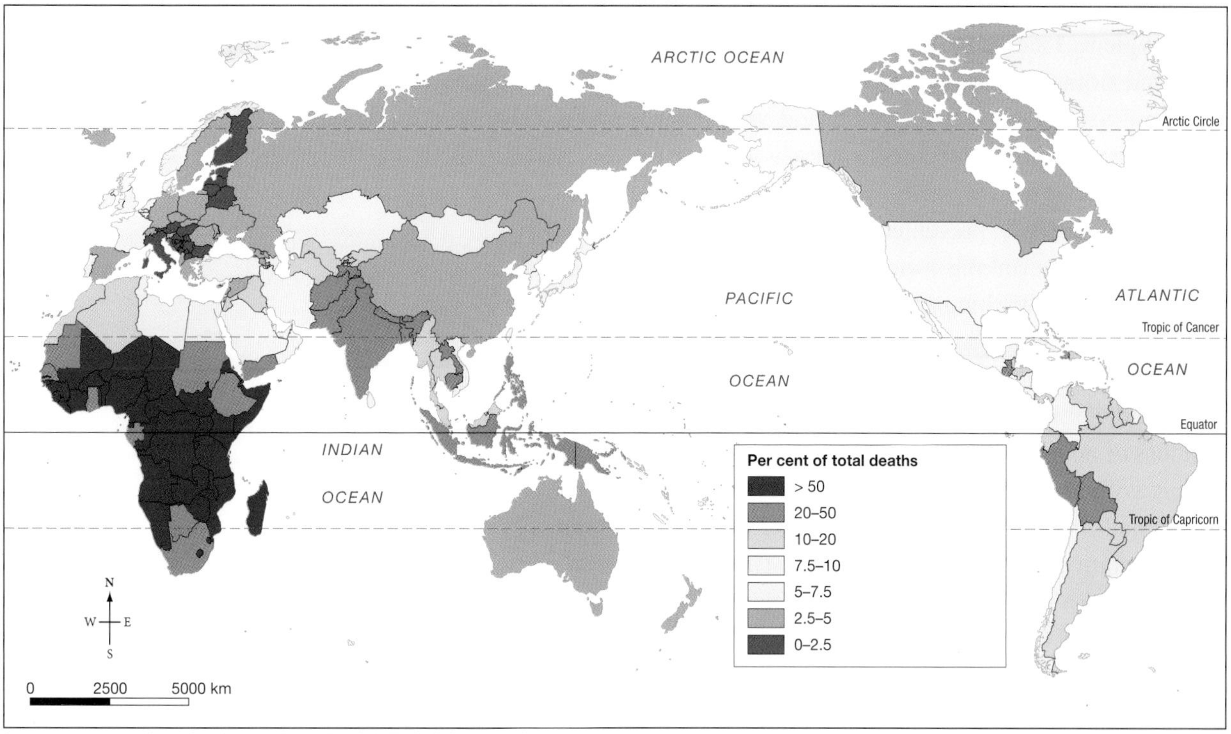

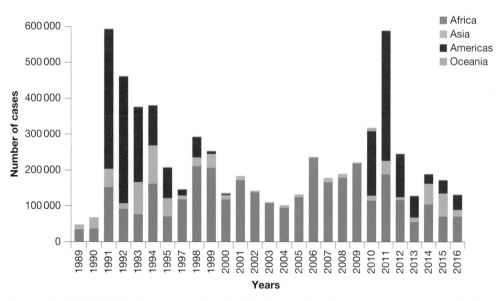

Source: Institute for Health Metrics and Evaluation (IHME). GBD Compare. Seattle, WA: IHME, University of Washington, 2017.
Available from http://vizhub.healthdata.org/gbd-compare. (Accessed 19 May, 2018.)

FIGURE 2.59 Cholera cases reported to WHO by year and by continent, 1989–2016

Source: © 2018 WHO http://www.who.int/healthinfo/global_burden_disease/estimates/en/index1.html

The World Health Organization's cholera prevention strategies include:
- surveillance: quickly detecting cholera cases through medical testing (to properly diagnose cases) and monitoring spread and controls
- water and sanitation health: funding and facilitating access to clean sources of drinking water and good hygiene practices, and the promotion of breast feeding to prevent young children from drinking infected water

- treatment: making oral rehydration treatments readily accessible and rapidly available, along with antibiotics and intravenous rehydration for severely affected patients
- hygiene: local education programs to develop sound hygiene (handwashing with soap, safe food handling and storage, sewage disposal, etc), and awareness of the disease and its symptoms
- oral vaccines: there are three oral vaccines for cholera, which are recommended during outbreaks in areas with **endemic** cholera (the disease can be contracted in the local area, rather than being carried in from other areas).

Cholera is easily prevented if people have access to clean drinking water and do not use dirty water for washing fruit and vegetables or clothing. Children and malnourished adults who catch cholera suffer badly from acute diarrhoea and vomiting, which leads to dehydration and loss of appetite. If not treated quickly, most people with acute cholera die, but rapid administration of oral rehydration therapy (ORT) will reduce fatalities to less than 1 per cent. Unfortunately, most cholera treatment in African is undertaken on a country-by-country basis and often as a response to an outbreak. While emergency action is necessary to save lives, a broader regional approach that focuses on prevention and preparedness would enable international health authorities to be more successful in helping communities at risk. Local education about personal hygiene and water standards, better surveillance of water holes and sanitation disposal areas, along with district alert systems would reduce the impact of future outbreaks, particularly when most occur during the rainy season. One of the major hurdles in achieving these goals is a lack of political cooperation between districts and neighbouring countries.

FIGURE 2.60 Oral rehydration therapy is used in developing countries to combat diarrhoea.

8 teaspoons of sugar   1 teaspoon of salt   1 litre of water

Activity 2.8a: Cholera's impact

Using the information you have learned about cholera, answer the following questions.

Explain cholera's impact
1. Which continents are most prone to outbreaks of cholera? (see figure 2.59)
2. What factors make the population of a country or region more vulnerable to cholera?
3. Why is cholera almost non-existent in developed countries like Australia? What factors contribute to this?

Analyse the spread of cholera and propose ways to manage the risk
4. If you were planning to travel overseas, what personal steps might you have to take before departing to reduce your risk of getting cholera?
5. In 2016 and 2017 there were several cholera outbreaks in Yemen, with up to 1 million suspected cases among the population of approximately 27 million. Locate Yemen on a map and research the political, economic and social situation in Yemen at that time. Write a short paragraph to explain why the population of Yemen might have been more vulnerable to a cholera outbreak at that time.
6. There have been seven significant cholera pandemics in the last 200 years. Research the seven pandemics and construct a map to show the advance of each one.
   (a) Using a different colour for each pandemic, shade the effected countries on a blank world map.
   (b) Show your colour legend.
   (c) Use arrows to show the direction of advance in each pandemic.
7. What patterns of distribution can you identify on your cholera map?

8. The 1991 cholera outbreak in Peru killed more than 10 000 people and was caused by cholera strains living in the ballast water that was dumped from ships.
    (a) How would you propose preventing and managing such an outbreak in a community where cholera was not endemic?
    (b) Construct a management plan to respond to the outbreak.
    (c) Evaluate what difficulties you might encounter in implementing this plan.

2.8.3 Climate change and infectious diseases

Even before people understood the causes of infectious disease, they were aware of the connection between climate and disease outbreaks. In Ancient Rome, wealthy people migrated to surrounding hills during summer to avoid outbreaks of malaria. Like malaria, many diseases tend to be seasonal – as we know, influenza is most common in winter.

The biological causes of infectious disease can be influenced by temperature, precipitation and humidity. It is likely that some infectious agents (bacteria, viruses and parasites) multiply more rapidly in warmer temperatures. The numbers of vectors, especially mosquitoes, are also sensitive to temperatures. Malaria and dengue fever, both mosquito-borne diseases, are much more prevalent in higher temperatures and humidity. Increased rainfall often leads to an increase in water-borne diseases such as cholera, especially when flooding occurs. If climate change has an impact on temperatures and on precipitation patterns around the world, it is likely that patterns of infectious diseases will also change, both temporally (related to time) and spatially (the areas affected). Higher temperatures and longer wet seasons in tropical areas, for example, provide ideal breeding conditions for mosquitoes, which will increase the incidence of diseases such as malaria. Warmer temperatures may also reduce outbreaks of influenza and other diseases that thrive in colder climates.

Spatially, we may see a growth in the number of places susceptible to a variety of tropical infectious diseases. Climate change may also have indirect or secondary effects on infectious diseases. It seems probable that, if sea temperatures increase, we will see an increase in the severity of tropical storms. This may, in turn, increase the danger of natural disasters caused by cyclones. Already, especially in poorer countries, natural disasters trigger outbreaks of disease. If cyclones become more destructive because of climate change, it is likely that the threat of infectious disease outbreaks will increase too. Climate change might also have tertiary effects on disease, if it has an impact of agriculture and food supplies, for example. Reduced food supplies may in turn lead to poor nutrition, making people more vulnerable to infectious diseases.

Malaria

Malaria is one of the world's most common and ubiquitous parasitic diseases. It is also one of the oldest known afflictions to humans. The term *malaria* is an Italian word meaning 'bad air'. It is believed that malaria was rampant for centuries around Rome, probably due to the surrounding swamps. Affecting up to 40 per cent of the world's population and endemic in at least 160 countries, malaria is one of the world's biggest killers. The World Health Organization has estimated that malaria kills more than one million people annually, many of whom are children. An unusual feature of malaria is that it is specific to humans and that mosquitoes are the only known vector to carry its parasites. In 2016, 216 million cases of malaria were reported worldwide, with approximately 445 000 people dying from the disease.

Malaria is caused when the female *Anopheles* mosquito injects a small number of single-celled (protozoa) blood parasites called plasmodium below the human skin. The mosquito bite is felt as its drill-like proboscis enters the skin when about to gorge itself on blood. To prevent coagulation and ensure blood flows easily, the mosquito squirts the skin of its victim with an oily film of saliva containing malaria plasmodia. This ensures that some intrusion of parasites occurs into the body through the surface blood vessels. There are four types of plasmodia parasite known to cause malaria. The most common strain is the *Plasmodium vivax*, which accounts for up to 80 per cent of all cases; however, the *p. falciparum* strain is the most lethal.

FIGURE 2.61 The life cycle of malaria parasites

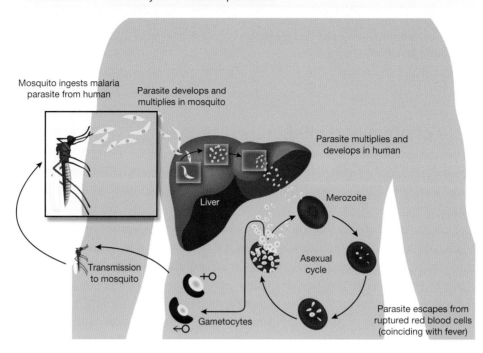

Within days, plasmodia work their way to the liver, feeding on red blood cells and multiplying to the point where they explode into the bloodstream. As the body's immune system attempts to respond to the invaders, fevers and headaches occur, but it is too late. With the body literally boiling from fever and many red blood cells destroyed, vital organs are deprived of oxygen and gradually shut down. Continuing acidification of blood causes brain cells to die; at this stage children often lapse into coma (see figure 2.61).

Malaria hazard zones

Malaria is endemic in many tropical and subtropical regions of the world. It is most prevalent in countries of southern Asia, Africa and South America. In the past, the majority of cases (up to 80 per cent) existed in the east African nations of Zambia, Kenya, Mozambique, Uganda, Tanzania and Ethiopia, as well as the sub-Saharan nations of central Africa (Democratic Republic of Congo, Republic of Congo, Central African Republic, Rwanda and Burundi), and west Africa (Benin, Ghana, Nigeria, Niger, Senegal, Equatorial Guinea and Gabon). Malaria is so prevalent in Africa that WHO doctors claim a child can expect to have up to five malarial episodes per year and that two children will die every minute from it.

The next largest remaining cluster is in southern Asia, mainly in India, Afghanistan, Sri Lanka, Thailand, Indonesia, Vietnam, Cambodia and China. There are also large pockets in Brazil and Central America. Because malaria victims are often unable to work or attend school, or afford healthcare, they become trapped in a cycle of poverty. This also creates a significant economic burden on many countries with a high prevalence of malaria.

In these regions, new strains of the parasite that are showing a resistance to conventional drug treatments are beginning to emerge. The WHO also closely monitors the efficacy of anti-malarial drugs, used to prevent contraction of the disease, to ensure that early signs of resistance in the parasite are recognised and can be managed.

Prevention and treatment

Countries most affected by malaria generally have a low GDP per capita, with governments spending much of the healthcare budget trying to prevent or treat the disease. The WHO reports that in 2016, approximately US$2.7 billion was spent on malaria control, eradication and prevention strategies around the world. The international investment in malaria prevention and treatment strategies has seen a 29 per cent reduction in malaria mortality rates between 2010 and 2016, with a 35 per cent decrease in mortality rates for children under five.

Key strategies in this progress have been:
- early diagnosis and treatment, which help to prevent the disease being transmitted to others and mild cases from becoming severe
- combination drug treatments for newly emerging drug-resistant strains of the parasite
- distribution of insecticide-treated mosquito nets in high transmisison areas
- indoor insecticide sprays to kill mosquito vectors in high transmission areas.

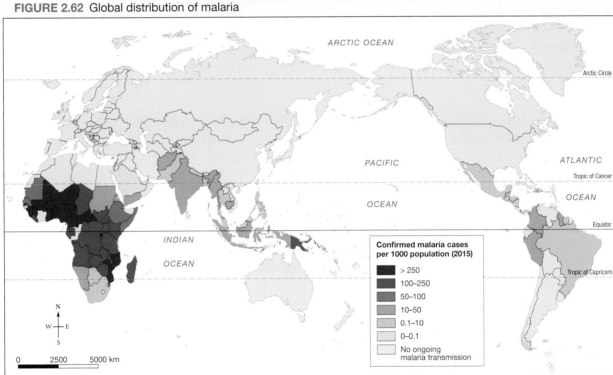

FIGURE 2.62 Global distribution of malaria

Source: Reprinted from WHO, World Malaria Report 2015 — Map — Projected Changes in Malaria incidence rates, by country, 2000–15.

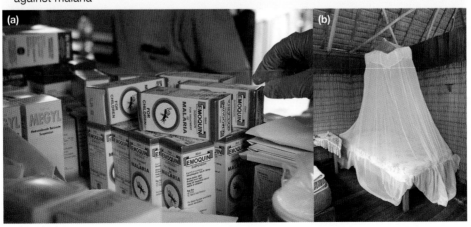

FIGURE 2.63 (a) Malaria medication and (b) mosquito nets, key strategies in the fight against malaria

Malaria patients are treated with hospitalisation and drugs when symptoms appear. Because of the complex life cycle of the malaria parasite, it is very difficult to prevent outbreaks using vaccination. Drugs such as quinine and artemisinin, which are effective at first, have limitations because of the ability of the parasite to develop resistance over time.

Another effective prevention is to provide people in malaria prone areas with durable insecticide-treated netting (ITNs); however, this has limitations if people are not living in permanent housing or cannot repair them if they are damaged. Improved education so that people might better understand weather conditions conducive to malaria outbreaks, as well as prevention and diagnostic tools, will particularly help pregnant women and children. However, this too only works when people are living in circumstances that support this kind of community education – without the threat of war, public disorder or famine.

At a broader community level, steps are being taken to eradicate mosquitoes and reduce habitats and breeding grounds by using insecticide sprays such as DDT (dichlorodiphenyl trichloroethane). Some communities have even gone so far as to drain swamps and nearby wetlands. While reducing the extent of mosquito habitats may reduce local malaria outbreaks in the short-term, these methods have an adverse effect on the environment and non-targeted species.

Is malaria preventable?

While there are many social and economic obstacles that make cure and prevention difficult, it is possible to prevent malaria. Some of the constraints on complete eradication include:
- having access to clothing, repellents and netting that prevent people being bitten in the first place, then a reluctance by some people to use them properly
- having access to preventative anti-malarial treatments
- a lack of knowledge at identifying weather conditions and recognising early symptoms
- a lack of community resources to eradicate mosquito breeding grounds
- a lack of health and hospital facilities
- a chronic shortage of funding to promote further research and development of vaccination programs, particularly when poorer countries are already receiving considerable funding for other food and social projects.

Private donors like the Bill & Melinda Gates Foundation are playing a leading role in malaria research and the development of a vaccine, but many vaccine trials fail before one eventually succeeds. The complex life cycle of malaria parasites makes the disease difficult to treat or prevent by vaccine. Yet, the declining effectiveness of anti-malarial drugs have created an urgent need to discover an effective vaccine. Even though vaccines are highly effective against bacteria and viruses, a vaccine against the malaria parasite has yet to be developed. Viruses such as polio (with 11 genes) are genetically simple compared to the complex structure of parasites like *Plasmodium falciparum* (with over 5000 genes). Also, the rapid movement of Plasmodia from the gut of a mosquito to the human blood vessels, then the liver, and then to red blood cells makes it difficult to target with a specific vaccination. The World Health Organization Malaria Vaccine Technology Roadmap, set in 2006 and updated in 2013, established the goal of developing a malaria vaccine with 75 per cent efficacy by 2030.

Resources

Weblink: Current malaria data
Weblink: Reducing malaria cases

Activity 2.8b: Global malaria spread

Using the information you have learned about malaria, answer the following questions.

Comprehend and explain the spread of malaria

1. List the challenges that exist in combatting malaria in developing nations that do not present such a challenge in developed nations.
2. Explain why mosquito nets have been an effective strategy in combatting the spread of malaria. In your answer, explain at which stage the nets interrupt the transmission cycle.

Apply your understanding and propose response strategies

3. Examine figure 2.64, which shows the change in and sources of funding for malaria prevention around the world.

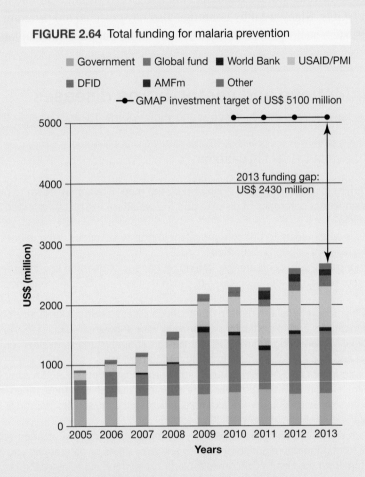

FIGURE 2.64 Total funding for malaria prevention

(a) Describe how funding for malaria prevention changed between 2005 and 2013.
(b) The WHO reported a decline in malaria between 2010 and 2016. How might the patterns in funding for malaria shown in this graph help to account for that change?
4. There was a US$2 430 million gap between what was spent fighting malaria and the 2013 investment target.
   (a) Suggest reasons why this target was not met.
   (b) Propose how these additional funds might be raised.

2.8.4 Reducing the risk of infectious and vector-borne diseases

As the examples of cholera and malaria demonstrate, the impact of infectious diseases can range in scale from individuals and families, to local communities, countries and even worldwide. Reducing the risk of disease also occurs at a number of levels. Individuals and families often choose to be vaccinated against diseases, and local communities can ensure that school children are educated about the importance of hygiene and have community-based healthcare centres. Most often, though, it is state or national governments that play a leading role in managing the risk of infectious diseases.

Risk management may involve prevention, treatment (including emergency responses to major outbreaks of disease) and mitigation. Vaccination programs are designed to prevent or mitigate the effect of infectious diseases, and the provision of treated water and sanitation helps mitigate outbreaks of water-borne diseases. Governments are responsible for health infrastructure (hospitals, healthcare clinics) and healthcare systems (doctors and nurses).

International cooperation is important in managing the spread of infectious diseases across country borders. The WHO is an important body for managing diseases worldwide, as are international non-government organisations (NGOs), such as Doctors without Borders (*Medecins sans Frontieres*). Poorer countries often find it difficult to manage infectious disease outbreaks and may rely on international assistance, both from other countries and from NGOs. Many wealthy countries also have official aid programs that are often used for health-related projects, such as hospitals and clinics, fresh water supplies and sanitation.

Activity 2.8c: Infectious and vector–borne diseases

Using the information you have learned about infectious and vector-borne diseases, answer the following questions.

Explain and analyse the data to identify patterns

1. Refer to figure 2.56.
   (a) Describe the pattern of disease outbreaks illustrated by the map. Does there seem to be a relationship between recent outbreaks of disease and the wealth of the places affected? Why might this be the case?
   (b) Identify any diseases which seem to be more prevalent in poorer countries. Why might these diseases be more prevalent in poorer countries?
2. Refer to figure 2.57. In a paragraph, compare the 10 leading causes of death in low-income and high-income countries. As part of your response, identify causes of death that are significant in one income group, but not the other.
3. Refer to figure 2.58. Describe in detail the pattern of infectious disease outbreaks illustrated by this map. Refer both to general locations and to particular countries in your answer.
4. The United Nations Development Program (UNDP) releases an annual Human Development Report. Access the latest report online and find the table showing health outcomes.
   (a) Select a sample of countries (five or six) with very high human development and a sample with low human development. Construct a table showing the countries' names and three different health and disease indicators for each country (for example, infants lacking measles immunisation, deaths due to malaria and physicians per 10 000 people).
   (b) Write a paragraph in which you analyse the data in your table. What conclusions can you draw from your analysis? What factors might contribute to any similarities or differences you found?
   (c) Select two indicators from the Health Outcomes table that you think are related. Explain how and why you think they are related.
   (d) Select a range of countries from the table from each of the four levels of human development. Draw a scattergraph to illustrate any relationship between the two indicators chosen in part c. Does your completed graph confirm that the two indicators seem to be related?

Resources

Video eLesson: SkillBuilder: Constructing and interpreting a scattergraph (eles-1756)
Interactivity: SkillBuilder: Constructing and interpreting a scattergraph (int-3374)

CHAPTER 3
Challenges for Australian places

3.1 Overview

3.1.1 Introduction

The places we come from and grow up in shape us and live with us forever. We explore them, understand them and their history; we get to know them, their patterns and routines. They become part of us. Likewise, the place we first move out of home to, the place we first settle with a family or the place we will retire to one day can be important places that affect us in deep, personal ways.

Even though places constantly change, individuals and communities can influence that change. People who live in and use the places around us can influence decision makers and take their own actions to ensure those changes are positive and sustainable. By understanding how the places around us function we can make better decisions about their use and future.

In this chapter, you will examine the characteristics of rural, remote and urban places in Australia and the challenges they face. You will explore how places in Australia are defined, patterns of settlement and the factors that help to shape a community's identity. You will explain and identify factors and processes that influence patterns in settlement, such as the physical environment and access to resources, employment opportunities, affordability and access to services. You will also conduct fieldwork to assess and propose solutions to a challenge faced in your community.

FIGURE 3.1 Aerial view of farmlands near Scenic Rim, Queensland

3.1.2 Key questions

- What defines a place as 'rural', 'remote' or 'urban'?
- What factors affect the identity of Australian places?
- How have Australian places changed?
- How have the population distribution and location of settlements in Australia changed?
- What physical, social and economic factors affect settlement patterns in Australia?
- What challenges do people face living in rural, remote or urban areas?
- What are some of the key challenges where you live, and how might you manage these challenges?

Activity 3.1: Reflecting on your knowledge of Australian places

Reflect on your own experience and knowledge of the different types of places in Australia to answer the following questions.

1. What features would define a place as rural, remote or urban? Consider population, location, facilities and any other characteristics that might help to define or differentiate between the three.
2. What makes on place more 'liveable' than another? Consider a range of physical, economic and social factors.
3. What factors, both historically and geographically, have affected where people settle in Australia?
4. What factors do you think will drive population change in Australia in the future? Consider what might affect different rural, remote and urban areas.
5. What specific challenges (economic, social or physical) is your community facing? (Consider whether different groups within your community might list different challenges to the ones you have identified.)
6. Are the challenges you listed long-term challenges that your community has been faced with for a long time or are they relatively new?
7. What do you think is the most significant challenge facing your community? How do you think this challenge should be managed?
8. Is your community 'sustainable'?

3.2 Places in Australia

3.2.1 Defining 'place'

The world is made up of places, both natural and constructed. Places may be:
- sites of biodiversity
- locations for economic activity
- centres for decision-making and administration
- sites for the transmission and exchange of knowledge and/or ideas
- meeting places for social interaction
- sources of identity, belonging and enjoyment
- areas of natural beauty and wonder
- areas of spiritual significance.

A **community** is a system of interacting and interdependent social groups occupying a particular area. They are characterised by patterns of demography, ethnicity, income, family structure, religion or culture. Community groups reside in settlements that vary in size from small hamlets and villages to giant megacities. Their location, size and structure are often dependent on the physical features of an area and its economic and social development.

One thing we can say with certainty is that the concept of 'places' is a cultural construct. They are places because humans define them as such. Every part of earth exists in a 'space', which means it can be represented numerically, using latitude and longitude for instance. However, not every part of earth is considered a 'place', it is only those spaces that humans have defined and attached meaning to that become 'places'.

The importance of Country/Place to Indigenous peoples is an example of the interaction between culture and identity, and shows how places can be invested with spiritual and other significance. In the context of the study of places it is important to consider historically how humans have created places to live and work, and how those places are changing over time.

3.2.2 Historic urbanisation

Human settlement is not a new phenomenon, but its current size, complexity and rate of growth in the world certainly all are. Around 12 500 years ago, humans began to transition from a hunter-gatherer lifestyle to one more settled and sedentary. This period is known as the Neolithic revolution. This fundamental change in human lifestyle was brought about by the domestication of animals (for food and as a source of power) and the rise of agriculture. This change is still impacting on our species and the planet.

Most urban development as we know it has occurred over the past 150 years; it has largely been driven by one or more of the following:
- population increase and a need for living space
- industrial growth, and an expansion and concentration of the workforce
- economic growth and consumer demand for goods and services
- government requirements to make living space more cost-efficient.

Settlements today, whether large cities or small towns, are dynamic places that reflect a diversity of human functions. They are not just collections of concrete, brick and tarmac, but exciting places where people live, work, learn, grow, play, relax and sometimes showcase their enterprise, technology and culture.

Site and situation of early Australian settlements

Most of Australia's major urban areas were established soon after European colonisation of the continent. The locations chosen for these cities have common features: they are on the coastline, they are at or near the juncture of an inland waterway and the sea, they are in or near relatively fertile land and they have plentiful natural resources in the vicinity. (Figure 3.2 shows the expansion of towns and cities in Australia since 1911.)

The factors that most impacted the location of early settlements, colonial settlements in Australia particularly, were:
- water supply, including flood avoidance
- resources such as building materials, food and fuel supply
- geographical situation, including topography, defence, shelter and aspect
- interconnections, such as nodal points and bridging points, for example road crossings, river crossings and connection points on trade routes – any natural or human connection point
- harbours – ocean going transport was critical to survival of the colonies as goods and people had to be transported in by sea from England or other colonies
- location of specific resources for trade, for example areas rich in ore, such as Mount Isa, or areas that once conducted whaling operations, such as Eden in New South Wales.

Specific settlements also had very particular reasons for their location. A year after settlement, the site of the new colony of Brisbane, in what became Queensland, was moved from what is now Redcliffe to counter a hazardous and annoying threat: the mosquito. The new, and current, location of Brisbane provided less of a mosquito problem and offered better transport connections to Sydney and the rest of the world. Colonisers took advantage of the fresh water from the springs in what is now Spring Hill, plentiful resources in the local area and great potential for agricultural production.

FIGURE 3.2 Australian towns by population (a) 1911 (b) 1961 and (c) 2016

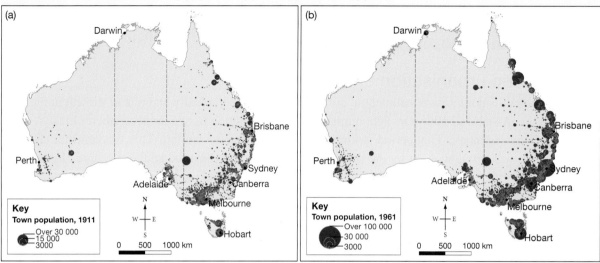

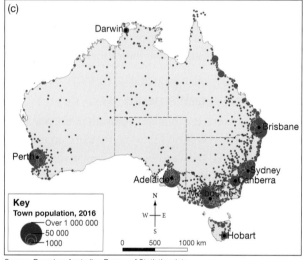

Source: Based on Australian Bureau of Statistics data.

Activity 3.2: Population and place

Considering the different factors that affect the population of a place, answer the following questions.

Explain and comprehend Australian places

1. Explain what makes a place different from a space.
2. Identify two factors that affected the location of early settlements in Australia, and explain why these factors were important.
3. Use the **Australian Bureau of Statistics (ABS) QuickStats** weblink in the Resources tab to find statistics on the population of your nearest town or suburb. Create a table to show how that place compares with Queensland and Australia with regards to median age, average number of people per household, full-time workers and median weekly household income.
4. Create a multiple line graph of the changing population of your nearest major urban area over time to visualise how the population has changed.
5. Describe the factors that make the area where you live an important place. Consider cultural, economic and physical features.
6. Write a paragraph explaining one of the significant changes you have seen in your nearest major urban centre during your lifetime.

Analyse the data and apply your understanding

7. Based on the ABS data for your local area, suggest reasons for population changes in your local area: are the causes social, economic, environmental or a combination? Present your ideas as a Venn diagram or table.
8. Does giving something a name make it a place or is there more to a place than that? Create a list of criteria for what defines a 'place'.
9. Think of three places that were once important in Australia but have lost their significance? Suggest reasons why these places are no longer important.
10. If you could live in any place in Australia, where would it be and why? What benefits would living in this location afford you?
11. With reference to figure 3.2, describe the changes that occur to the distribution and size of Australia's places between 1961 and 2016. Research the events that occurred during this period that might explain the changes you have described?

---

on Resources

Video eLesson: SkillBuilder: Constructing multiple line and cumulative line graphs (eles-1740)

Interactivity: SkillBuilder: Constructing multiple line and cumulative line graphs (int-3358)

Weblink: ABS QuickStats

3.3 Hierarchy of settlements

Settlements vary in size and complexity according to their location, political-cultural background, population, levels of infrastructure and functions (services). When ranked according to size and 'importance', this order of settlements is known as a hierarchy. Smaller settlements such as villages and towns provide residents with low-order functions such as grocery shops, service stations and primary schools, while larger settlements such as cities or metropolises offer high-order functions such as universities, hospitals or television stations.

3.3.1 Broad categories of settlement

Conurbation

A **conurbation** is an extensive urban area that forms when there is an amalgamation of several cities. Tokyo-Yokohama in Japan or Gold Coast-Brisbane-Caboolture in south-east Queensland are conurbations. Although cities may merge together, they tend to retain their separate identities.

FIGURE 3.3 A conurbation: south-east Queensland

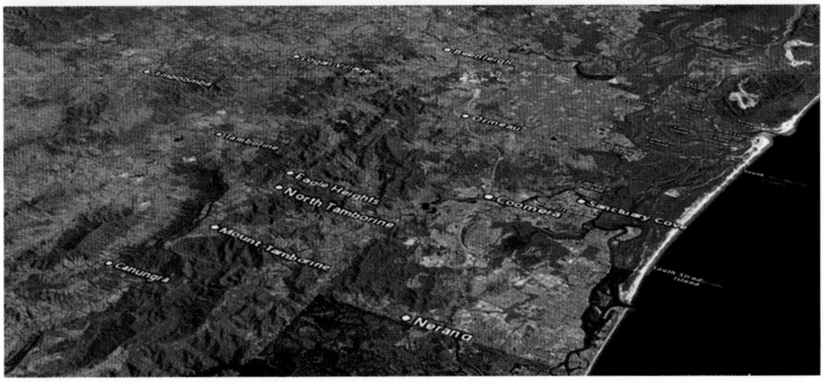

Source: © State of Queensland 2018, Qld Globe

Metropolis

A **metropolis** is the largest single urban settlement in a state (often the capital) or district: Brisbane, for example. They not only have large populations but also offer a broad range of high-order functions such as state government, and legal and administrative services.

FIGURE 3.4 A metropolis: Brisbane

City

A **city** is a large urban settlement with clearly defined boundaries and municipal functions. Toowoomba (pop. 2016: 100 032), Gladstone (pop. 2016: 55 616), Townsville (pop. 2016: 229 031) and Mount Isa (pop. 2016: 18 342) are all cities. Cities are often the centre of local government and regional services.

FIGURE 3.5 A city: Townsville

Town

A **town** is an urban settlement that is smaller than a city but larger than a village, for example, Maleny (pop. 2016: 3734), Clermont (pop. 2016: 3031) and Weipa (pop. 2016: 3095). Towns can be large or small, and generally offer residents low-order functions such as grocery shops, cafes, hardware stores, service stations and schools.

FIGURE 3.6 A town: Maleny

Village

A **village** is a small settlement, such as Burketown (pop. 2016: 238) or Aratula (pop. 2016: 453), with a residential population and a small number of low-order services. Small towns can also be called villages. In recent times, the term 'village' has also been used to describe suburban satellite areas, such as Samford, west of the central business district (CBD) of Brisbane.

FIGURE 3.7 A village: Aratula

Source: © State of Queensland 2018, Qld Globe

Hamlet

A **hamlet** is a tiny settlement consisting of a small number of residential and work buildings, with possibly some other low-order functions, such as a general store, service station or sports oval. On many farm properties, a similar cluster of buildings is referred to as a homestead, for example, Tarong Homestead.

FIGURE 3.8 A hamlet: Cooyar

Source: © State of Queensland 2018, Qld Globe

Activity 3.3a: Examining your local area

Examine demographic statistics in your local area using the online **ABS QuickStats** tool (see the weblink in the Resources tab).

Explain the demographics
1. What is its area?
2. What is its population?
3. What category of settlement would you classify your area of residence?
4. What are the functions found in your local area that justify the classification you have chosen? Justify your decision with statistics.

3.3.2 ABS definitions of settlement sizes

The ABS defines places in Australia in a number of ways. **Mesh Blocks** (MBs) are the smallest geographical area defined by the ABS. They are designed as geographic building blocks rather than as areas for the release of statistics themselves. All statistical areas in the Australian Statistical Geography Standard (ASGS), both ABS and non-ABS structures, are built up from Mesh Blocks. There are 358 122 Mesh Blocks covering all of Australia.

Through the ASGS, any settlement of more than 1000 people is considered an urban area. Settlements of more than 100 000 people are considered 'Major Urban Areas'. Settlements with between 1000 and 99 999 people are considered 'Other Urban Areas'. Settlements with fewer than 1000 people are considered rural areas.

An **Urban Mesh Block** is defined as a Mesh Block with a population density of 200 persons or more per square kilometre. The ABS defines **urban centres** as areas that either:
- have an Urban Mesh Block population greater or equal to 45 per cent of the total population and a dwelling density greater or equal to 45 dwellings per square kilometre or
- have a population density greater or equal to 100 persons per square kilometre and a dwelling density greater or equal to 50 dwellings per square kilometre or
- have a population density greater or equal to 200 person per square kilometre.

FIGURE 3.9 The Gold Coast is an urban centre and locality, but also forms part of a conurbation.

Many rural areas have a central settlement that serves as a hub for the wider community. Under certain conditions, the ABS considers small settlements with a population over 200 people to be an **Urban Centre and Locality** (UCL) within a rural area or locality. These areas are identified and classified using population density data from the most recent census.

TABLE 3.1 Australian Statistical Geography Standard (ASGS)

Name	Definition	Urban/Rural
Major urban	A combination of all Urban Centres with a population of 100 000 or more	Urban
Other urban	A combination of all Urban Centres with a population between 1000 and 99 999	Urban
Bounded locality	A combination of all localities in a region/area	Rural – may contain UCLs as part of a regional centre (e.g. Cunnamulla)
Rural balance	The remainder of state/territory	Rural

Source: Australian Bureau of Statistics

3.3.3 Urban and rural places

There are a number of physical differences between urban and rural areas that can help us understand these places and the challenges they face. There can also be differences in the types of services or industries found in urban and rural areas.

The three-sector theory divides industry into three types:
- **primary sector**: industry producing raw materials such as agriculture or mining
- **secondary sector**: manufacturing industries such as food processing plants or paper milling
- **tertiary sector**: services such as retail, hospitality, healthcare or education.

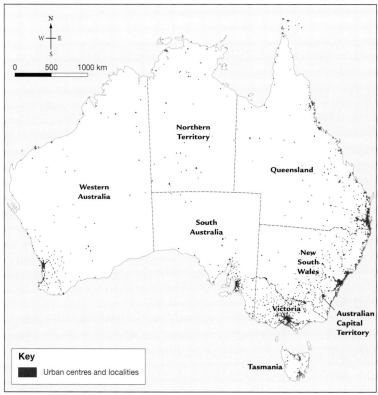

FIGURE 3.10 Australia's urban areas, 2016

Population size

Generally, urban centres have populations of a larger size and higher density than rural areas – although some rural areas have large populations, they are spread over a greater area. In Australia, the ABS defines a settlement with a population of more than 1000 people to be an urban area (see table 3.1). This number can vary in other countries.

Economic activity

Rural settlements have traditionally been defined as areas where most of the workforce are farmers or work in primary production, that is, those who farm or harvest and produce other primary resources. Traditionally, the economies of urban areas are predominantly in secondary industries, such as manufacturing and service, or tertiary industries, such as health, education, retail or hospitality. Of course, these sectors exist in rural areas but in urban areas they are found in higher numbers and as a higher proportion of overall jobs.

Services available

Rural areas have significantly fewer services, such as tertiary institutions or specialist medical services, and less infrastructure, such as public and private transport options, than urban areas due to the higher costs associated with servicing a dispersed and small population.

Development density

Urban settlements have higher density of development and a mix of land use types. Rural areas tend to have much lower densities, more space available and less variety in the types of land use. Rural land use is mainly agricultural but in rural settlements there may be some residential and small-scale industrial land use associated with servicing the primary industry in the area. In contrast, urban areas feature higher density residential areas and heavily industrialised zones.

FIGURE 3.11 Satellite image of rural areas pattern: Torrens Creek

Source: © State of Queensland 2018, Qld Globe

FIGURE 3.12 Satellite image of urban areas pattern: Toowoomba

Source: © State of Queensland 2018, Qld Globe

Demography – social

Urban areas are more likely to have a higher proportion of younger population while rural areas are more likely to have a higher proportion of people over 65. Younger people tend to leave rural areas to look for work and other opportunities in cities thus impacting the demography of these places (see figure 3.13 and table 3.2). One measure of the social demographics in a community is the total dependent population, or proportion of the community not of typical working age (between 15 and 65).

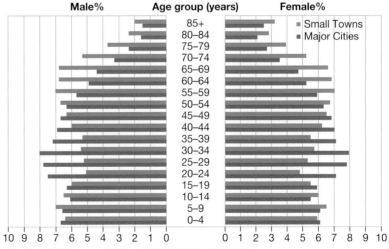

FIGURE 3.13 Age and sex distribution of people living in small towns and major cities in Australia, 2016

Source: ABS Census of Population and Housing, 2016

TABLE 3.2 Age dependency in two cities and two towns in Queensland

	% under 15	% over 65	Total % dependent population	Population 2016
Brisbane	19.6	12.8	32.4	2 270 800
Toowoomba	19.1	18.3	37.4	100 032
Murgon	19.9	24.8	44.7	2058
Barcaldine	20.3	19.6	39.9	1287

Source: Australian Bureau of Statistics

Activity 3.3b: Understanding urban and rural places

Examine table 3.2 and figures 3.13 and 3.14 to answer the following questions.

FIGURE 3.14 Examples of rural and urban areas

Source: Mike Swaine / Above Photography

Explain and comprehend the demographics of urban and rural areas

1. Do urban or rural areas tend to have a higher proportion of people over 65? Why might this be the case? Give evidence to support your answer.
2. Do urban or rural areas tend to have a higher proportion of under 15s? Why might this be the case? Give evidence to support your answer.
3. Explain why urban areas have a higher proportion of people in their 20s.
4. Categorise and describe the features of each of the places in figure 3.14.

Analyse the age-profile data

5. Consider the different challenges faced in providing services and facilities to people in remote, rural and urban areas in Queensland. How might a better understanding of the age profiles of urban and rural areas impact service provision in Queensland?
6. How can knowledge of this age profile change how services are provided in different parts of the state?

3.3.4 Remote places

Remote is a term used to describe parts of our world that are not typically accessible to most people. Remote areas are those that are away from built up areas but pinning down a clear definition can be hard. For instance, the Australian Government uses three different scales to measure remoteness in different parts of the country. This can have huge impacts on how much money is allocated to different areas as well as the services provided to the people who live there.

The **Rural Remote and Metropolitan Areas classification** (RRMA) uses population size and distance from the nearest service centre to classify places into seven categories: capital cities, other metropolitan centres, large rural centres, small rural centres, other rural areas, remote centres and other remote areas.

The **Accessibility/Remoteness Index of Australia** (ARIA) uses Geographic Information System (GIS) data to measure road distance to service centres to produce a sliding scale of remoteness. ARIA includes five categories: highly accessible, accessible, moderately accessible, remote and very remote.

The **Australian Standard Geographical Classification** (ASGC-RA) defines remoteness by Census Collection Districts on the basis of the average ARIA score within the district. The remoteness of local areas is then assessed and classified by the ARIA categories.

Certain areas of Australia do not have many people engaged in primary industries but these areas can be hundreds of kilometres from the nearest service centre. For instance, Kakadu, The Pilbara or the desert areas of inland Australia would be considered remote rather than rural.

3.3.5 Land use

Land use describes how the land is being used, either by humans or naturally, how people use the land or the functional dimension of land for different human purposes.

How the land is used can demonstrate an area's urban or rural qualities. Most people in most places around the world use the land in relatively standard ways and most cities contains streets, ports, bus and train stations, building sites, power stations, houses, apartments and shops.

Similarly, in rural areas across the globe land is commonly used for farms, dams, roads and highways, rail lines and/or ports, and patches of native vegetation. There are a range of standard land use categories that can be found across different settlements and communities. Examples of these are shown in table 3.3 and figure 3.16.

FIGURE 3.15 Aerial images of the Pittsworth area show the transition to large scale farm lands.

Source: © State of Queensland 2018, Qld Globe

TABLE 3.3 Satellite imagery showing examples of different land use

Land use and example	Land use and example	Land use and example
Residential: land where people live; can be high, medium or low density (Gladstone) 	Commercial: land used for commercial purposes, such as a shopping complex or strip (Toowoomba) 	Water: land for water delivery and storage in communities (Advancetown Lake and Hinze Dam) 
Light industrial: land for industrial activity that uses light-weight materials and produces light goods, such as foodstuffs, arts and crafts, clothing or small electronic items (Townsville)	Recreational: land used for recreational purposes, such as a green space, a park, a playground, sporting field or skate park (Glass House Mountains) 	Primary industry/agricultural: land used for primary industries (farming and cropping) (between Cecil Plains and Millmerran) 
Heavy industrial: land for industrial activity that uses heavy materials or heavy equipment, such as automotive, aeronautical or ship building, chemical and electronic manufacture, and machine tooling (Rocklea)	Administrative: land used for governmental purposes, such as council offices or courts (Brisbane City Hall) 	Military: land used for defence purposes (Gallipoli Barracks, Enoggera) 
Educational: land devoted to educational use, such as a kindergarten, school or university (Craigslea State School) 	Transport: land that is used to transport people or things, such as airports, ferry terminals, ports, bus corridors, or train lines and stations (Roma airport) 	Natural/protected: undeveloped or rehabilitated land in a natural or close to natural state (Daintree National Park) 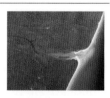
Medical: land used for the medical industry, such as a hospital (Royal Brisbane and Women's Hospital) 	Utility: land for utility services, such as electricity, telephone and internet, power and sewage removal (Stanwell Power Station) 	Mining: land used for mining (Mount Isa) 

Source (images): © State of Queensland 2018, Qld Globe

FIGURE 3.16 Examples of rural and urban areas

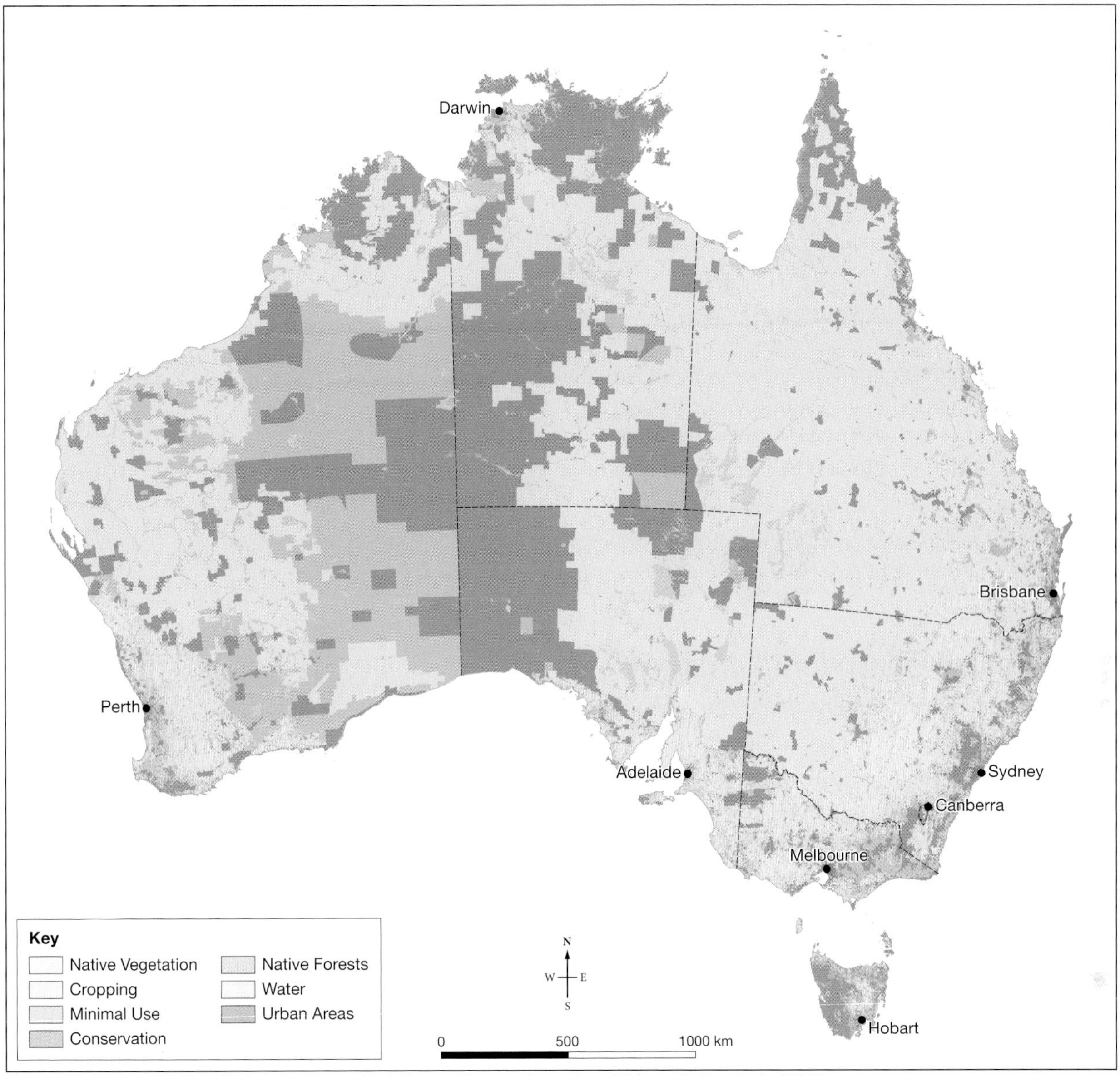

Key
- Native Vegetation
- Cropping
- Minimal Use
- Conservation
- Native Forests
- Water
- Urban Areas

Source: © Commonwealth of Australia Australian Bureau of Agriculture and Resource Economics and Sciences Redrawn by Spatial Vision. Used under a CC Attribution BY 4.0 International Licence.

Activity 3.3c: Map land use in your area

Use an online interactive mapping tool (such as the Queensland Globe, ScribbleMaps or Google Maps) that allows annotation. Individually or in small groups examine a specific suburb or area in your region. Interpret the satellite image layer of data.

Explain and comprehend the land use

1. Identify and mark the different types of land use evident on the map. (Ensure you select appropriate symbols, patterns or colours to represent different land use types.)
2. Adjust transparency to better view the image and the colours.
3. Annotate your map or create a key.

Analyse the map and apply your understanding to consider the opportunities and challenges

4. Analyse your map to identify the land use patterns. Describe the pattern of land use evident in your area and explain what factors might contribute to those patterns. (Consider a range of factors, including social, physical and economic.)
5. Predict how your area might look in 20 years. Give justifications for your prediction.
6. What does a well-planned area look like? Write a list of factors that should be taken into account when planning land use in an area.
7. What types of land use is in the area you examined that could lead to potential negative outcomes? Suggest how these negative outcomes might be avoided.

---

on Resources

Video eLesson: SkillBuilder: Constructing a land use map (eles-1755)

Interactivity: SkillBuilder: Constructing a land use map (int-3373)

Video eLesson: SkillBuilder: Describing change over time (eles-1753)

Interactivity: SkillBuilder: Describing change over time (int-3371)

3.3.6 Interpreting aerial imagery

Interpreting aerial and satellite imagery is an important part of examining the features of a specific area. NASA recommends the following five strategies as a starting point for anyone wanting to interpret a satellite image.
- Look for a scale
- Look for patterns, shapes and textures
- Define the colours (including shadows)
- Find north
- Consider your prior knowledge.

Look for a scale

Identifying the scale of your image gives you clues that can help you interpret what you see on your screen. Most maps use a traditional scale bar to help you understand the map's scale. Satellite images refer to their resolution to identify scale and this refers to the size of the earth represented within each pixel of the image.

All digital images are made up of pixels; the more pixels, the higher the resolution of the image. A pixel is also a square, so if someone refers to a 2 cm pixel size, it means that each pixel covers 2 cm x 2 cm of the earth, allowing you to see very high detail. A 30 m pixel would cover a larger area, 30 m x 30 m, but the resolution would be lower.

Satellite images can be taken at different scales for different purposes. Large scale imagery has pixels that cover small parts of the earth, such as 2 cm pixels. These images would show things, such as a car, in very high detail. In small

FIGURE 3.17 Satellite image of Mossman and surrounding area

Source: © State of Queensland 2018, Qld Globe

scale imagery, the pixels could be kilometres in size but the image would cover a much larger part of the earth. Depending on the resolution of the image you are viewing, a town could be a dot on the image or it could take up the entire image with high levels of detail.

Large-scale high-resolution images can be used to monitor crop health, record changes in vegetation cover, monitor waterway health or manage emergency service resources in real time. Small scale low resolution images can be used to view weather conditions, monitor region, state or nation-wide vegetation conditions, examine state-wide flood impacts or view drought impacts across the state.

Look for patterns, shapes and textures

There are many patterns that identify how humans use the land. Look for the distinctive patterns that farms demonstrate with their regular rectangular shapes of different fields. Some circles may be evident if circular irrigation is being used.

Transportation and human infrastructure, such as roads and railways, can be very obvious. Cities usually have a number of transport routes going into and out of them. Urban areas have lots of areas in regular shapes that are grey or other lighter colours than the natural world around them.

Look for landforms and natural features such as mountains, canyons, islands, volcanoes or clouds. Clouds in a particular pattern may indicate an area of land moving to a higher altitude.

Define the colours (including shadows)

The colours in an image tell you something about the world below. What this is depends on the wavelength on the sensor of the satellite you are viewing data from. For this example we will focus on true colour images, that is, images that reflect the visible spectrum and thus show us what our own eyes would see. There are also satellite images that work on infra-red through to ultraviolet on the electromagnetic spectrum.

- Green generally refers to vegetation: darker greens usually indicate thicker and taller areas of vegetation while paler greens indicate grasslands. The colours of some vegetation are seasonal so additional knowledge can be helpful when interpreting imagery.
- Blues and blacks tend to indicate water, either oceans, rivers, lakes, creeks, ponds or other waterways. Sometimes algae or sediment can be seen in water as green or brown stains.
- Grey or white indicate built up areas, such as urban areas, or areas with a lot of ice.

The colours that show agricultural land will depend on what is being grown, but lighter greens, yellows and some browns indicate farmland. Browns or oranges indicate a lack of vegetation or very dry areas. These can be seen closer to arid areas.

Interpretation will often require the viewer to combine information. For example, in figure 3.17 there are a lot of brown and earthy colours to the east of the township that suggest farmlands. The regular rows of buildings show the placement of streets in the town. A golf course is also visible to the east of town, identifiable by the green grass lines of the fairways.

Find north

Most imagery will be oriented with north pointing to the top of the image but if not, it is useful to be aware of where north is. This can help to work out the direction of prevailing winds, a mountain range or other natural features.

Consider your prior knowledge

What you know about the area can be useful in working out what you are looking at and when the image was taken. Prior knowledge can help you work out features such as recent areas of bushfire or new buildings or roads.

Activity 3.3d: Comparing aerial images with topographical maps

Complete this activity by either referring to figures 3.18 and 3.19, or by using an online interactive mapping tool that allows annotation (such as the Queensland Globe, QTopo, ScribbleMaps or Google Maps) to examine an online aerial image or satellite image and topographical map of a place you are familiar with.

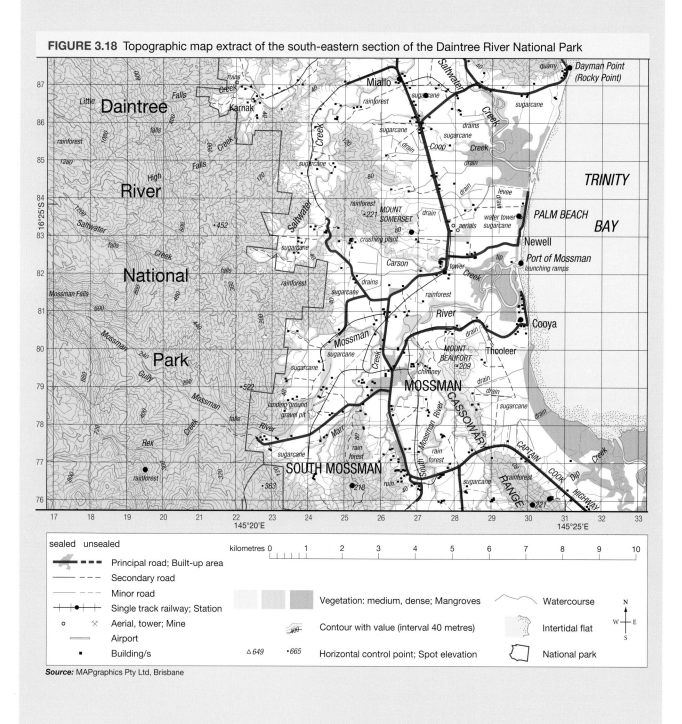

FIGURE 3.18 Topographic map extract of the south-eastern section of the Daintree River National Park

Source: MAPgraphics Pty Ltd, Brisbane

FIGURE 3.19 Examples of rural and urban areas

Source: © State of Queensland 2018, Qld Globe

Explain and comprehend the images

1. Look for any evidence of natural processes at work. Label waterways, mountains or hills. Is there any erosion in the image? Any islands? (These are some examples of evidence of natural processes – label others you see.)
2. The Queensland Globe and Google Earth Pro both provide access to banks of past imagery. View the changes that have occurred in this area since the early 2000s. What are the main natural and built changes that can be seen in this record?

Analyse the information and apply your understanding

3. Can you deduce any change over time using this evidence?
4. Compare the topographical map with the satellite image. Are there any differences? Give examples of how the two data sources show changes to the land use.

Resources

- **Video eLesson:** SkillBuilder: Interpreting an aerial photo (eles-1654)
- **Interactivity:** SkillBuilder: Interpreting an aerial photo (int-3150)
- **Video eLesson:** SkillBuilder: Comparing an aerial photograph and a topographical map (eles-1751)
- **Interactivity:** SkillBuilder: Comparing an aerial photograph and a topographical map (int-3369)
- **Weblink:** QTopo
- **Weblink:** Queensland Globe

3.4 Where are Australian places?

3.4.1 Indigenous places in Australia

It is estimated that between 300 000 and 750 000 people inhabited the Australian continent prior to European contact in the 18th century. These people were separated into around 250 individual nations (see figure 3.20). Most nations spoke their own language and had a defined geographic area in which they lived.

The lack of written records means that it is difficult to say with certainty exactly how the population of Australia was distributed at that time. Indigenous Australians lived a primarily nomadic lifestyle, although some cultural groups were more settled than others. They developed ways of managing the natural environment for sustainable use, including sophisticated land management using fire and other agricultural techniques. The Gunditjmara people are famous for creating the world's oldest aquaculture infrastructure, traps that corral and catch eels, found in Budj Bim, Victoria.

The Indigenous population, although covering most of the continent, was most likely more populous on the east coast, in what is now Queensland, New South Wales and Victoria. This was due to a favourable biophysical environment, food availability, climate and plentiful water sources from the Great Dividing Range and the Murray–Darling Basin.

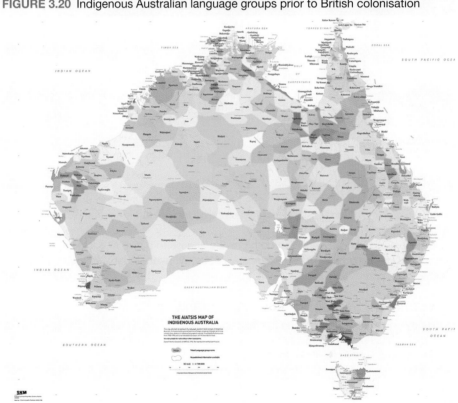

FIGURE 3.20 Indigenous Australian language groups prior to British colonisation

Note: A larger version of this map appears in the eBookPLUS.
Source: This map attempts to represent the language, social or nation groups of Aboriginal Australia. It shows only the general locations of larger groupings of people which may include clans, dialects or individual languages in a group. It used published resources from 1988–1994 and is not intended to be exact, nor the boundaries fixed. It is not suitable for native title or other land claims. David R Horton creator, © AIATSIS, 1996. No reproduction without permission. To purchase a print version visit: www.aiatsis.ashop.com.au/

Indigenous Australia now

In 2016, Indigenous Australians made up 2.8 per cent of the nation's population, with one-third of Indigenous Australians living in capital cities. However, the overwhelming majority, 79 per cent, of Indigenous

Australians now live in urban areas, including those outside of the capital cities. This is compared to 68 per cent of non-Indigenous Australians who live in capital cities.

Statistically, Indigenous Australians are significantly younger than the general population, with a median age of 23 compared to Australia's overall median age of 38. Only 4.8 per cent of Indigenous people are over the age of 65, which is a considerably smaller proportion than for non-Indigenous people at 16 per cent.

Figure 3.21 shows 60 per cent of Indigenous Australians live in Queensland or New South Wales but make up only 4 per cent and 2.9 per cent of each state's total population respectively.

TABLE 3.4 Indigenous population in Australia by state and territory

State	Urban (%)	Urban pop.	Rural (%)	Rural pop.	Total
New South Wales	85.5	184 830	14.1	30 480	216 176
Victoria	86.8	41 479	12.6	6021	47 788
Queensland	81.2	151 423	18.4	34 312	186 482
South Australia	80.7	27 586	18.5	6324	34 184
Western Australia	72.6	55 160	26.6	20 210	75 978
Tasmania	72.3	17 042	27.5	6482	23 572
Northern Territory	50.0	29 124	48.8	28 425	58 248
Australian Capital Territory	99.3	6462	0.3	19	6508
Total (Australia)	79.0	512 844	20.4	132 430	649 171

Source: Australian Bureau of Statistics

Activity 3.4a: The distribution of Indigenous Australians

Refer to figure 3.2c, 3.21 and table 3.4 to answer the following questions.

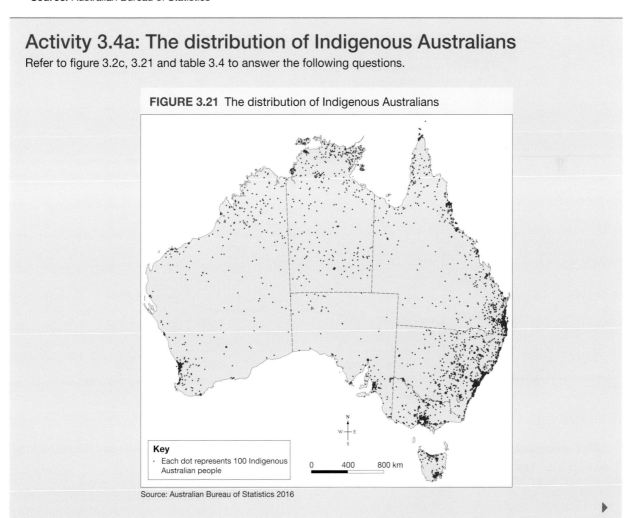

FIGURE 3.21 The distribution of Indigenous Australians

Key
- Each dot represents 100 Indigenous Australian people

Source: Australian Bureau of Statistics 2016

> Explain and comprehend Indigenous population distribution
> 1. Describe the distribution of Indigenous people in Australia.
> 2. Describe the similarities and differences between the distribution of Indigenous Australians and the general population of Australia. (Begin by referring to figure 3.2c.)
> 3. Note that the number of Indigenous Australians residing in urban areas plus the number residing in rural areas doesn't always add up to the total. Why might this be?
> 4. Draw a graph to show the percentage of Indigenous Australians living in urban areas in each state and territory.
> 5. Why might the ratio of Indigenous Australians in urban areas in the ACT be so high?

3.4.2 Australia's population distribution

Australia's population is growing, as shown in figure 3.22. Despite the fact that Australian identity heavily references the bush and the outback, our population distribution is overwhelmingly urban, with nearly two-thirds of the population living in a capital city. This distribution is heavily influenced by our geography. Most of the major urban settlements can be found on the coastline and, like the pre-European Indigenous population, most of us can be found in the east, in Victoria, New South Wales and Queensland (see table 3.5), although there is also a smaller population cluster in south-west Western Australia. Historically, these major urban settlements, usually each state's capital city, required access to the open seas as the only way of connecting with the outside world.

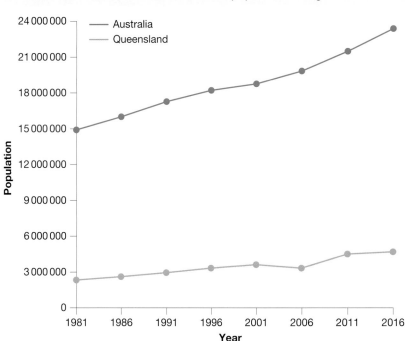

FIGURE 3.22 Australia and Queensland's population change

Source: ABS, © The State of Queensland 2009

Other urban areas can be also located primarily on the coastline but some urban centres are located inland. Table 3.6 lists Australia's 30 most populous urban areas.

TABLE 3.5 Australian state populations

State	Population (June 2016)
New South Wales	7 725 900
Victoria	6 068 000
Queensland	4 844 500
Western Australia	2 617 200
South Australia	1 708 200
Tasmania	519 100
Australian Capital Territory	396 100
Northern Territory	244 900
Total (Australia)	24 127 200

Source: Australian Bureau of Statistics, ABS, 3101.0 — Australian Demographic Statistics, Sep 2017

Remote and rural settlements are also located inland along the western and northern coastline between Perth and Darwin and Darwin and Cairns. These settlements tend to follow one of the following patterns:
- **isolated**: smaller, individual dwellings (usually not close to settlements) as natural resources may not be plentiful enough to sustain a large population (see figure 3.23a)
- **dispersed**: settlements that tend to be spread out across a large area with very few examples of nucleation evident (see figure 3.23b)
- **nucleated**: settlements where some buildings have come together due to economic or social/cultural reasons, or sometimes as a means of security (see figure 3.23c)
- **linear** or **ribbon**: settlements in which buildings are strung out along lines of transport or communication. This could include villages that exist along roads or at intersections, or buildings that congregate along rivers and other waterways. (see figure 3.23d)
- **ring** or **green villages**: settlements found in more remote parts of the world, these villages, such as Aurukun, are built around a central meeting area and point to the communal aspect of many traditional lifestyles. (see figure 3.23e)

FIGURE 3.23 Settlement patterns: (a) isolated (b) dispersed (c) nucleated (d) linear or ribbon (e) ring or green villages

Source: © State of Queensland 2018, Qld Globe

Activity 3.4b: Population distribution in Australia

Refer to figure 3.24 and an online mapping tool with annotation function (such as the Queensland Globe, ScribbleMaps or Google MyMaps) to complete the following questions.

Explain and comprehend settlement patterns

1. Examine figure 3.24. Describe the pattern of population distribution and density in Brisbane.
2. Find an online topographical map of Brisbane (such as at QTopo). What physical features that are not shown in figure 3.24 might have influenced population distribution patterns?
3. Examine the spatial organisation of Brisbane's and Canberra's Central Business Districts using an online mapping application. Describe the differences you see.
4. Suggest reasons why the spatial patterns in Brisbane and Canberra are so different.
5. Use the satellite image view from an online mapping tool to examine different rural and remote settlements in Australia. Find an example of each of the settlement patterns displayed in figure 3.23 and record its latitude and longitude, and draw a sketch of the settlement pattern in a table like the one below.

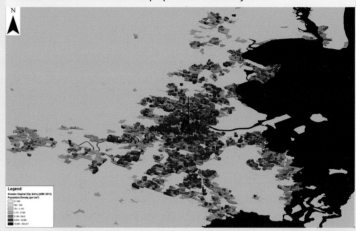

FIGURE 3.24 Brisbane's population density in 3D

Source: This material was compiled and presented by .id, the population experts. www.id.com.au. This material is a derivative of ABS Data that can be accessed from the website of the Australian Bureau of Statistics at www.abs.gov.au, and which data can be licensed on terms published on the ABS website.

Settlement pattern	Latitude	Longitude	Sketch of pattern
Isolated			
Dispersed			
Nucleated			
Linear or ribbon			
Ring or green villages			

TABLE 3.6 The most populous places in Australia

Rank	Name	State/territory	Population (June 2016)	Proportion of population (%)
1	Sydney	New South Wales	5 029 768	20.74
2	Melbourne	Victoria	4 725 316	19.24
3	Brisbane	Queensland	2 360 241	9.74
4	Perth	Western Australia	2 022 044	8.56
5	Adelaide	South Australia	1 324 279	5.50
6	Gold Coast-Tweed Heads	Queensland/New South Wales	646 983	2.64

(continued)

TABLE 3.6 The most populous places in Australia (*continued*)

Rank	Name	State/territory	Population (June 2016)	Proportion of population (%)
7	Newcastle–Maitland	New South Wales	436 171	1.81
8	Canberra–Queanbeyan	Australian Capital Territory/New South Wales	435 019	1.80
9	Sunshine Coast	Queensland	317 404	1.27
10	Wollongong	New South Wales	295 669	1.23
11	Hobart	Tasmania	224 462	0.92
12	Geelong	Victoria	192 393	0.79
13	Townsville	Queensland	178 864	0.76
14	Cairns	Queensland	150 041	0.62
15	Darwin	Northern Territory	145 916	0.60
16	Toowoomba	Queensland	114 024	0.48
17	Ballarat	Victoria	101 588	0.42
18	Bendigo	Victoria	95 587	0.39
19	Albury-Wodonga	New South Wales/Victoria	90 576	0.37
20	Launceston	Tasmania	86 335	0.36
21	Mackay	Queensland	80 755	0.35
22	Rockhampton	Queensland	78 795	0.33
23	Bunbury	Western Australia	74 113	0.32
24	Bundaberg	Queensland	70 310	0.29
25	Coffs Harbour	New South Wales	70 134	0.29
26	Wagga Wagga	New South Wales	55 960	0.23
27	Hervey Bay	Queensland	52 806	0.22
28	Mildura-Wentworth	Victoria/New South Wales	50 998	0.21
29	Shepparton-Mooroopna	Victoria	50 693	0.21
30	Port Macquarie	New South Wales	46 247	0.19

Source: Australian Bureau of Statistics, 3101.0 — Australian Demographic Statistics, Sep 2017

6. Using the data in table 3.6, calculate one in how many people in Australia live in:
   - Sydney
   - Brisbane
   - Hervey Bay.
7. Imagine you have been asked to create a map of Australia's 30 most populous places using the data in table 3.6. You have been asked to use four classes of data for your city populations. Describe how you would break the data up into four classes and the parameters you would use.

8. What symbols or shapes would be most appropriate for a map that illustrates the data for the whole of Australia? Create a key using your chosen symbols.
9. Would the symbols you chose for question 8 change if you were only mapping the population of suburbs in a large city like Townsville? Explain why your symbols might change.

3.4.3 Australia's natural places

As a continent, Australia has many places that haven't been developed or influenced significantly by humans. Australia is known for its natural beauty and much of this has been officially recognised and deemed special or significant by being officially named parklands, national parks, state forests or recognised as important through other state, national and international registers. Figure 3.25 shows the main gazetted places of natural and cultural significance in Australia.

Many natural places have special significance to Australia's Indigenous population. Some of these places are recognised formally, while knowledge of others is passed from generation to generation through spoken word, stories and songs.

Some natural places are deemed important because of their economic potential, either as forests, dams or areas of resource extraction.

Australia's biomes and climate zones are two more ways to classify Australia's natural places. Figure 3.29 shows the main climate zones that influence Australia's landmass.

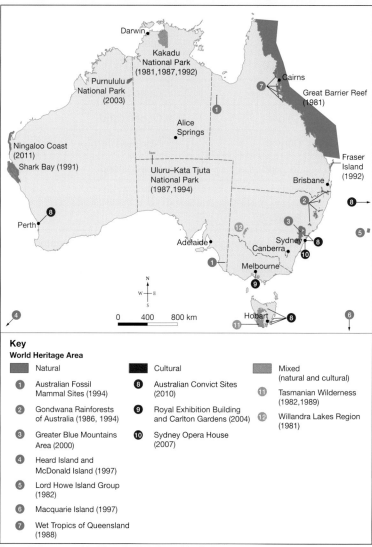

FIGURE 3.25 Australia's sites of natural and cultural significance

Source: © Commonwealth of Australia 2018. State of the Environment 2011 Committee. Australia state of the environment 2011. Independent report to the Australian Government Minister for Sustainability, Environment, Water, Population and Communities. Canberra: DSEWPaC, 2011.

Activity 3.4c: Natural places in Australia

Refer to figure 3.25 to answer the following questions.

Explain and comprehend significant places

1. What places shown in figure 3.25 have you heard about? Would you like to visit any of these places? What features of a place would make it 'significant'?

2. The places shown in figure 3.25 are important to Australia but what about the places that are important in your local area? List some local places of natural or cultural significance.
3. What places might we value into the future for their natural beauty? What places will be valued for their cultural importance in the future?

3.5 Factors affecting Australia's population distribution

Population distribution illustrates the way in which a population is spread across an area. If we think about Queensland, people are not distributed evenly across the state. More people are found along the coastline and in the south-east corner, and these people live in higher densities than in other parts of the state. This pattern is evident in figures 3.26 and 3.27.

Population distribution can be represented visually using a number of methods. Most common are dot density maps where each dot represents a certain number of people, and choropleth maps, where each region is shaded or coloured according to the value in that area, in this case population. Figures 3.26 and 3.27 both represent the same population data but in two different ways.

The dot density map of Queensland, figure 3.26, shows more dots towards the coastline and in the south-east corner of the state. It also shows some areas with no dot symbols at all, which doesn't necessarily mean those areas are empty, just that they don't have a large enough population to warrant a dot. Dot density maps should have a key that tells you how many people are represented by each dot. In figure 3.26 it is 1 dot per 100 people.

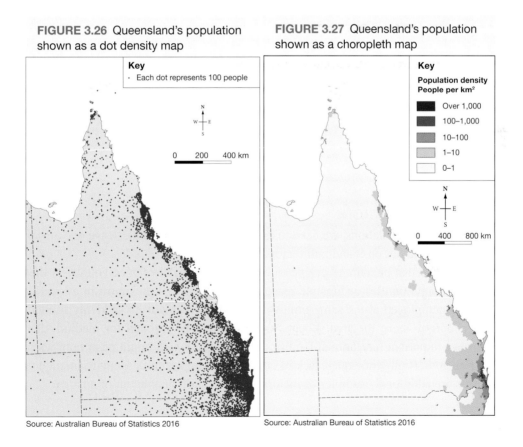

FIGURE 3.26 Queensland's population shown as a dot density map

Source: Australian Bureau of Statistics 2016

FIGURE 3.27 Queensland's population shown as a choropleth map

Source: Australian Bureau of Statistics 2016

Figure 3.27 shows a choropleth map of Queensland. Shades of orange are used to represent different population densities with lighter shades indicating lower values (lower density) and darker shades indicating higher values (higher density).

Population density for a place can be calculated easily by dividing the total population by the total area in question.

$$\frac{\text{total population}}{\text{area }(\text{km}^2)} = \text{population density}$$

Activity 3.5a: Comparing population data

TABLE 3.7 Population and area of Queensland and Australia, 2016 census

	Population	Area
Queensland	4 948 700	1 730 648
Australia	24 702 900	7 692 024

Source: Australian Bureau of Statistics

Explain and comprehend population data

1. Using table 3.7, calculate the population densities of Queensland and Australia.
   (a) Which is higher?
   (b) Which is lower?
   (c) What is the percentage difference?
   (d) What factors might contribute to the difference between the two densities?
2. Research the population and area of the following places and work out their population density. Compare them to Queensland's population density and describe the differences you see.
   (a) Your local area, town or suburb
   (b) An inner-city Brisbane suburb such as Milton
   (c) An outer Brisbane suburb such as Springfield
   (d) A regional centre such as Chinchilla
   (e) Tokyo or Mumbai.

3.5.1 Push and pull factors

When examining changing populations, we need to examine why people might leave one area and move into another. Geographers talk about push and pull factors.

Push factors are those that encourage people to leave an area. In developed countries, these factors are usually related to job opportunities or lifestyle reasons. For instance, a downturn in a local industry could leave many people unemployed at the same time so job prospects might be suddenly limited. This would encourage some people to leave the area. Local parkland might be rezoned for medium density housing, which would remove important recreational facilities for an area, which might push families out. These are examples of push factors. High crime rates, lack of employment opportunities, environmental degradation or removal of services are additional examples of factors that may push people out of an area.

Conversely, **pull factors** are those that attract people to a location. The potential for employment and better job prospects, or the provision of recreational facilities that could encourage families to move into an area would be considered pull factors. Other pull factors might include transport connections in an area, service provision, educational and medical facilities or the value of the natural environment.

3.5.2 Physical factors affecting population distribution

Geography

More physically remote, extreme and harsh areas such as mountains or deserts generally have fewer people. Coastlines and other water sources can boost human settlement as people are attracted to living by the water for a variety of reasons.

Physical connectivity is important, too. Post-European contact, most settlements in Australia were founded on the coastline because shipping was vital to maintaining a connection to the other colonies, Britain and the rest of the world. Indigenous Australians had well-established trade and transport routes that served as the basis for many of our current highways and roads as they were built upon by European settlers. Settlements would be more likely to spring up in these places as people moved through them regularly.

Water

People need steady and stable access to fresh water for drinking, cleaning, bathing, industry and agriculture. As early settlement populations increased, more areas were set aside for permanent water storage, such as Wivenhoe Dam near Brisbane. The provision of water is one of the fundamental drivers of population – populations will naturally adjust according to the capacity of water available for use. Pipelines and desalination plants can also help to provide water to areas that do not have sufficient natural sources of potable water.

Resources

Resources for building, eating and trading are all predictors of where people will live. Early settlers in particular required significant resources to establish and grow their communities. In more recent times it has been possible to 'manufacture' some of these characteristics to establish communities. Fly In-Fly Out (FiFo) communities in mining areas are small, temporary camps that have no local access to all of the required resources to properly establish a community from scratch. Water, food and other resources are brought into the area to allow workers to live near their mine sites. As more and larger areas of mineral production were discovered in central Queensland, populations in those areas increased significantly in order to service the extraction of ore and secondary industries.

Mount Isa, a city located 1825 km north-west of Brisbane (see figure 3.28), is an example of a settlement that has emerged solely because of the resources found there. Lead, silver, copper and zinc are found in abundance the region, which drove mining companies to push into the area in the early 20th century. These companies need staff and can offer attractive salary packages, so many people began to reside in Mount Isa and a community has slowly established itself there. Now, nearly 22 000 people call Mount Isa home and the city has most of the services and features of any other large Australian city. Like many mining towns, however, the number of people working in certain industries is very different to the proportion in the wider community (see table 3.8).

FIGURE 3.28 Mount Isa is located 1825 km north-west of Brisbane

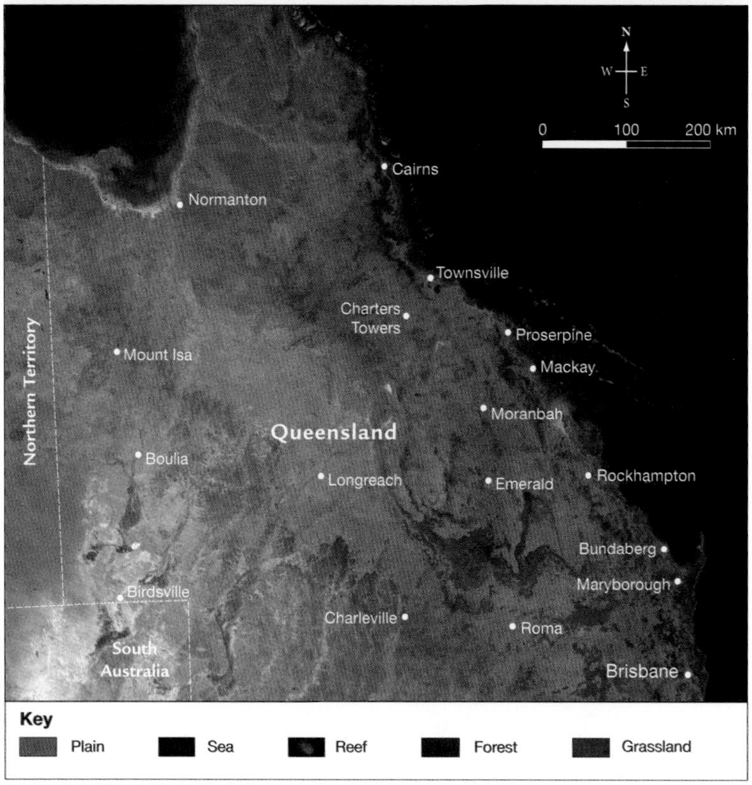

Source: Reto Stöckli, NASA Earth Observatory

TABLE 3.8 Australian census statistics for Mount Isa region employment, 2016

Industry of employment, top reasons (employed people aged 15 years and over)	Mount Isa	%	Queensland	%	Australia	%
Copper ore mining	1950	12.8	2867	0.1	6829	0.1
Beef cattle farming (specialised)	1407	9.2	16 552	0.8	44 309	0.4
Local government administration	876	5.8	30 286	1.4	142 724	1.3
Silver-lead-zinc ore mining	854	5.6	1605	0.1	3153	0.0
Hospitals (except psychiatric hospitals)	607	4.0	91 756	4.3	411 808	3.9

Source: Australian Bureau of Statistics, ABS: 2016 Census QuickStats Mount Isa Code SED30057 SED

Soil

The quality of soil and potential for growing crops attracts people. Higher quality soils maintain environmental quality, make healthier animals and people, and generate greater crop yields. Because of the potential for communities to develop around agricultural practices, areas with high quality soil attract people, while areas with poor soil quality do not.

The Darling Downs is a region in southern Queensland with excellent soil quality, which makes it the most fertile farming area in the state. The landscape is dominated by low, undulating hills used to grow different agricultural products including cotton, wheat, barley, sorghum and different vegetables. Sheep and cattle stations can also be found in the region. A number of settlements can be found on the Darling Downs including Toowoomba, Dalby, Warwick, Stanthorpe, Goondiwindi, Oakey, Miles, Pittsworth, Wallangarra, Allora, Clifton, Cecil Plains, Drayton, Millmerran and Chinchilla.

Climate

People are attracted to areas that experience temperate weather conditions as the temperature is moderate and bearable. Temperate zones are found between the Tropic of Capricorn and the Antarctic Circle in the southern hemisphere, and between the Tropic of Cancer and the Arctic Circle in the northern hemisphere. Most of the world's population is found in these zones, particularly in the northern hemisphere as it has more landmass than the temperate zones of the southern hemisphere (see figure 3.29 showing Australia's climate zones).

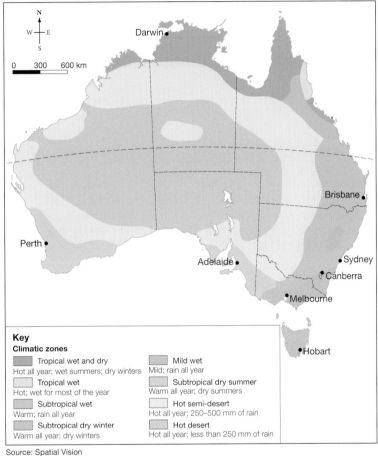

FIGURE 3.29 Australia's climate zones

Source: Spatial Vision

3.5.3 Social factors affecting population distribution

Australia's natural beauty has long been attractive to international tourists and encouragement of the tourism industry has meant viable communities have developed in regions that may not have any other opportunities. A higher proportion of people in Cairns are employed in the accommodation and recreation industries (3.5 per cent and 2 per cent respectively) than in Queensland as a whole (1.4 per cent and 1.6 per cent).

Another social factor that influences where people move to is the provision of health and educational services. Better educational opportunities mean that more families will live in an area. In some towns in regional Australia, most students finish primary school and then go to boarding school because there is no high school in their town or area. Tertiary institutions also tend to attract or keep younger people in an area. Older people will gravitate to areas with more comprehensive health services and medical facilities.

There is a greater cost associated with providing these services to rural, regional and remote areas as they are further away from population centres and other medical facilities. It is also much harder to get medical and educational professionals to move to and work in these areas as larger urban areas are often more attractive to professionals.

People who move to an area will also be more likely to settle near those with similar cultural and religious backgrounds to be close to familiar leisure, social and cultural experiences. This has led to some places being associated with different national groups. Brisbane's West End historically has a larger Greek community than other parts of Brisbane, many Italians have settled in Far North Queensland in the cane fields near Cairns since WWII, and the Sunnybank area of Brisbane has a history of Asian immigration.

Activity 3.5b: Social factors affecting population

Use the ABS QuickStats tool and other online data sources such as Google Maps, Google Earth and/or the Queensland Globe to complete the following questions.

Explain and analyse population data

1. Gather demographic data for these settlements:
   - Chillagoe
   - Burketown

   Collect data about:
   - dependent population
   - total population
   - health services in each settlement
   - educational services in each settlement
   - land use in each settlement.
2. Calculate the proportion of dependent people in each place (aged 0–15 and over 65).
3. Use an online mapping tool to examine different land use and services in each area. Describe the link between the land uses, services and facilities in each location and its demography.
4. What other factors may influence the different demographic characteristics of each location?

Resources

Video eLesson: SkillBuilder: Creating and describing complex overlay maps (eles-1656)

Interactivity: SkillBuilder: Creating and describing complex overlay maps (int-3152)

3.5.4 Economic factors affecting population distribution

Resource exploitation and trade

Political decisions play a large role in where a population is distributed. Governments provide services and incentives for certain types of business and in certain regions. For instance, the mining industry has seen significant support from local, state and federal governments in Australia over recent years. Mining companies are among Australia's most consistently profitable stocks in international trade, and infrastructure and services in mining areas, such as those in central Queensland, reflect this.

Resources are a significant input into any settlement and for that settlement to be sustainable it requires adequate resources. Early settlements need local resources for housing, food and water. As settlements grow and mature they may begin to trade their resources for other things that they need but cannot access in their region.

If an area doesn't have many resources it will not be able to sustain a large population unless it can import much of what it needs. This is why there are very few major cities in desert environments but those that are, Dubai for instance, import nearly everything.

Employment

Economic conditions influence population distribution significantly. During times of favourable economic conditions, areas can experience a massive influx of people who are looking to profit. However, during times of economic downturn people will move on to areas that offer better employment opportunities. Australia famously attracted many international migrants in the period after WWII as many significant infrastructure projects were undertaken thus increasing skilled and unskilled employment opportunities. Many current Australians can trace their ancestry back to these post-war migrants. The recent mining boom has also led to the influx of people to mining communities, both as residents and as FiFo workers.

Affordability

Affordability is probably the most significant factor in determining where people live. There are two factors to affordability: wages and the cost of living. The benefits of a well-paid job will be undercut if the cost of living is very high. This relationship between wages and living costs can vary considerably across the country.

Figure 3.30 shows the relative socioeconomic disadvantage in Australia. The values are generated from a combination of indicators that take into account the cost of living and wealth of different areas. Figure 3.31 shows the relative median weekly income for households in Queensland — two measures of the economic advantage in a specific area.

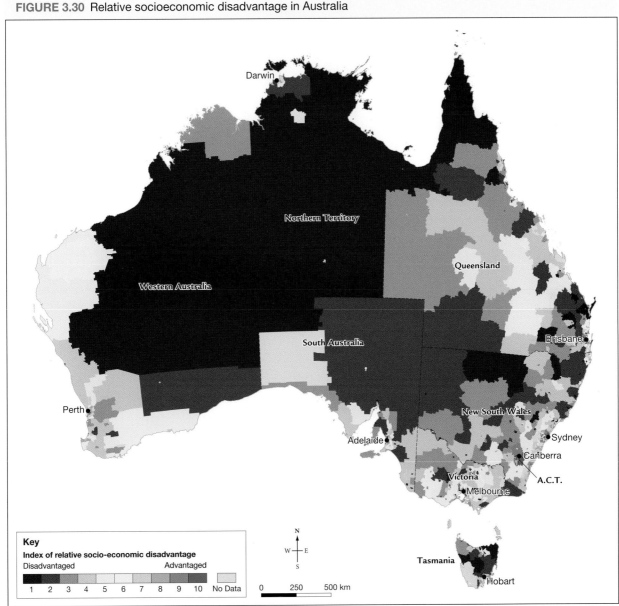

FIGURE 3.30 Relative socioeconomic disadvantage in Australia

Source: Based on Australian Bureau of Statistics data.

FIGURE 3.31 Median household income per week in south-east Queensland

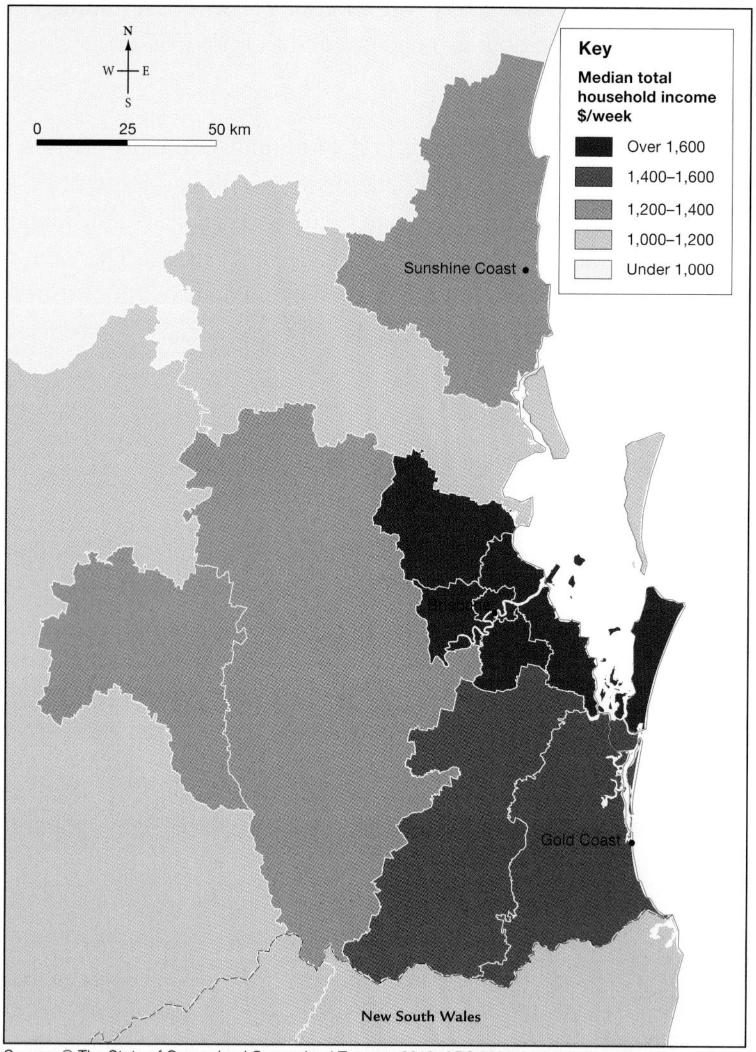
N
W
E
S
0
25
50 km
Key
Median total household income $/week
Over 1,600
1,400–1,600
1,200–1,400
1,000–1,200
Under 1,000
Sunshine Coast
Gold Coast
New South Wales

Source: © The State of Queensland Queensland Treasury 2018. ABS 2033.0.55.001 Census of Population and Housing: Socio-Economic Indexes for Areas SEIFA, Australia, 2016; GBRMPA; Google; ZENRIN.

Emergency and key services at risk due to property market boom

5 February 2018

Key Worker Housing Affordability in Sydney report released

Sydney's property market is pricing the city's key workers out of metropolitan areas, driving them potentially hours away from their workplaces, and threatening the viability of the key services they provide to our city.

This was the finding of a wide-ranging study, Key Worker Housing Affordability in Sydney, commissioned by member-owned Teachers Mutual Bank, Firefighters Mutual Bank and Police Bank, and undertaken by the University of Sydney's Urban Housing Lab led by Professor Nicole Gurran and Professor Peter Phibbs.

The report provides detailed analysis of declining levels of housing affordability across greater and metropolitan Sydney for key workers – the people we all depend on. These include teachers, firefighters, nurses, police, ambulance drivers and paramedics.

Based on current trends, the outlook for housing affordability in Sydney's Greater Metropolitan Area is grim.

'Sydney's over stretched housing market is locking teachers, firefighters, nurses, police, ambulance drivers, and paramedics out of home ownership. Our key workers are increasingly being forced to outer metropolitan areas in search of an affordable place to live,' said Professor Gurran.

The report found that in the ten years leading up to 2016, key areas in Sydney lost between 10 and 20 per cent of teachers, nurses, police and emergency service workers to outer and regional areas. Sydney's Inner South West (–14.6%), Inner West (–11.3%), Eastern suburbs (–15.2%), Ryde (–14.2%) and Parramatta (–21.4%) all experienced a net loss of key workers, while areas including the Illawarra (+10.5%), Southern Highlands (+17%) and Hunter Valley (+13.6%) all had net gains.

The nature of this group's shift work combined with living a distance away in areas with unsuitable public transport meant 77.4 per cent of key workers drove their private motor vehicle to work in 2016, compared with just over 43 per cent for the general population.

Only five per cent of key workers used public transport to get to work, compared with 12.7 per cent for the general population.

Teachers Mutual Bank CEO Steve James says that the pressure this situation puts on people already working in high pressure jobs is unfair, whether they're looking to rent or buy.

'Longer commute times, especially in private vehicles, lead to significantly higher financial costs and serious social consequences for key workers and their families, disrupting work life balance and impacting their lifestyle. Critically, lengthy commute times are also associated with lower rates of workforce participation.'

The report reveals that between 2003 and 2016, the median price of established homes in Sydney more than doubled from $400 000 to around $900 000 – well beyond the reach of many key workers, especially those who are single. Soaring rents have heightened the crisis, making a 20 per cent home loan deposit unattainable for many key workers.

For instance, a single key worker eyeing a property in Sydney's inner ring at the 2016 median price of just over $1 million would need 13 years to save for a deposit. This is a sharp increase from the 8.4 years needed to save a 20 per cent deposit in 2006.

'For a key worker, finding somewhere affordable to live in reasonable proximity to their work is becoming impossible for those not already in the property market,' said Steve James, CEO of Teachers Mutual Bank Limited.

'The report has found that the closest local government area with an affordable median rental price for an entry level enrolled nurse is Cessnock in the Hunter Valley. That's about 150 km from any hospital in Sydney city, making it a 300 km round trip per day,' explains Mr James.

The report also looks at urgent solutions that can be implemented to help key workers buy their own homes both close to their place of work and their established support networks such as family, friends, and the wider community including local groups, clubs, and schools.

It identifies five key priorities for policy makers and private sector stakeholders to consider. Professor Gurran emphasised that these approaches can be implemented largely within the current policy framework.

'There are a number of strategies that would improve housing affordability for key workers and their families, with government support. These include reducing the 'deposit gap' to home ownership that key workers now face despite their stable employment; boosting the supply of affordable 'starter homes' for first home buyers; reducing construction costs through design innovation; and looking at alternative forms of housing tenure,' said Professor Gurran.

Teacher's Mutual Bank CEO Steve James continued: 'This study shows that without urgent and genuine intervention on the part of policy makers and other institutions, a growing number of key workers in NSW and Sydney may never be able to afford to own their own home within reasonable distance of Sydney.'

Tony Taylor, Police Bank CEO, said that the five key measures identified in the report to improve housing affordability for key workers were worth serious consideration.

'Without urgent action by all levels of government, in conjunction with the finance and building industries, we could see more key workers pushed further out of the Sydney metropolitan area. The effects of that will be felt by each and every one of us,' he said.

Professor Phibbs added: 'By addressing the key barriers to affordability, residential housing in metropolitan Sydney could be as much as 20 per cent cheaper than today's prices, with vastly more manageable deposits for key workers looking to buy a home.'

In Sydney, 20% of key workers such as teachers, firefighters, nurses, police, ambulance drivers and paramedics left the city in the ten years leading up to 2016. Housing affordability was cited as a key reason. 77% of key workers drove their motor vehicle to work while only 5% of key workers took public transport, these are higher and lower than the Australian average respectively.

Source: The University of Sydney, School of Architecture, Design and Planning

Activity 3.5c: Key worker housing affordability
Answer the following questions based on your understanding of the article 'Emergency and key services at risk due to property market boom'.

Comprehend the information and identify the impacts
1. In general terms, where did key workers move from and to in the ten years up to 2016?
2. What are some impacts to the workers of this shift in living conditions? (Consider whether there might be both positive and negative impacts.)
3. What are the impacts of such changes on society more broadly? (Consider whether there might be both positive and negative impacts.)
4. What can and should be done to change this trend with key workers? Suggest two ways that might change the trend and justify your response.

3.5.5 Demographic factors affecting population distribution

Australia's national population is measured every five years through the census, but our population is steadily changing as people are born and die, and as they move to or from different places. Populations change constantly and we can represent this change over a set period of time as:

Total population growth: population + (births - deaths) + (immigration - emigration)

The rate of natural increase is calculated by subtracting the number of deaths from the number of births. The graph in figure 3.32 shows how natural increase, and interstate and overseas migration have affected Queensland's population, and are projected to affect the population in the future.

The state government of Queensland predicts: 'The current trend of population concentration in south-east Queensland is projected to continue, albeit at a slower pace than is currently occurring. The larger regional cities of Cairns, Townsville and Toowoomba will continue to grow and regional centres such as Rockhampton, Hervey Bay, Mackay and Gladstone are also projected to attract population, as a result of the lifestyle and employment opportunities they offer.'

Other likely changes include the expansion of the mining and resources extraction industries in the south and west, which are predicted to bring economic and other benefits to those areas. The related jobs are likely to be a significant pull factor for those areas.

Generally, people are choosing to live in established major population centres such as Brisbane, the Gold Coast, Ipswich, Townsville, Mackay and Toowoomba. There are several processes that can determine how our cities, towns and settlements will change as people move.

Urbanisation

Urbanisation occurs when people move from rural areas to urban areas. It is the growth in the proportion of a population living in urban environments. This phenomenon is occurring across the world with more people and a greater proportion of the global population living in cities. Currently, 54 per cent of people on earth live in an urban place and this proportion has been steadily increasing over the last century. The United Nations predicts that, from 2015 to 2030, the global urban population will grow between 1.44 per cent and 1.84 per cent per year. Rates of urbanisation are highest in developing countries, with African and South American cities predicted to see the highest rates of urbanisation in the near future.

In Australia, we have always had a high proportion of our population living in cities. European settlements that began as towns eventually grew and grew, thus attracting more and more people. Some would move into rural areas as the country was opened up to farming but the trend over the past 100 years has clearly been one of increasing urbanisation. Figure 3.32 shows the components of population change in Queensland since 1981, and how this is projected to continue.

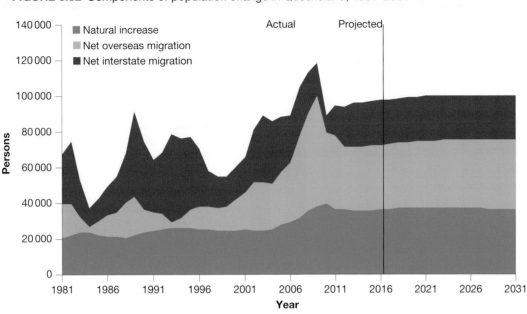

FIGURE 3.32 Components of population change in Queensland, 1981–2031

Source: © The State of Queensland Queensland Treasury 2011; Queensland Government population projections to 2031: local government areas, 2011 edition, Office of Economic and Statistical Research, Queensland Treasury.

Suburbanisation

Suburbanisation is the development and outward spread of new suburban areas. It includes a population shift from inner city areas to outer suburban areas, often on the rural fringe, that are accessible to the city centre. In Australian cities, there was significant population growth post-WWII and our cities expanded and encroached upon their rural edges. Many suburban areas across Australia's larger cities were farms or native vegetation not long ago.

Suburbanisation is the key process that leads to **urban sprawl**. Figure 3.33 shows population change at different distances from the Brisbane CBD between 2001 and 2011. The largest proportion of this change has been an increase between 10 and 25 km from the CBD.

Counterurbanisation

Counterurbanisation is a term that describes the migration of people from urban to rural areas. It is a loosely defined term and can occur for a range of reasons.

Counterurbanisation is generally related to push factors in urban areas that are suffering some sort of decline, such as the movement of jobs out of an area, increasing crime rates or poor-quality housing or environmental conditions. In Australia, counterurbanisation could apply to the migration of those seeking a 'tree change' or 'sea change'. With adequate transport connections, many people can live more than 100 km from where they work and make the trip, usually to a major city, daily or weekly.

The city of Brisbane is surrounded by several satellite centres including Ipswich and Caboolture, which are close enough that many people commute from one place to the other each day for work (see figure 3.34).

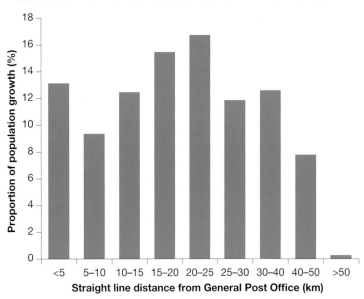

FIGURE 3.33 Comparison of proportion of population change at various distances from Central Business District, Brisbane, 2001–11

Source: © Commonwealth of Australia 2013, Bureau of Infrastructure, Transport and Regional Economics BITRE, 2013, Population growth, jobs growth and commuting flows in South East Queensland, Report 134, Canberra ACT.

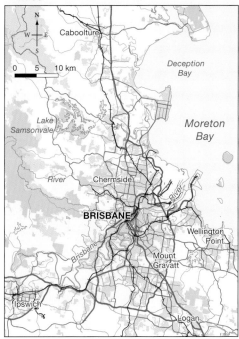

FIGURE 3.34 Brisbane and its satellite cities

Source: © Commonwealth of Australia Geoscience Australia 2018 / © Openstreetmap contributors

Periurbanisation

Periurbanisation is the process of urban growth that creates hybrid landscapes with dispersed rural and urban characteristics. This type of growth is logically found on the edges of cities and can create places with both urban and rural features.

FIGURE 3.35 Examples of urban settlement patterns

Source: © State of Queensland 2018, Qld Globe

174 Jacaranda Senior Geography for Queensland 1 Units 1 & 2 Third Edition

These areas typically contain agriculture as well as land uses that have purposefully been moved out of more populous areas such as:
- highways
- rubbish tips, recycling centres and landfill sites
- airports
- utilities
- hospitals and other large-scale services
- factories and manufacturing
- larger shopping facilities.

Activity 3.5d: Population distribution changes

Use the Queensland Government resource **QImagery** weblink in the Resources tab and the historic aerial imagery that many local councils have available on their online interactive maps to complete the following questions.

Analyse data and apply your understanding of population changes

Go to an online repository of historic maps and examine the following areas from 1980 to today.
- Your local area
- Logan
- Gold Coast
- Gladstone
- Beerburrum State Forest
- Weipa

1. Describe the settlement patterns of each location. Can you find evidence to support the patterns covered in the section 3.5.5?
2. Describe the changes you see in both the built and natural environment at each location from 1980 to today.
3. Find data on the number of government services available at each site. Can you see a pattern in the distribution of services to different areas of the state?
4. Identify what challenges and opportunities residents in each area might have experienced because of the changes you described. Choose the challenge and the opportunity that you think would have the most impact on your life if you lived in these areas, and explain how they would affect you.

Resources

- **Weblink:** QImagery
- **Weblink:** The cost of living in Queensland
- **Weblink:** Queensland thematic maps

3.6 Challenges facing remote and rural Australia

Australia's outback contains some of the most beautiful, diverse, remote and harsh landscapes in the world. As you have seen, many people live in rural and remote Australia, and these parts of our country are important sources of resources, such as food and minerals. There are unique challenges that come with living in this part of the world that are not seen in cities; these challenges have different impacts and solutions.

For many rural and remote Australian places, it is their isolation that defines them. Figures 3.26 and 3.27 show the distribution of Queensland's population, and the areas that are habitable but remote are evident. Being so far away from other places can help a community to build a sense of camaraderie in an area but it brings with it specific challenges.

FIGURE 3.36 There are benefits and challenges of living in a remote community.

3.6.1 Employment and industry

A strong local economy is vital to the survival of a small town. In good economic times, there is more money in the area, which encourages employment and drives spending – good things for the economy of a rural town. When times are difficult there is less money being spent and businesses are less likely to employ people as spending drops. These times can see many people leave areas to look for employment elsewhere.

Australia experienced a mining boom in the 2000s that added billions to the economy and created jobs and wealth for hundreds of thousands of Australians who work in the industry or associated industries. Populations expanded across areas of remote, rural and regional Australia, including in Queensland, where mining activities were based.

Table 3.9 shows the population change between 2001 and 2016 for the central Queensland mining town of Moranbah, which is part of the Isaac region. Although Moranbah doesn't have a large population, and many mine site workers actually live elsewhere and 'Fly-In-Fly-Out' (FiFo), the population still expanded by more than 25 per cent from 2001 to 2011. From 2010, the mining boom started to wind down and Moranbah's population slightly contracted, as the effects of the mining boom slowed. As mining in the area slowed, fewer mine site employees were required and this had a knock-on effect for those secondary businesses that fed into the mining industry, whose businesses slowed as well.

Activity 3.6a: Comparing population growth

Refer to table 3.9 and the ABS QuickStats tool to answer the following questions.

TABLE 3.9 Population change in Moranbah and Queensland, 2001–16

	2001	2006	2011	2016
Moranbah	6124	7133	8626	8333
Queensland	3 585 639	3 904 532	4 332 739	4 703 193

Source: Australian Bureau of Statistics

Explain the data

1. What is the percentage change in population in Moranbah and Queensland between 2001 and 2011?
2. What is the percentage change in population in Moranbah and Queensland between 2011 and 2016?
3. Create a line graph with a dual scale to show the changing populations of Moranbah and Queensland.
4. Write a paragraph describing the rate of change for each location. What factors might contribute to this difference?
5. Find out the change in population over the same period in smaller centres from the same region, such as Copabella and Glenden. (Use the ABS QuickStats tool to help.)

Analyse and apply your understanding
6. What do your findings suggest about the relationship between size and vulnerability to economic downturn in these regions?
7. List the different ways the population change seen in the Isaac region, where Moranbah is located, would affect a community. Consider social and economic impacts.

3.6.2 Transportation connections

Better transportation connections will ensure more people can access a place and be willing to live there. Rural and remote areas do not generally have good public transport options although many of these places exist, survive and thrive because of a nearby train line, maritime port or airport.

FIGURE 3.37 Rural transport defines communities

Source: Photo News Corp / Brenda Strong / News Regional Media

3.6.3 Housing availability and affordability

Encouraging people to buy into an area is a good way to get them to stay. The mining boom of the early 21st century saw many rural and remote areas change dramatically as property values often skyrocketed as soon as mining or mineral projects were announced. This brought with it developers and investors, and soon housing prices and rents were so unaffordable in the area that those who had resided there before the boom were suddenly priced out of the market.

In contrast, when mining and other projects finish, the reverse can happen: many property investors and buyers pull out of an area, which suddenly guts the market and leaves many people with overpriced property and no interested buyers.

Managing the transition between different phases of the economic cycle is important for economic stability and sustainability in these areas.

FIGURE 3.38 News headlines from the end of the mining boom

Trapped by a mining town mortgage after the boom

Property market drops by 50 per cent in Queensland mining towns

Gladstone families stuck in a one-time real estate hotspot where most homes now sell for a loss

3.6.4 Waste management

Managing waste and wastewater in small, regional areas can be challenging depending on the needs of the community. All towns and cities in Queensland have some form of wastewater treatment facility in the town. Smaller or more dispersed areas may still be seeking an option.

Remote and rural areas often have their own landfill or waste management processes.

In some rural and regional areas that rely on one industry, for instance mining or agriculture, waste management is controlled by that industry. This can lead to issues where residents may be left out of the decision-making process and waste is managed in an inappropriate or unsustainable manner for that area.

3.6.5 Fresh water quality and availability

Water is the key resource required in any settlement. If there is no regular supply of water, there will be no settlement. Careful management of fresh water is crucial, especially in smaller towns and settlements, to ensuring a stable and healthy population.

In the mid-2000s, south-east Queensland experienced a prolonged drought that heavily impacted on the region's ability to supply water to the population. Severe water restrictions were put in place on residents as water storage dropped to critical levels. Water delivery infrastructure was also upgraded, which allowed water to be moved around the region more easily.

3.6.6 Access to communication technology

The National Broadband Network (NBN) is a nation-wide piece of infrastructure designed to modernise Australians' access to the internet. Faster internet speeds will allow users to do more online, including watching more television shows or movies, creating more digital content such as videos, 3D models and games, conducting medical procedures in real time using remote access, and educating people in rural and remote areas who rely on distance education more effectively.

Over time, the scope and technology proposed by the NBN has been reduced by the government. Critics say this has been done to minimise the financial impacts on private corporations offering products that rely on downloads, such as pay television channels and online movie distributors, although the official reason is saving costs.

Many parts of rural and regional Australia have very poor telephone and internet coverage and this absolutely hampers the ability of people in those areas to participate in online communities and the online economy. When initially announced, the NBN was promoted as a 'nation building infrastructure project' by then-prime minister Kevin Rudd. Sadly, the NBN has been used as a political football by successive governments and its technological scope has been eroded over time. Those charged with delivering the NBN would argue this has been done to curb its cost.

Activity 3.6b: Researching Emerald

Research the history of Emerald, a mining town in central Queensland.

Explain and comprehend
1. When was Emerald established?
2. Why was it established in that location?
3. What industries were the initial drivers of the economy of the area at that time?
4. What industries are now driving the economy of the area?

Apply your understanding of the challenges
5. What challenges would Emerald face from a declining population?
6. What could the people of Emerald do to mitigate any negative impacts of a contracting economy? (For some possible strategies, research how Newcastle in New South Wales has managed their changing economy after the mining boom.)

 Resources

 **Weblink:** Australian National Waste Report

3.7 FiFo communities in central Queensland

3.7.1 Defining FiFo

FiFo stands for 'fly in-fly out' and relates to people who work and live in very different places. Most of these workers are employed in the mining industry in remote parts of the country but live in more populated areas, such as south-east Queensland. Their employer pays very well due to the remote location in which they work, and flies the employees to work and provides temporary accommodation.

The Queensland Government Statistician's Office uses the following definition: 'Non-resident workers are people who fly-in/fly-out or drive-in/drive-out (FiFo/DiDo) to work and live in the area temporarily while rostered on, and who have their usual place of residence elsewhere. Non-resident workers include FiFo/DiDo mining and gas industry employees and contractors, construction workers and associated sub-contractors.'

The practice has positive impacts on people in terms of the money they can earn, but it also has negative impacts on communities as they manage increasing costs and an ever-changing community. Individuals suffer due to regular movement in their lives and the impacts of constant travel and change. This makes it an increasingly controversial practice and one that the state government has seen fit to address through a Parliamentary Inquiry.

3.7.2 FiFo in Queensland

Essentially, FiFo communities can be found in most areas of resource extraction and sometimes in surrounding centres as secondary industries set up to support the main mining operations.

In Queensland these include:
- Bowen Basin
- Surat Basin
- Gladstone region

3.7.3 Factors driving FiFo work

Economic drivers in the wider community play a key role in bringing the mining industry to an area. As resources are discovered and a mining operation begins, a lot of money is poured into local infrastructure. This starts with the initial development of the mine site and then upgrades to nearby transport infrastructure, including roads, airports, ports and rail lines, to cope with the mining operation. These infrastructure projects alone can be massive undertakings and usually require a significant influx of skilled workers as they may not be readily available in a small country region.

FIGURE 3.39 FiFo workers on a mine site

Secondary industries start to benefit right away as they support those parts of the mine that contract work out. As more people move into the area, real estate prices rise. These areas do not usually have a lot of dwellings, so prices rise quickly and they are hard to control.

Demographic changes occur as workers move in and out of a community, often on short-term contracts. Many of these workers do not require housing in nearby settlements or they can't afford nearby housing due to the very high cost of real estate in mining areas. They often end up in temporary accommodation that can lack some basic features of a regular community.

3.7.4 Impacts of FiFo on communities

Economic

The mining industry is one of Queensland's highest performing in terms of income. In 2016 mining was the fourth largest industry sector contributing 6.9 per cent to Queensland's Gross State Product.

TABLE 3.10 Comparison of Queensland resources industry, 2004–14

Measure	2004	2014
Employment	24 000 people	80 500 people
Gross State Product	$19.5 billion	$27.3 billion
Royalties	$1 billion	$2.4 billion
Coal production	156 million tonnes	250 million tonnes
Coal value of production	$9.6 billion	$23.6 billion
Mineral value of production	$5.6 billion	$9.6 billion
Coal seam gas production	30 petajoules	250 petajoules
Petroleum wells	217	1553
Exploration expenditure	$244 million	$1.2 billion

Source: © The State of Queensland Department of State Development, Manufacturing, Infrastructure and Planning 2018

FIGURE 3.40 Community facilities in rural and regional Australia based around resource extraction (a) Emerald (b) Quilpie

Activity 3.7a: Impacts of FiFo workers

Answer the following questions based on your understanding of FiFo employment in Queensland.

Apply your understanding and propose justified actions

1. What are the positive and negative impacts of FiFo on the environment and on society? Consider the demographic, cultural and political issues around FiFo.

2. Explain the benefits and challenges of the FiFo issue from the perspective of a different stakeholder in the industry. Choose one of the following:
   - community associations – 'We want what's best for the community.'
   - universities – 'Academic research is our focus.'
   - local government – 'We look after our local ratepayers' interests.'
   - resource industry – 'We want to take the ore out of the ground and move it on for profit.'
   - members of parliament – 'We support whichever group is most convincing.'
   - peak bodies – 'We look after our members, mostly smaller industry groups that orbit the mining and resources industry.'
   - agricultural groups – 'We supply much of Australia's food for consumption and export, and we're already suffering as cities expand.'
   - economic development groups – 'We want to generate money at all costs.' Groups come with local, state and federal scope.
3. Suggest what changes, if any, each stakeholder group might want to see, and which of the other stakeholders might be opposed to these changes.

Resources

- **Weblink:** FiFo industry
- **Weblink:** FiFo review
- **Weblink:** FiFo community concerns
- **Weblink:** Queensland Government FiFo review
- **Weblink:** Strong and Sustainable Resource Communities Act 2017

3.7.5 Responding to the challenges of FiFo

The Queensland state government undertook a Parliamentary Inquiry into FiFo that handed down its recommendations in 2016. The report detailed a number of recommended action points, including:
- improved social impact assessments by mining companies
- workforce plans that maximise the opportunity for local workers to get jobs
- workers to live in local existing housing, or in purpose-built villages, where there is community support
- accommodation that provides a safe, clean and healthy environment for workers.

Activity 3.7b: Examining the Parliamentary Inquiry

Read the Queensland Government's **FiFo review** (weblink in the Resources tab) to answer the following questions.

Analyse the information and propose actions

1. Analyse each recommendation in the report using the PMI method. PMI = Plus, Minus, Interesting. What are the positives and negatives, and interesting or unanswered points?
2. Develop ways to evaluate the effectiveness of each recommendation. Come up with some ways to measure how each recommendation might be successful.
3. Consider one of the locations within the Bowen or Surat basins or the Gladstone region. Assess that location to see how well it meets the recommendations made in the Parliamentary Inquiry based on the available data.
4. What could be done at your chosen place to help it meet the recommendations that were suggested? Justify your suggested response with examples and evidence.

3.8 Challenges facing urban Australia

3.8.1 Urban sprawl

As urban populations in Australia steadily increase, so does the demand for residential space. At times, this may be at a rate faster than that of natural population increase and immigration. Many cities progressively grow outwards, spreading into the surrounding countryside and invading adjacent towns, regions and undeveloped land.

This process of outward expansion of urban areas from the CBD is known as urban sprawl. The term is imprecise but it is accepted that urban sprawl:

- lacks planning (in that each housing estate is clearly planned but how they tend to work together is generally not)
- encourages **monofunctional development**
- usually consists of large areas of residential development
- is low density
- includes commercial strip development
- has less public transportation
- is dominated by motor vehicles.

FIGURE 3.41 Urban sprawl between Brisbane and the Gold Coast

Source: © State of Queensland 2018, Qld Globe

Urban sprawl occurs as a direct result of the increasing population in urban areas. The 'Great Australian Dream' of a quarter acre block (roughly 1000 m^2) has been a driving factor as urban populations and the demand for large residential blocks of land have increased while the availability of land has decreased. This has pushed developers to build large, low density residential developments on the edges of our cities, causing these urban areas to expand rapidly (see figure 3.42). In 1990 there was almost no development in the area – it was effectively farmland. Over the next 25 years, development slowly expanded from Brisbane's northern suburbs and, in 2018, North Lakes is a vibrant suburb housing more than 22 000 people.

FIGURE 3.42 North Lakes in (a) 1997 and (b) 2017

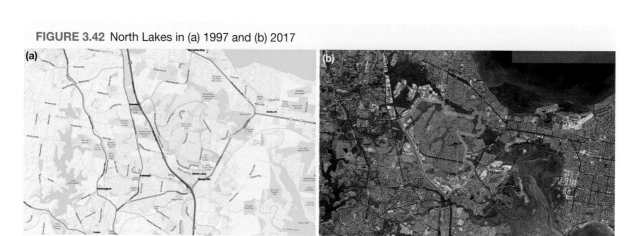

Source: © State of Queensland 2018, QImagery Brisbane 1997, QAP5562 Frame 25 and Queensland Globe

Impacts of urban sprawl

It can be argued from a positive standpoint, that urbanisation increases population density, which increases efficiency in the economy and also increases revenue for local governments. Urban areas are centres of economic growth and development that bring many people, skills, industries and ideas together which positively affects education outcomes in urban areas and economic rates. Urban residents also tend to have better health outcomes than rural residents, with higher life expectancies, and lower rates of obesity, infant mortality and fertility. This is probably as a result of better access to recreational and medical facilities.

However, an increasing population in a sprawling urban environment also increases levels of consumption (of food, water and energy, for example). Rising levels of consumption increase the output of the city — pollution, waste and heat (see figure 3.43).

FIGURE 3.43 Inputs and outputs of cities

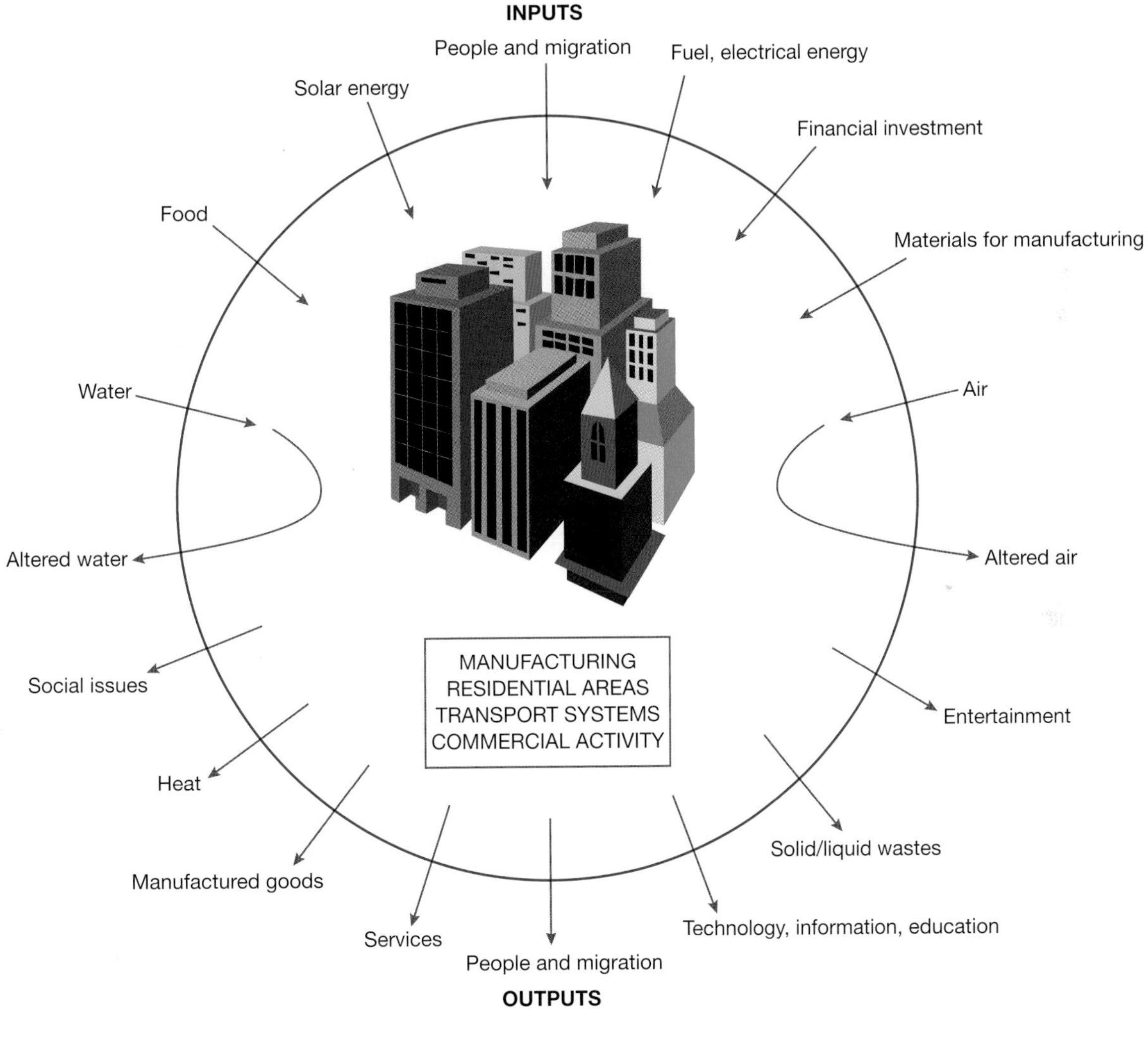

INPUTS: Solar energy; People and migration; Fuel, electrical energy; Financial investment; Materials for manufacturing; Air; Food; Water

Central processes: MANUFACTURING, RESIDENTIAL AREAS, TRANSPORT SYSTEMS, COMMERCIAL ACTIVITY

OUTPUTS: Altered water; Social issues; Heat; Manufactured goods; Services; People and migration; Technology, information, education; Solid/liquid wastes; Entertainment; Altered air

Loss of biodiversity and reduction of agricultural land

Environmentally, urban sprawl leads to a number of negative impacts. Environmental degradation occurs as land is taken for residential use. In some cities, native vegetation is replaced with housing, roads and the utilities required to sustain a modern community. This results in a loss of habitat, such as the loss of koala habitat in Brisbane's southern suburbs, a loss of animal species and an increase in fragmentation that results in the loss of wildlife corridors that enable easy movement and migration of different animal species.

Biodiversity is lost as habitat is lost. Many native species of plants and animals are displaced as their habitat is replaced by new suburbs. Sometimes species can adapt to these new environments but generally the losses are significant. It isn't uncommon to have native species, such as possums, mingle closely with people living in residential areas but it is very uncommon to see other types of species, koalas for instance, living so close to humans.

Agricultural land on the edges of cities is slowly replaced by residential development. This reduces the amount of available arable land and increases the transportation costs associated with moving agricultural products to our cities.

Waterways and flooding

When natural vegetation and land cover is replaced with hard and impervious surfaces, like concrete, that don't allow water to soak into the ground as readily, this means rainfall and other flowing water runs faster across the surface and picks up more pollutants (such as oils, pesticides and metals). This fast-flowing water is more likely to cause erosion when it does come into contact with a natural surface, due to its increased velocity. This, in turn, can cause increased sediment run-off and all of these materials end up in waterways and water supplies.

Flooding in built-up areas is also increased as more moving water is channelled via hard surfaces and there are limited opportunities for it to disperse into groundwater or waterways, which may have been built over.

Energy use

Increased energy use, increased waste generation, increased air pollution and warmer microclimates are all additional negative side effects of urbanisation. For example, **urban heat islands** occur when urban areas are hotter than rural areas surrounding them. This occurs due to the hard, dark surfaces found in cities, usually buildings, concrete or tarmac on roads, which absorb heat that is slowly released across the day and night. Urban areas in Australia can experience temperatures up to 4°C higher than surrounding areas due to this effect. (This is explored further in chapter 4.)

Access to facilities and vehicle dependency

Socially, the impacts of urban sprawl are complex and far-reaching. Many perceive that these areas lack functional public spaces and effective public transport or road infrastructure. Areas of urban sprawl are seen as mono-characteristic and devoid of the life and atmosphere that can be found in more established inner city areas. Strip development leads to commercial zones along major roads that take up valuable open space and limit public access to residential areas, and they usually contain the same or similar stores. Fast food outlets, service stations, grocery shops, pet supply warehouses, car yards, and large-scale international retail chains can all be found in areas that have experienced urban sprawl. These areas tend to be indistinguishable across the country.

Urban sprawl also encourages increased motor vehicle dependency. For most residents in these areas, grocery and convenience stores, schools, retail and hospitality are not found within walking distance of housing. Public transport is often lacking on the edges of cities as these areas are lower in density and take up much more space than inner city suburbs, thus requiring more public transport investment per person. Residents need to use private transport to get around and car use can be seen to increase further away from the CBD. This in turn can have health implications for residents of these areas as well as increasing atmospheric pollution in and around the city.

Activity 3.8a: Challenges facing urban Australia

Refer to figures 3.43 and 3.44 to answer the following questions.

Explain and comprehend the information

1. For each input and output of cities in figure 3.43, think of an actual example from your nearest city.
2. Suggest ways that the water and air that is input into our urban cities might be 'altered'?
3. What are some positive and negative social issues that come out of cities? Create a list of opportunities and challenges and choose one of each to explain in detail.
4. Figure 3.44 shows the variation in property values relative to the CBD. Describe the general relationship between distance from the CBD and property values.
5. In some outer suburban areas, property values are higher than other areas, sometimes significantly. Explain two factors that might increase the value of land in outer suburbs.

Factors affecting urban sprawl

As urban sprawl increases and gets further from the CBD, it becomes more inefficient and costly to provide services to these areas, which can lead to increasing inequality.

The factors that influence urban sprawl are complex and will differ from place to place, but it can generally be attributed to a combination of:

- cost of land
- desire for low density living
- transport
- safety
- population increase
- development guidelines
- housing policy.

Affordability

The price of land usually drops the further it is from a city's CBD. There can be exceptions to this rule but generally, those on lower incomes or who are less wealthy will be attracted to these places where they can afford to buy or rent a home (see figure 3.44).

The Great Australian Dream has pushed many people into these areas as they look for affordable, low density living. Families with children would prefer to have their own backyard, especially if they are moving from a part of the city with little green space or infrastructure for children, such as playgrounds or skate parks. Urban sprawl can be driven, in part, by this consumer attitude that prefers the quarter acre block over inner-city, apartment living.

As more households drive more motor vehicles, the ability for people to travel long distances, especially for work, has also increased. This has meant that suburban development, which used to only occur along transport corridors (such as rail lines or highways), can move further away from these corridors and more land can be developed.

Safety

Outer suburbs are often perceived as being safer than inner city areas in terms of individual safety from personal crime but also in terms of quality of air and extent of the natural environment. This perception is not always accurate but it is a factor in driving people into these outer suburban areas.

FIGURE 3.44 Land–rent relationship

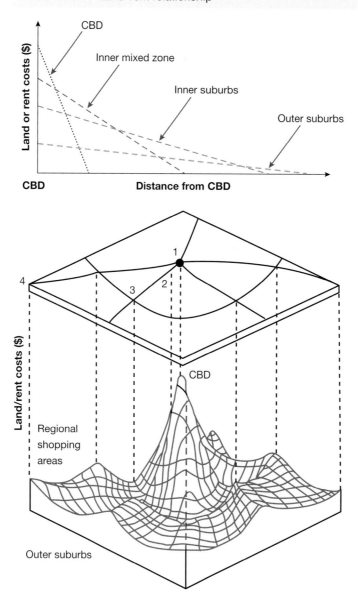

Population increase

Interstate and overseas migration helps to increase the population in many urban areas and this also contributes to urban sprawl. Families that expand tend to stay within the same city but as children move out of home, more homes are required.

Development restrictions

In the past, outer fringe subdivisions were allowed to occur with few restrictions. Ill-considered land development and marketing practices on the rural–urban fringe resulted in large areas of land being cleared and quickly developed for commercial gain by real estate developers – sometimes before sufficient infrastructure could be put in place to serve the future residents. Councils also benefit financially from these new suburbs of ratepayers – approving new developments boosts their income.

Housing policy

In the past, government housing policies have favoured ownership of detached single dwellings as opposed to government-controlled social housing. Direct government assistance, such as home saving grants and tax incentives, are designed to encourage private ownership while the lending policies of financial institutions have encouraged new home ownership. Australia's recent political history has seen house prices skyrocket in the past 20 years but personal income has not grown at the same rate, which has made housing, particularly in urban areas, less affordable for those who are yet to participate in the housing market.

Activity 3.8b: Car use in outer and inner suburban areas

Refer to table 3.11 and use an atlas or online map to answer the following questions.

Explain and analyse the data

TABLE 3.11 Car ownership and distance from CBD in Brisbane

	Average motor vehicles per dwelling (2016)	Straight line distance from Brisbane CBD (km)
Spring Hill	1	1
Kedron	1.6	7
North Lakes	1.9	27
Wamuran	2.6	50

1. Describe the relationship between vehicle ownership and distance from the CBD in Brisbane.
2. The suburbs selected in table 3.11 are all north of the CBD. How would this relationship look if we examined suburbs south of Brisbane's CBD? Use an atlas or online map to examine this area.
3. What result would you expect for places that are 75 km and 100 km south of the CBD? Justify your hypothesis.

3.8.2 Land use zoning

Zoning can be a cause of problems in urban areas but also a solution. Zoning is one way we can manage our growing urban environments. Local governments have a system of zoning in place to ensure that only appropriate development occurs in particular places. For instance, it would be inappropriate to place a waste management facility next to a hospital or in the middle of a residential area. Appropriate zoning ensures this doesn't happen.

Although this system can sometimes be taken advantage of, generally it keeps our built-up areas ordered and logical. The main types of zoning generally correspond to the different land use types outlined in section 3.3.3.

Problems can arise when authorities adapt zoning as a means of solving density problems. This can lead to conflict in these areas as residents object to the changes and their negative impacts. For more detail, see subtopic 3.9.

FIGURE 3.45 Brisbane zoning

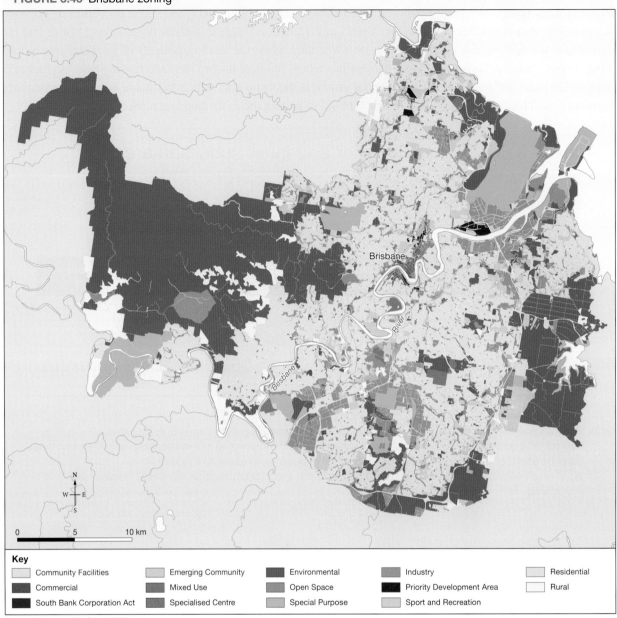

Source: © Brisbane City Council 2018

Resources

- Weblink: Sorry, we're closed
- Weblink: Cities can still grow without getting us stuck in traffic
- Weblink: Brisbane City Plan 2014
- Weblink: Planning and building in Brisbane
- Weblink: Queensland's planning system

3.8.3 Housing availability and affordability

Housing affordability is of particular concern to those in urban areas as populations and demand increase. Everyone needs a place to live but this goal is becoming harder to achieve for younger Australians as housing costs have increased markedly since the late 1990s and demand has also increased with population increases.

Successive governments have been unable to curb the rapid decrease in housing affordability (see figure 3.46). Over time, both income and house prices have increased but incomes have not kept up with the rise in house prices. This means that a house that took an average of 7.9 times the annual income to purchase in the 1970s is now 12.6 times the average income.

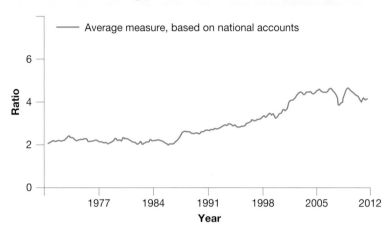

FIGURE 3.46 Dwelling price to income ratio over time in Australia

Source: Yates 2011, ABS 2017c and ABS 2017d

Activity 3.8c: Mortgage stress and land use

Refer to figures 3.45 and 3.47, and Queensland Globe, or another online mapping tool, to answer the following questions.

Explain and comprehend the data

1. Describe the difference in land use between central Brisbane and outer Brisbane.
2. Based on the differences you have identified, suggest and explain the reasons for one challenge that may arise with further expansion of the outer suburbs.
3. Describe the extent and intensity of mortgage stress across Queensland shown in figure 3.45.
4. Describe the extent and intensity of mortgage stress across south-east Queensland.
5. Identify the areas that are more likely to be under mortgage stress. Can you explain why this may be so? What might these places have in common?
6. Describe the extent of industrial land use in Brisbane shown in figure 3.47.
7. Identify the land categorised as 'Environmental' in Brisbane. Describe where this type of land is generally located.
8. What could 'Special Purpose' land use entail?
9. Explain how the Brisbane River impacts land use in Brisbane.
10. Using the Queensland Globe or another online mapping tool, examine the area around the 'Priority Development Area' located approximately 8 km downstream from the mouth of the Brisbane River in Hamilton. What sort of development is located there now?
11. Research the area and find out what development is planned or has gone ahead in the area. Suggest what infrastructure will be needed to support the development in the area.

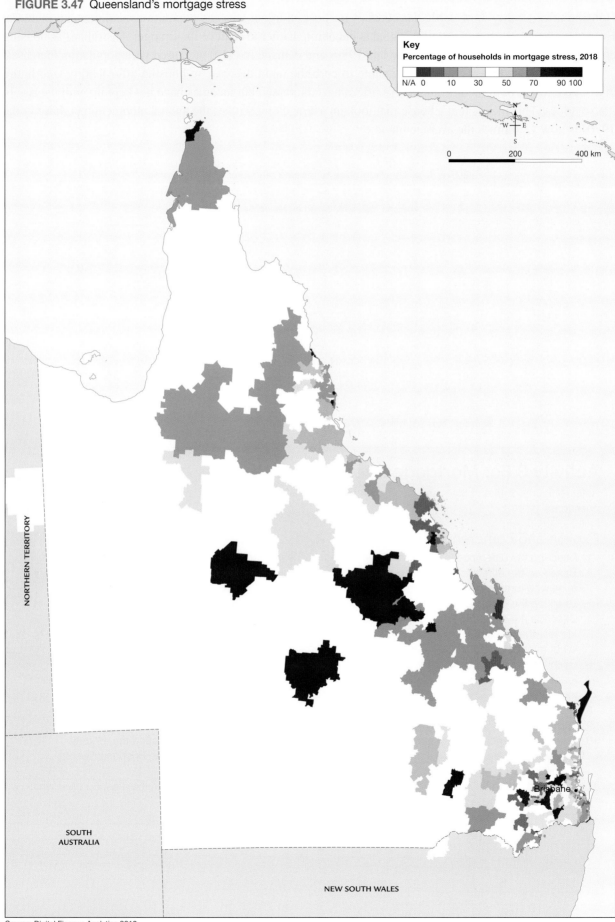
FIGURE 3.47 Queensland's mortgage stress

Source: Digital Finance Analytics 2018

3.8.4 Gentrification

Gentrification is the process of older areas, usually of lower socioeconomic status, being slowly bought out and renovated by wealthier people. This is a controversial topic as gentrification usually improves property prices in an area but it is at the expense of many of the neighbourhood's original inhabitants and character.

Usually, poorer groups, such as students or artists, inhabit rundown areas of urban spaces due to the more affordable rent. They bring character and flair to these areas, which increases the demand for housing and causes property prices to rise. Eventually these 'early gentrifiers' are pushed out of the area as they cannot afford to stay and the character of the area subsequently changes.

Paddington in Brisbane is a great example of an area that has been gentrified. The suburb sits only 3 km west of Brisbane's CBD. In the years following WWII, the area was mainly working class but in the 1980s the demographics shifted as more people bought into the area looking for a home near the Brisbane CBD. These increasingly wealthier investors drove up property prices and now Paddington is one of Queensland's most desirable, and expensive, suburbs.

FIGURE 3.48 Paddington (a) in 1949 and (b) now

Source: Paddington Kindergarten viewed from the playground, Paddington, Brisbane, ca. 1949. Brisbane John Oxley Library, State Library of Queensland; Anon 2008

Resources

Weblink: Gentrification

3.8.5 Transport options

Transport options in urban Australia are very different to those in rural and remote Australia.

As Australia's cities have grown, and continue to grow, higher population densities lead to more traffic congestion. Traffic in these areas is a constant problem and getting worse, particularly during peak times. Brisbane has witnessed significant investment in roads and tunnels since the early 2000s in an effort to curb increasing traffic congestion.

FIGURE 3.49 Brisbane's road infrastructure

Source: © State of Queensland 2018, Qld Globe

Public transport has also undergone huge investment. Often, the challenge for those providing public transport in urban areas involves how to manage more people using the system at peak times of the day. Active, sustainable transport options, such as cycling and walking, have also led to changes in city's infrastructure.

3.8.6 Service provision and management

Just like in rural and regional Australia, equitable and appropriate service provision is important to urban residents. Traditionally, areas with lower incomes and more marginalised people have struggled to attract the sort of service provision that wealthier areas have. This has impacts on the quality of life of people in these areas as well as discouraging people from moving into these areas.

3.8.7 Waste management

Urban areas have lots of people, which means lots of waste. Recycling of metropolitan waste is a very common thing: we know that yellow bins mean recycling and most of us have one at home. What many people don't know is that most of that recyclable material is sent overseas for processing.

As cities expand, they are encroaching on places that used to be waste dumps and landfill, which leads to land management issues. In many cases, older landfills are turned into common use areas such as sporting grounds or parks.

Waste management is a local council issue in Queensland. For many years, most councils would export their recycling to China as it was cheaper than dealing with it here. Recently, however, China has stopped taking in foreign recycling waste and this will be a challenge for Queensland councils as they will have to weigh up the costs associated with managing the waste versus the damage inflicted to the environment if the waste is not recycled.

FIGURE 3.50 Queensland's recycling challenge

Recycling will be dumped by councils nationwide as costs blow out, government association says

In 2014–15, each Australian produced, on average:
- 565 kg of municipal waste
- 831 kg of construction and demolition waste
- 459 kg of fly ash (coal production by-product)
- 849 kg of other, and commercial and industrial waste.

Activity 3.8d: The characteristics of Brisbane

Absolute change refers to the change in the number of people living in a suburb. To calculate the absolute change, subtract the earliest population data from the most recent population data. A negative result tells you the population has dropped.

Relative change is the percentage difference in population over the same time period. To calculate this, divide the absolute change by the earliest population data, and then multiply by 100. If the population has grown, the result will be positive; if the population has declined, the result will be negative.

Refer to table 3.12 to complete this activity.

TABLE 3.12 Population change in Brisbane's inner suburbs

Suburb	Population		
	1996	2006	2016
Bowen Hills	736	1585	3226
Brisbane City	1789	NA	9460
Dutton Park	1496	1363	2024
East Brisbane	5092	5230	5934
Fortitude Valley	1709	5082	6978
Herston	1921	1795	2215
Highgate Hill	5227	5428	6194
Kangaroo Point	4717	6868	8063
Kelvin Grove	3763	4246	7927
Milton	1629	1733	2274
New Farm	9200	11 245	12 542
Newstead	1182	4818	2193
Paddington	7222	7625	8562
Red Hill	4680	5403	5560
South Brisbane	2572	4285	7196
Spring Hill	3144	4835	5974
West End	5870	6206	9474
Woolloongabba	4173	3917	5631

Source: Australian Bureau of Statistics

Explain and comprehend population patterns

1. Calculate the absolute and relative (percentage) change in population between 1996 and 2016 for each of the suburbs shown in table 3.12.
   (a) Draw up a table of your results.
   (b) Which suburbs experienced the greatest increase in population in relative and absolute terms? Which suburbs experienced decline and by how much?
   (c) Identify suburbs where population growth fluctuated over the period. Suggest some possible reasons for such fluctuations in population growth.
   (d) The table provides evidence that it is sometimes necessary to treat percentage growth figures of population with some caution. Explain why this may be necessary, given the evidence of the data in the table. You might, for example, compare the figures of New Farm with Fortitude Valley.

Refer to figure 3.51 and your answers to question 1.

2. Compare the various suburbs' population growth rates with the characteristics shown in figure 3.51. Which of the characteristics seems to correspond most closely with population growth and which corresponds the least? Suggest reasons why this might be the case.

FIGURE 3.51 Selected characteristics of Brisbane's inner suburbs

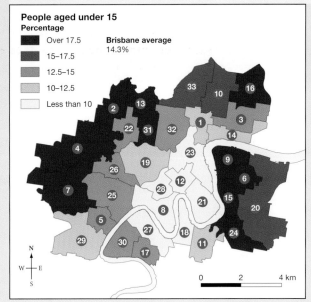
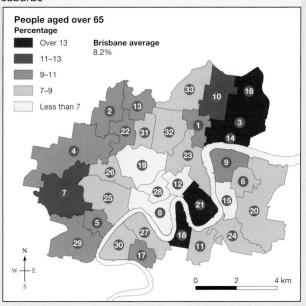
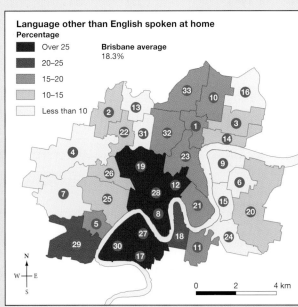
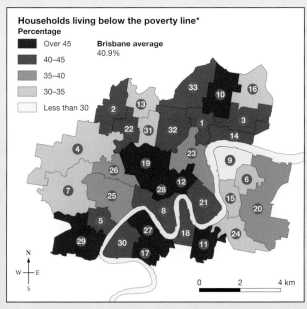
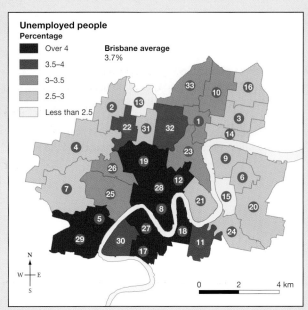
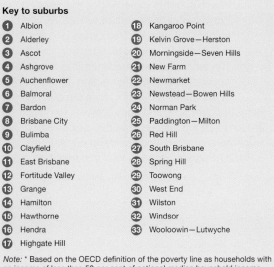

Key to suburbs

1. Albion
2. Alderley
3. Ascot
4. Ashgrove
5. Auchenflower
6. Balmoral
7. Bardon
8. Brisbane City
9. Bulimba
10. Clayfield
11. East Brisbane
12. Fortitude Valley
13. Grange
14. Hamilton
15. Hawthorne
16. Hendra
17. Highgate Hill
18. Kangaroo Point
19. Kelvin Grove—Herston
20. Morningside—Seven Hills
21. New Farm
22. Newmarket
23. Newstead—Bowen Hills
24. Norman Park
25. Paddington—Milton
26. Red Hill
27. South Brisbane
28. Spring Hill
29. Toowong
30. West End
31. Wilston
32. Windsor
33. Wooloowin—Lutwyche

Note: * Based on the OECD definition of the poverty line as households with an income of less than 50 per cent of national median household income. ABS data from 2015–16 shows the median weekly gross income in Australia as $1616.

Source: Australian Bureau of Statistics, Source Brisbane Inner City SA4 305

Analyse the data and apply your understanding

3. Write a paragraph to compare the demographic, social and economic characteristics of Brisbane's inner suburban communities, as illustrated by figure 3.51, with those of Brisbane as a whole. The average for Brisbane in each characteristic is provided in each map key.
4. Select two inner suburban Brisbane communities that exhibit distinct contrasts in their demographic, social and economic characteristics. Write a response that compares and contrasts these communities, making reference to the location of each suburb and the communities' various characteristics.
5. Is there evidence to suggest that any of Brisbane's inner suburban communities may face the risk of high unemployment and population loss in the future? Why or why not? What circumstances would need to exist for this to happen?

Propose and justify strategies for inner-urban sustainability

6. The Brisbane City Council wants to maintain the social and cultural character of the city's inner-suburban communities, as well as enhance their demographic and economic sustainability. Suggest some strategies that might help to achieve the council's aims for these communities. Explain how your suggested strategies would help to achieve the council's aims.

3.9 West End, Brisbane

3.9.1 Development history

The area now known as West End was originally called Kurilpa and was inhabited for many thousands of years by the Turrbal and Jagera nations. Early British colonisers renamed the area after the London borough of the same name.

The Kurilpa peninsula, which covers what is now South Brisbane, West End and Highgate Hill, was once covered in dense rainforest.

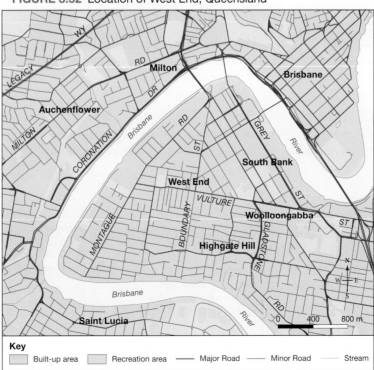

FIGURE 3.52 Location of West End, Queensland

Source: Natural Resources, Mines and Energy, Queensland Government, various maps and spatial data sets, licensed under Creative Commons Attribution 4.0 sourced on 31 May 2018

By the 1850s and 1860s, West End was a district of mainly farming families, with the native rainforest having been slowly removed. In 1865, the first bridge crossing into Brisbane was built and was later washed away. Another, more permanent, crossing was built in 1874.

In the 1880s the dry dock at South Brisbane was opened, which expanded the shipping industry in the area. This increased the residential population in South Brisbane and West End. In 1884, a train line to South Brisbane was opened.

From the 1890s until the 1930s, there was a steady increase in industrial development in West End, including sawmills, a steam joinery, gas works, a brewery and farms. Most of this development occurred around the edge of peninsula to take advantage of the river for transportation.

FIGURE 3.53 West End brewery, 1896

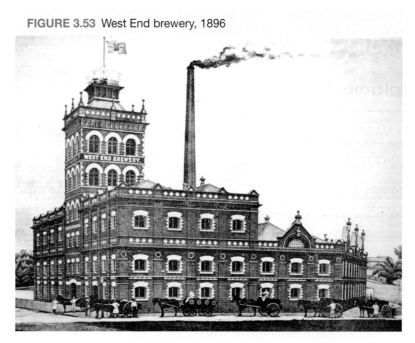

Source: Brisbane John Oxley Library, State Library of Queensland; 2003

In 1910 the Mater Hospital opened and high schools came soon after. In 1923 the shipping industry and other related businesses moved to Hamilton Wharves, which reduced the industrial character of the area. The 1938 opening of the Story Bridge continued the gradual decline of local industry as connections were opened to other parts of Brisbane.

The residential property market was relatively inexpensive and suited the new immigrants who had moved into the area post-WWII. A significant Greek community emerged in the 1950s around the Greek Evangelical Church and a Vietnamese community followed – these communities are still a visible part of the community and local area today.

FIGURE 3.54 Celebrations at the Paniyiri Greek Festival

Source: Andrew Porfyri for Paniyiri Greek Festival. Reprinted with permission.

By the 1980s the cultural precinct comprising of the Queensland Art Gallery, the Performing Arts Complex, the Queensland Museum and the State Library had opened. The Brisbane Expo in 1988 led to the creation of South Bank Parklands as a huge public space where the former wharves and associated industries were located.

This sparked gentrification of the area in the 1990s. Units began to replace houses and real estate prices began to rise substantially. Mass development occurred in the 2000s and the process of gentrification was in full swing as younger people moved into the area and the character began to change. Now, cafes and restaurants sit alongside niche bookstores and art galleries.

West End is currently changing as the older, bohemian, artistic and creative character of the area faces off against the 'new money' that has recently moved in, ostensibly to take advantage of that character and charm.

3.9.2 Demographics

According to the 2016 census, 9474 people lived in West End. Tables 3.13 to 3.16 summarise some of the main data available about West End and compares it to Australian data. Figure 3.55 shows the population pyramid for the area based on 2016 census data.

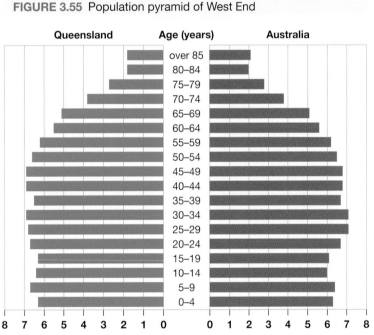

FIGURE 3.55 Population pyramid of West End

Source: Australian Bureau of Statistics, ABS, 2016 Census QuickStats West End Brisbane - Qld Code SSC33063 SSC

TABLE 3.13 General data

	West End	Australia
Population	9474	23 401 892
Median age	34	38
Average people per household	2.3	2.6
Average number of children (in households with children)	1.7	1.8
Average motor vehicles per dwelling	1.3	1.8

Source: Australian Bureau of Statistics, ABS, 2016 Census QuickStats West End Brisbane - Qld Code SSC33063 SSC

TABLE 3.14 Demographic data

	West End (%)	Australia (%)
Dependent population (% of total under 15 and over 65)	21.7	34.5
Not married (% of total population)	51.2	41.9
Population with at least a university degree level of education	45.6	22.0
Australian-born population	55.7	66.7
Religious affiliation – No religion or not stated	57.9	39.2
Language other than English – Greek	4.2	1.0
Language other than English – Vietnamese	2.9	1.2
English only spoken at home	64.4	72.7

Source: Australian Bureau of Statistics, ABS, 2016 Census QuickStats West End Brisbane - Qld Code SSC33063 SSC

TABLE 3.15 Employment data

	West End (%)	Australia (%)
Occupation – Professionals	42.4	22.2
Occupation – Managers	14.2	13
Industry – Higher Education	6.7	1.5
Industry – Medical	5.4	3.9
Industry – Cafes and restaurants	5.1	2.4
Travel to work – public transport	23.7	11.5
Travel to work – car	42.3	68.4

Source: Australian Bureau of Statistics, ABS, 2016 Census QuickStats West End Brisbane - Qld Code SSC33063 SSC

TABLE 3.16 Family and home data

	West End (%)	Australia (%)
Family composition – couple family without children	46.7	37.8
Family composition – couple family with children	37.1	44.7
Dwelling structure – house/townhouse	36.0	85.6
Dwelling structure – flat/apartment	63.3	13.1
Property tenure – rented	57.5	30.9
Property tenure – owned (outright or mortgage)	38.8	65.5

Source: Australian Bureau of Statistics, ABS, 2016 Census QuickStats West End Brisbane - Qld Code SSC33063 SSC

Activity 3.9a: Comparing census data

Use the data about West End in tables 3.13, 3.14, 3.15 and 3.16, and figure 3.55 to answer the following questions.

Analyse the information

1. Comment on the characteristics of West End from the 2016 census and compare and contrast those with the data for Australia. Write an extended paragraph response that refers to specific data to support your observations.
    - Survey a part of the suburb to understand the types of activity that occurs in the area. (Perhaps examine the types of businesses using the historic imagery data in Google Street View and compare that to what exists today.)
    - Use the ABS QuickStats tool to gather the equivalent data from earlier census for comparison.
2. Select two or three variables that you find most interesting and graph them using a spreadsheet tool such as Microsoft Excel or Google Sheets. Create a multiple bar graph that shows the rate of your selected data sets for West End and for Australia on the same graph. If you have enough data over a longer period of time you could use a comparative line graph.

Using ICT to graph data

For Microsoft Excel

Open Excel and set up data. Each row contains a different attribute, each column a different census year. In this example, we will create a line chart showing how different types of dwellings in West End changed in number from 2001–16. This data was used to create the line graph in figure 3.56.

Use ABS QuickStats to find the data and enter it into your table in Microsoft Excel. Enter the suburb and year and scroll down to find data on dwellings.

Start your table in cell A1.

	A	B	C	D	E
1		**2001**	**2006**	**2011**	**2016**
2	**Separate house**	1209	1229	1219	1148
3	**Semi-detached or terrace or townhouse**	141	115	260	200
4	**Flat or apartment**	1239	1435	1801	2411

Select all of the data in your table, including column and row headings.
In the Insert menu, select the Insert line graphic option.

From the Chart Tools menu, you can customise the look and feel of your chart.

For Google Sheets

Sign in to Google Drive using your Google Account credentials. Open a new Google Sheet in the relevant folder for this activity.

Enter data with dwelling types as rows and a different year for each column:

	A	B	C	D	E
1		2001	2006	2011	2016
2	Separate house	1209	1229	1219	1148
3	Semi-detached or terrace or townhouse	141	115	260	200
4	Flat or apartment	1239	1435	1801	2411

Select the entire table, including the top row and the first column.
Go to the Insert drop down menu and select Chart.
　Here you can adjust the chart type and use the Chart Editor to refine your data, how your chart is set up and how it looks. Change the chart type to Line chart.
　You can link the chart to your table so that as you refine or add data in the table it updates in the chart itself. Copy the newly created chart and insert it into any Google Documents you are using or save the chart as an image for use elsewhere.

3.9.3 Liveability in the West End

West End is routinely considered one of Brisbane's most liveable suburbs due to its proximity to the CBD, multiple transportation connections, strong sense of community, safety, recreational facilities, advanced economy, quality local amenities and its proximity to excellent services.

Tensions

Since the inception of West End there has been objection to the extent and manner of development in the area. An article from 1930 describes in detail the native rainforest that was removed to create the modern roads and areas for industry in the 19th century while lamenting the lack of preservation and foresight of residents and policymakers of the time.

　'This jungle was a tangled mass of trees, vines, flowering creepers, staghorns, elkhorns, towering scrub palms, giant ferns and hundreds of other varieties of the fern family, beautiful and rare orchids and the wild passion flower ... Here at our very door we had a wealth, a profusion, of botanical beauty which can never be replaced by the hand of man [sic]. Too late have we recognised the desirableness of conserving these glorious works of Nature.' *The Brisbane Courier, March 1930*

By the 1990s no native vegetation was left in West End and most dwellings were stand-alone houses. More apartments were developed and this picked up pace in the 2000s, such that by 2001 there were more apartments in West End than houses. The ratio of apartments to houses has continued to grow ever since, as shown in figure 3.56.

The Brisbane City Council had been encouraging infill development for most of the 2000s and formalised this with the 2006 'Brisbane 2026 CityShape' initiative. This called for new homes and development to be built around Brisbane's major shopping centres or along major growth corridors, and for jobs to be

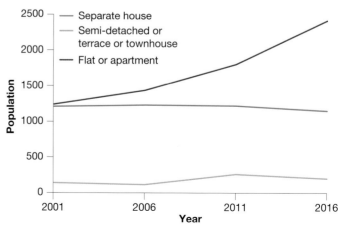

FIGURE 3.56 Dwelling types in West End over time

Source: Australian Bureau of Statistics

located close to where residents lived, with better local services and facilities. The Brisbane City Plan 2014 also maintains this outlook.

As the processes of gentrification and urban consolidation have become more and more visible in West End, with more apartments, building sites, towers and traffic, there has been increased opposition to developments, particularly those that are large in scale and visible in impact. One example of such a development is the West Village development on the former Asboe furniture site. The site is significant as it is very large and is also part of the commercial stretch of Boundary Road, the shopping, eating and drinking hub of the area.

FIGURE 3.57 Development in West End occurred at pace in the 2000s.

The initial proposal included:
- a 2.6-hectare site
- a 1532 square metre common area
- 1250 apartments
- 7 buildings, each 8–22 storeys high.

In addition, 30 per cent of the site was proposed as a 24-hour accessible open space of laneways and arcades.

The development attracted much controversy in the local community and resulted in many letters to the editor, submissions to Brisbane City Council and the Queensland government over the public consultation period, protests, public meetings, news reports and court challenges. The Queensland government stepped into the planning process to modify the initial proposal. The project was approved in 2016 and will be developed in stages between 2018 and 2027.

Right to the City

One notable result of the planning process around the West Village development was the rise of community groups as a vehicle of opposition to significant local developments. A progressive candidate had recently been elected to the Brisbane City Council representing the area and another progressive candidate was considered a genuine possibility to win the seat at state government level during the planning process. This meant the government had to take note of community reaction to the proposed development.

FIGURE 3.58 Grassroots action

Source: Natalie Osborne Right to the City

These candidates drove much of the opposition and the Right to the City movement was able to influence the planning process. Right to the City was able to reverse or alter some of the more negative aspects of the development, such as increasing the amount of open space on the site and limiting the height and visual impacts of some of the buildings in the development. However, the developer still managed to force the state government to approve the development, which did not meet the Brisbane City Council planning limits for building height at the time.

FIGURE 3.59 Alternative West Village development plan

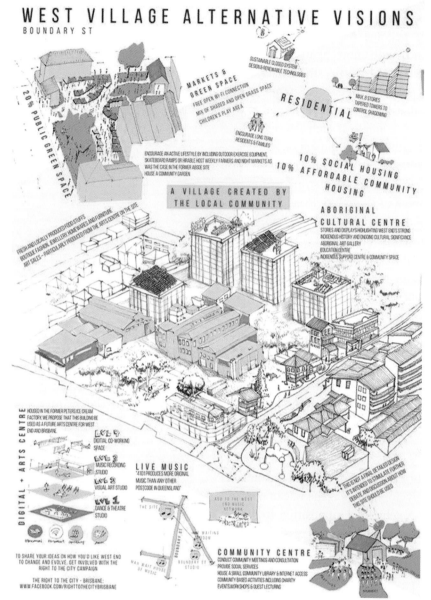

Source: Giselle Penny, Luke van de Vorst & Marilena Hewitt of Right to the City. Reprinted with permission.

Activity 3.9b: West End development
Using what you have learned about development in the West End, answer the following questions.

Explain and analyse urban development
1. What are some of the negative consequences of infill development, particularly those that might be evident in West End?
2. What could be some of the impacts of having an oversupply of apartments or units in an area? How might this impact the character of the suburb of West End?
3. Develop a set of criteria that should be used for future West End developments. Ensure that your criteria consider all aspects of sustainability including environmental, economic and social aspects.

4. Evaluate the current and alternative plans that have been provided as part of the planning process. Use the criteria you developed in question 3. If you had to choose between the two options, which would you choose? Why? What would need to change in the proposal you rejected to meet your criteria?
5. The concept of liveability is fluid and can be defined in multiple ways. Research one of the commonly cited measures of liveability and think about what additional measures you would use.
6. Consider what criteria you would develop to assess the liveability of an area. What factors would you consider important?

3.10 Responding to the challenges facing places in Australia

There is no doubt that the places we live, work and play in have a huge impact on our communities and on us as individuals. They give us identity, impact our health, let us relax, give us opportunity and drive our economy. As you have seen, the way our places are designed is of vital importance to communities and individuals. So how do we respond to the challenges that arise in the places we live?

3.10.1 Green belts — Garden cities

Historically, the first real effort to curtail unhealthy development came in the form of the garden city.

In the early twentieth century, governments began to show greater interest in improving the physical and social conditions of cities beyond the bare necessities. This period marks the beginning of modern town planning in Australia and the influence of social reformers such as Ebenezer Howard.

FIGURE 3.60 Howard's garden city proposal

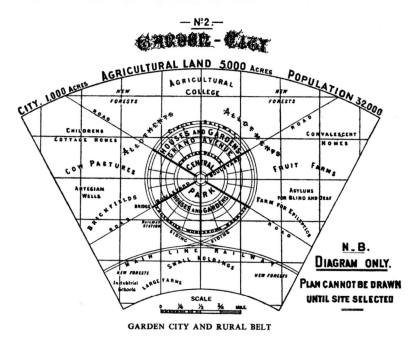

Source: Garden Cities of Tomorrow by Ebenezer Howard 1902

Ebenezer Howard was an English journalist who publicised his ideas in 1898 in a book entitled *Tomorrow: a Peaceful Path to Real Reform*. He had been influenced by factory towns and suburbs built by industrial philanthropists in the nineteenth century, and devised the concept of the garden city. He had seen how high

land values in cities encouraged high residential densities and discouraged the provision of social amenities such as open space. He therefore formulated principles for the construction of garden cities, in which the benefits of healthy rural living could be combined with the advantages of city life. These included:
- a population ceiling of 30 000 to ensure that a sense of community was not lost
- all land owned by a public trust run by 'honourable people not driven by the pursuit of profit'
- a green core surrounded by two concentric rings of public buildings and housing, both of which would be separated from industry by a green belt
- a self-sufficient population that worked in local factories and bought food from surrounding farms.

Resources

Weblink: Sydney's failed greenbelt plan

3.10.2 Inclusive planning

We know the environment shapes people's behaviour and their physical and mental health. We have created places, mainly metropolitan, that encourage a sedentary lifestyle where motor vehicles are almost mandatory for getting to work, shops and schools, and to engage with other members of the community. This reliance impacts rates of obesity and diabetes, incidence of loneliness and overall mental health. Well-designed places, though, can reduce inequality, encourage healthier transport options and impact positively on the physical and mental health of their inhabitants.

Essentially, we can respond to the challenges that face our rural, remote and metropolitan places by planning and designing them well. But 'planning' encompasses so much. What is planning? The planning we employ to shape and grow the places we live in needs to be inclusive.

Inclusive planning recognises the importance of the design of our places to all who currently inhabit them and will inhabit them in the future. In many instances, planning has meant that profit has been put before people, which has led to us creating places that are unhealthy and detrimental to the communities and individuals that inhabit them. Too often decisions are made that fundamentally change and shape places without consulting the people who live there.

Zoning is a planning tool that, as previously discussed, could be a potential solution to some of the problems that exist in our remote, rural and metropolitan places. As our cities undergo infill development, effective zoning that takes into account future populations' needs can guide appropriate development in communities.

3.10.3 Urban villages

Urban villages are attempts to enact some of the broadly agreed planning concepts that can positively impact communities. These encompass things like mixed use development (where residential, retail, commercial and sometimes industrial land use are intermixed), good public transport options, the encouragement of vegetation in the area, medium density housing, connectivity for pedestrians and ample public spaces for recreation and community building.

FIGURE 3.61 Central Adelaide is surrounded by a ring of parkland.

Kelvin Grove and Dutton Park in inner Brisbane are two Queensland examples of areas that have employed some or all of the strategies above to enhance their social and physical space.

3.10.4 Urban renewal and regeneration

Urban renewal is a philosophy that sees run down areas of cities repurposed in an effort to revitalise the area. The term has slightly different meanings in different parts of the world but in Australia this means increased density in urban areas and more appropriate use of vacant or underused land. This process is driven by local, state and federal governments who can use their legislative powers to encourage certain types of development in certain places.

FIGURE 3.62 The Gold Coast light rail network is a great example of infrastructure that will enable transport-oriented development.

Infill development is also a part of urban renewal and can be employed to help revitalise areas of cities that have lost their charm over time. Infill development is development that aims to increase the population density in an area while minimising impacts on the existing community. This often involves the organic growth of apartment blocks or increased density by way of building works.

FIGURE 3.63 Examples of infill development in Queensland

3.10.5 Smart cities

Smart cities are cities that encourage the collection, analysis and use of data to help make better decisions about development. Smart cities allow data to be collected from a range of sources including sensors, devices, citizens themselves and existing government sources. In theory, smart cities make better use of the information that exists in our cities. In practice, the idea of sharing so much personal and intimate data can be off-putting for some people.

FIGURE 3.64 Smart cities harvest data from many sources

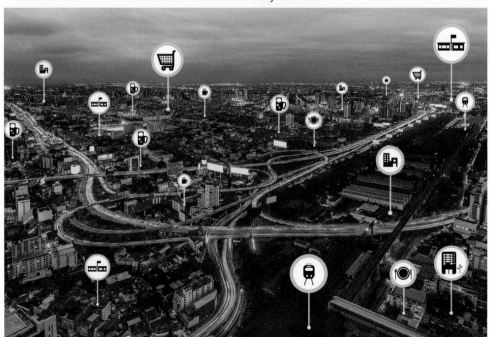

Activity 3.10: Liveable places

Consider your local area when answering these questions. That could be your suburb, your town or even your city. The extent that you consider might even change with each question.

Analyse the liveability of where you live

1. List the features that make a community a good place to live.
2. Categorise each idea on your list from question 1 as a social/cultural feature, environmental feature or economic feature. (If you can, compare your list to other groups or people's lists.) Adjust yours if you think you need to.
3. Use your ideas to develop criteria to assess what makes a local area a good place to live. Think of social/cultural, environmental and economic criteria.
4. What methods have you seen in your local area that have been designed to make the place a better place for people to live?

Propose strategies to improve your community

5. If you were given the power, what changes would you make to your local area to increase its sense of community and to make it a better place to live? Prioritise your ideas according to what would have the greatest positive impact.
6. Write an extended response to explain how you would achieve the change that you have given the highest priority and justify your response by outlining how this action will make your community more liveable.

Resources

Weblink: Smart Cities: Songdo, South Korea
Weblink: Critiquing smart cities

3.11 Preparing for your field study

3.11.1 Using the geographic inquiry model

The geographic inquiry model is a framework used by geographers to help structure an inquiry. It is a great way to help you organise your thinking as you undertake planning an investigation and can be used to address issues from local to global scales.

FIGURE 3.65 The geographic inquiry process

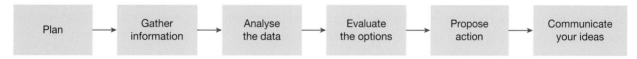

3.11.2 Plan

Every inquiry should start with a question or problem. Try and frame your question in such a way as to allow you scope to collect data, analyse it, make conclusions and respond. This might mean expanding your geographic area or the scope of your problem.

Develop key questions and focus questions that help you to organise your research and data collection. These key questions usually take the form of:
- What and where is the issue or problem at hand?
- How and why does it occur?
- What are the environmental, social and economic impacts?
- What are the solutions?

The key questions can then be narrowed down through the use of focus questions that focus your research. These focus questions can be constantly updated and can change as you gather more information through your research and field work.

Once you have established your key questions, write a plan outlining how you will undertake your research to find answers to these questions — your methodology.

Figure 3.66 outlines how you might plan an inquiry into development in a coastal, urban suburb. As you begin to research your focus questions you will find more information and uncover more questions to ask, which means your investigation should be reviewed and will be constantly evolving.

FIGURE 3.66 Planning a geographic inquiry

3.11.3 Gather information

Here you carry out the inquiry by collecting your **qualitative** and/or **quantitative** data and analysing it. This also includes conducting any research to prepare your understanding of the issue at hand.

Primary and secondary data

Primary data is data that is collected first-hand. In research, primary data holds the most weight as it is data collected by the author to help them answer specific research questions – it is data collected for that purpose. As you collect your data, reflect on whether you are collecting the right data to answer your questions, and whether there might be any ways you could improve your collection methods.

Secondary data is data collected by someone else and can include census data, government data and information, as well as research findings. Your analysis, and subsequently your solutions or proposals, will be evidence-based provided you have collected appropriate primary data, and sourced reliable and relevant secondary data.

Spatial technologies

Spatial technologies are one of the tools of geographers. They help us answer 'where' questions and can help us illuminate some of the issues that geographers study. They are used in a range of geographical and nongeographical ways, including environmental management, species tracking, weather and climate analysis, oceanic and inland water monitoring, and economics and history.

Geospatial technologies help us to measure, analyse and represent spatial or geographic information, usually digitally. For your field study, you could use spatial tools to:
- examine land use and/or land cover from a satellite image in the Queensland Globe
- investigate public transport and cycling infrastructure in the area around your place of study using the Queensland Globe
- visualise census and other demographic data using Google Maps or the ABS online mapping tools like TableBuilder
- visualise environmental data, such as geology, vegetation or species extent using online digital tools like the Queensland Globe
- map data you have collected in the field, such as water quality data or survey results using Google MyMaps or QGIS.

3.11.4 Analyse the data

Once you have collected your data you need to use it by analysing it. There are a number of methods used to analyse data. In the main, you are looking for patterns in the data that are related to the topic at hand. Sometimes these patterns are self-evident (e.g. poor water quality near an industrial area) but sometimes additional data needs to be collected or analysed in conjunction with your data to draw out the message in the data.

The primary data that you collect and secondary data that you select and interpret should be able to help you to explain the geographic processes at play in your issue and why they have led to the source of any geographical challenges. Your analysis should also focus on the different impacts that could occur as a result of your challenge continuing. You should be able to extrapolate from your analysis to generalise about the impacts of your issue.

At this stage, you should also consider whether the data you collected and the questions you asked have any flaws or limitations. Do you have the data you need to answer your research questions and suggest ways to combat the problem? Is all of the data you collected accurate?

3.11.5 Evaluating the options

Evaluating is the art of selecting or choosing between options. You should always back up any decision you make with evidence. Think about what your data is telling you about the issue and refer to the specific data that supports your position.

You can use a decision-making matrix, such as those used in chapters 1 and 2, to choose between options. First develop criteria and then score each option against the criteria to make your final decision.

3.11.6 Propose actions

Once you have analysed your data, made your decision and/or evaluated your options, you need to propose action(s) to address the problem. This could be selecting one option over another or many, proposing changes to a project, modifying existing suggestions or developing a proposal from scratch.

3.11.7 Communicate your ideas

The communication element of the inquiry process is the most important. Geography is a subject with change at its heart and encouraging change requires effective communication of ideas and prosecuting of arguments. Effective communication could be the difference between research going ahead or not, or a particular environmentally significant place getting more or less funding, a building being built, a park being revitalised or a patch of forest being cut down.

There are many methods of communicating information to an audience apart from speaking: using a presentation aid such as Microsoft PowerPoint, promotional videos, and visual methods such as billboards, posters or flyers. However, the field report is still the most common method of reporting findings in the professional and academic worlds as field reports are easy to catalogue and retrieve for later use.

3.11.8 Fieldwork example
Sustainable use of public space in Bowman Park, Brisbane

In this fieldwork example, you will consider how a prominent public space can be used to increase sustainability in an area. You will evaluate a public space and recommend changes that would increase the likelihood of positive sustainability in the area. You could examine any area in your local community that is publicly accessible and used by the community, such as a town square, local park or showground. In this example, we will examine Bowman Park in the Brisbane suburb of Bardon and ask the question:

How can Bowman Park in Bardon be more sustainable and improve the liveability of the local area?

You will:
- assess the current status of the park: How is it used? Is the site overused or underused by the local and wider community? Does it meet the needs of residents?
- research and make some recommendations as to what improvements could be made to make the space more sustainable and accessible
- communicate your findings using a report format.

The following section will take you through the whole process.

FIGURE 3.67 Preparing for your field study

Key questions

To develop your key and focus questions, see section 3.11.2.

Use a PMI extension activity to flesh out your ideas. See the following example key and focus questions. Remember that these focus questions should change as you begin to investigate your local place – you might find some interesting information that changes your perspective.

TABLE 3.17 Ideas for key and focus questions

	Focus questions	Data collection
What and where?		
What is the nature of the public space?	What is the history of the space? Who is it named after?	History
Where is the public space located?	Map of the area? Where is it in the region/Brisbane? What surrounds it?	Maps
How and why?		
How is the nature of the public space used?	What is the land use around the park? What are the public transport links? How do people use the park?	Land use survey. Count people using different features of the park (running, sport, playground, BMX track, basketball hoops, outdoor gym, creek tracks, BBQ areas)
Why is it used the way it is?	Who uses the public space? What are the facilities in the park?	Resident/user survey
How else is it used?	Types of use at different times of day, week and year?	Resident/user survey
What is the extent of the challenge?		
What are the social impacts?	How does the space impact the social, cultural and political aspects of the area?	Resident/user survey
What are the environmental impacts?	How does the space impact the local environment?	Resident/user survey. Water quality assessment

(continued)

TABLE 3.17 Ideas for key and focus questions (*continued*)

	Focus questions	Data collection
What are the economic impacts?	Does the space encourage or generate economic benefit to the entire surrounding community?	Resident/user survey Research – BCC
What can/should be done?		
What should be done?	Make some recommendations	
What are other places doing?	What works elsewhere?	Research

Develop criteria

Think about what might be an appropriate use of the space. You will need to develop criteria to determine the nature, location and extent of the challenge. Consider the role of each of these factors in determining your criteria (see table 3.18).

- social
- environmental
- economic
- political
- cultural
- historical

TABLE 3.18 Potential criteria to assess the nature of the challenge of how to improve the sustainability of Bowman Park, Bardon.

	Criteria
Social	How well does Bowman Park encourage social interaction and connectivity? OR How does the space impact the social, cultural and political aspects of the area?
Economic	Does Bowman Park encourage or generate economic benefit to the entire surrounding community?
Environmental	How does Bowman Park impact the local environment?

Collect data

Go into the field and collect data on:
- local history: notable events and changes in economy or demography that may be relevant. Don't forget to record reference details of any books, articles or websites that you use here as your report will need to reference them.

FIGURE 3.68 Local history

CHAPTER 3 Challenges for Australian places **211**

- survey land use: use spatial technologies to evaluate land use in the area and surrounds. Try the Queensland Globe from the Department of Natural Resources, Mines and Energy or Google Earth. Use the QImagery tool to examine past aerial photographs of the area. Refer to section 3.3.6 for advice on how to determine land use using an aerial or satellite image.
- survey public transport and cycling connections: include these on the same map as land use. Google Maps and the Queensland Globe has a public transport layer and a cycling layer.

Take a detailed field sketch of surrounding shopfronts and buildings. If you can't map or collect data for the entire area take a representative sample. Include information on what the building is used for (shops, type of shop, office, business, public service, etc) and the age of the building.

You could also take a photo with your mobile phone camera to annotate.

FIGURE 3.69 Nearby connections

 **Resources**

    Video eLesson: SkillBuilder: Constructing a field sketch (eles-1650)

    Interactivity: SkillBuilder: Constructing a field sketch (int-3146)

Survey residents on their views on how the space is used. It would also be useful to ask them about their suggested alternative uses of the space or to offer your proposals.

Estimate the economic value of the park, and then try to find economic data from the local council or newspaper that outlines the actual value of the space in some way. This might be hard and you may need to look at other means to determine the economic health of the area, such as the number of shops or spaces for lease, or pedestrian data for the area and surround.

FIGURE 3.70 Conduct a resident survey

FIGURE 3.71 Bowman Park

Source: Mick Law

CHAPTER 3 Challenges for Australian places 213

Make recommendations

Once you have analysed your data, you are ready to consider options for what needs to be changed in your local space to make it better. Find out what strategies could be employed by examining what is successful in other, similar areas, then propose and justify what actions should be taken to make the park compliant with the criteria you developed earlier. Use an interactive mapping tool to summarise your response and annotate where required.

FIGURE 3.72 Google My Map or ScribbleMap with annotations

Communicate

You should take some time to write your report. Use the following report structure to create your response to your teacher. Consider presenting your report in digital format. Videos, story maps, Google MyMaps and Google Tour Builder are all appropriate tools that could be used to present your findings and proposed actions in a compelling way.

- Introduction
- Body
- Method
- Processes
- Impacts – challenges
- Recommendations
- Conclusion
- Appendices
- References

FIGURE 3.73 The finished product

on Resources

Video eLesson: SkillBuilder: Creating a survey (eles-1764)

Interactivity: SkillBuilder: Creating a survey (int-3382)

CHAPTER 4
Challenges for megacities

4.1 Overview

4.1.1 Introduction

In this topic, you will explore urbanisation, particularly around the growth and challenges of **megacities**: cities with populations of more than 10 million people.

You will learn about where and why megacities have grown in specific places and the factors that have contributed to their development. You will also look at the challenges that growing urban environments present for people and the natural environment.

FIGURE 4.1 Japan's capital city, Tokyo, is the world's largest megacity. It is home to more than 13 million people, and more than 38 million people live in Tokyo's greater metropolitan region.

4.1.2 Key questions

- What factors led to the growth of megacities?
- What patterns or trends can be seen in the way that urban areas and megacities have developed?
- What patterns or trends can be seen in where urban areas and megacities have developed?
- What predictions can be made about future patterns of megacity development?
- What challenges does urbanisation present for people and the environment?
- How do countries manage the challenges presented by urban development? Do responses differ between developed and developing countries?
- Can megacities be planned in a way that promotes sustainability and the wellbeing of the people who live there?

Activity 4.1: Urbanisation and megacities

How much do you know about megacities? Reflect on your own experience and knowledge of urban environments to answer the following questions.

1. Based on your current knowledge, and without looking up an answer, list the cities that you think have a population of 10 million or more.
2. Have you ever visited a city with a population of more than 10 million? Describe the type of housing, transport system, services and public spaces you saw. If you have never visited a megacity, take a virtual visit online.
3. Why might the population of some cities grow very rapidly but not others? What challenges might rapid population growth present? Would these challenges be different for developing and developed nations?
4. How might climate change affect the severity of challenges faced by megacities?
5. What economic benefits/risks might life in a large city bring?
6. What wellbeing benefits/risks might life in a large city bring?
7. Do you think a megacity can be sustainable?

4.2 Global patterns of urbanisation

4.2.1 The rise of cities

Megacities are a product of both world population growth and **urbanisation** (increasing levels or proportions of people living in urban areas), which is now occurring at a rate faster than ever before in human history. Figure 4.2 shows the percentage of the world's population living in urban areas and the total number of people who reside in urban areas.

Urbanisation has its roots in the ability of early communities to produce reliable food surpluses – to feed the whole community throughout the year rather than each family or smaller group relying on producing their own food to survive. Working together to produce food for the broader community also led to the division of labour – different people completing different jobs to achieve a common purpose. This meant that not

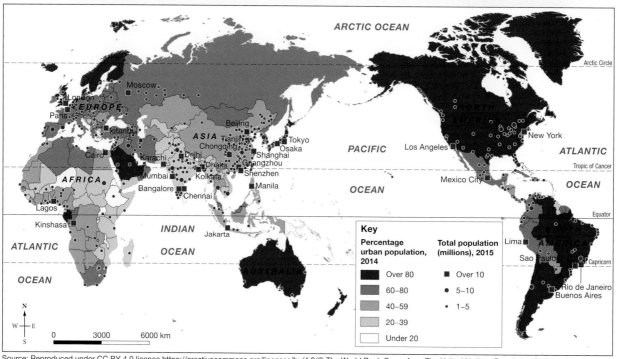

FIGURE 4.2 World urbanisation, 2014

Source: Reproduced under CC BY 4.0 licence https://creativecommons.org/licenses/by/4.0/© The World Bank Group from The United Nations Population Divisions World Urbanization Prospects.

everyone was tied to the land, working to produce food to survive. Instead, those with other skills, such as craftspeople, were able to work full-time in their own trade and were, in effect, paid by the food surplus.

Over time, skill specialisation increased and eventually trade took place thereby increasing the total wealth of the community and the individuals within it. The ancient Sumerian city of Ur, dating from 3800 BCE, was located in what is now southern Iraq and is a good example of how such communities quickly expanded and thrived. The Sumerians successfully controlled the Tigris and Euphrates rivers to supply water for farming and drinking in a very arid region.

Urbanisation has always been closely linked to economic development, but important advances in manufacturing and production techniques during The Industrial Revolution of the late 1700s and early 1800s spurred the rapid growth of existing cities in Europe and North America. New factories were generally built in or near existing urban areas, accelerating the growth of cities during the late 1800s as people flocked to them looking for work. New York, for example, grew from half a million people in 1850 to 3.5 million people in 1900.

FIGURE 4.3 Views of New York City, 1874 and 1907

Source: Library of Congress, Prints & Photographs Division, [LC-DIG-pga-02708]

Source: Library of Congress, Prints & Photographs Division, [LC-DIG-ds-00182]

New industrial cities also emerged, including Chicago. Developments in the transport network — such as canal construction and, later, railways that allowed for more efficient transportation of goods — made Chicago the fastest growing city in the world. In 1850, the population of Chicago was 29 963, but by 1890 it had risen to 1 099 850. During this period, the proportion of Americans living in towns and cities more than doubled as migrants from the surrounding countryside and abroad moved into the cities in search of work.

Most of these large cities in North America and Europe reached their peak by the mid-1900s; other parts of the world have experienced significant city growth since then. For example, the economies of Latin America transitioned from agriculture to industry and services as they adopted the import–substitution industrialisation strategy, making more products rather than importing them. In 1950, more than half of the total employment in Brazil was in agriculture but by 2000 this had fallen to just over one-fifth. Today, more than half of all Latin American countries' populations live in urban areas. By 2050, it is estimated that this figure will rise to 90 per cent.

This kind of economic structural change increased demand for labour, which led to large scale migration from rural to urban areas. This has been a major contributor to the total growth in urban populations. However, urbanisation has not only been stimulated by industrialisation and economic development; it can also lead to a phenomenon known as **pseudo-urbanisation**, a type of urbanisation without economic growth that results in large numbers of poverty-stricken residents living in informal settlements, such as the favelas in Rio de Janeiro and São Paulo, Brazil (*favela* is Portuguese for slum).

FIGURE 4.4 The Rocinha favela in Rio de Janeiro, Brazil

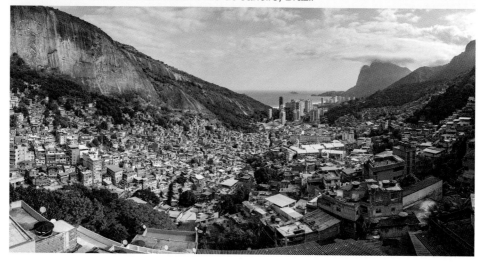

As shown in figure 4.5 and reported by the UN, the proportion of the world's population living in urban areas passed the 50 per cent mark in 2007. This proportion is expected to climb to two-thirds by 2050, largely due to increases in urbanisation in Africa and Asia. It is estimated that India, China and Nigeria alone will account for one-third of this growth in the world's urban population.

FIGURE 4.5 Change in global urban, rural and total population, 1950–2050

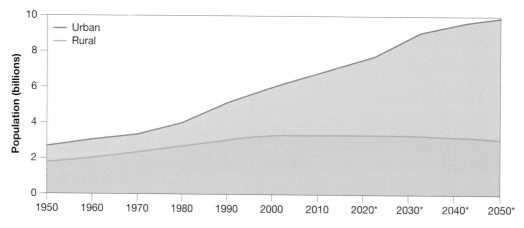

Note: * Projected
Source: United Nations (2018) *World Urbanization Prospects: The 2018 Revision*

4.2.2 Patterns in urban growth

Although urbanisation rates have been slowing in recent decades, the number of people living in urban areas has been increasing steadily. It is important to make the distinction between urbanisation and population growth in urban areas. Rates of urbanisation refer to the proportion of the total population of an area living in an urban environment. Urban population growth is the increase in numbers of people living in an urban area. Urban populations can grow without changing the overall rates of urbanisation if the rural populations are growing at the same rate. The rate or level of urbanisation measures levels that urbanisation is changing, usually expressed as a percentage. The United Nation's urban population projection for 2015–50 is for more than 2 billion urban dwellers in Africa and Asia alone. It is estimated that two-thirds of the world's population will be calling cities home by 2050. These population increases are the result of both natural increases (from births) and migration from rural areas.

Figure 4.6 shows that there are considerable differences in the population trends for developed and developing countries. While developed regions have seen a slight and consistently gradual rise in urban populations since the 1950s, the growth of urban centres in less developed regions has been far more rapid and is projected to continue.

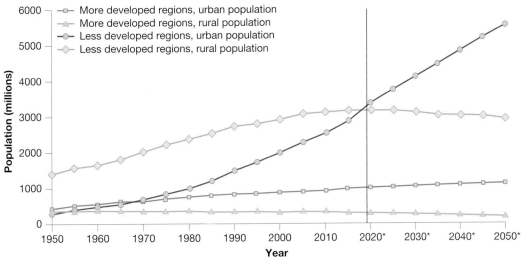

FIGURE 4.6 Global urban and rural population growth in developed and developing regions, 1950–2050

Note: * Projected
Source: Alirol E, Getaz L, Stoll B, Chappuis F, Loutan L, Urbanisation and infectious diseases in a globalised world, Lancet Infect Dis, 112, 131–241. Reprinted under STM guidelines.

Urban growth and economic growth

There is evidence to suggest a strong connection between economic growth and urbanisation, with higher levels of urbanisation associated with higher per capita incomes. This positive correlation can be explained by the fact that urban locations can have more economic advantages and opportunities than rural areas. Increased job opportunities and greater wealth expectations have long been major **pull factors** that have caused cities to grow from rural migration. According to the World Bank, cities now account for more than 80 per cent of the world's **Gross Domestic Product** (GDP).

FIGURE 4.7 The rise of globalisation has led to the widespread availability of some brands around the world.

The economic wealth of a country is most commonly measured by its GDP. This refers to the total value of all goods and services produced within a country, usually over a period of one year. To compare countries, GDP is converted to a common currency, and expressed as an average per person (or per capita); that is, how much, on average, each resident of a country produces. Currencies are usually converted to PPP (purchasing power parity) in US dollars. This method of converting each country's currency is calculated by finding the number of units of a country's currency required to purchase the same basket of goods and services that US$1 would buy in the United States.

There is also evidence to support the view that **globalisation** and urbanisation are inextricably connected. Globalisation is the process by which the world has become increasingly inter-connected through freer movement of capital, goods and services. This is reflected in the value of cross-border world trade expressed as a percentage of total global GDP. Since 1990 it has doubled to around 30 per cent. It manifests itself in the marketing and sale of identical goods and services simultaneously around the world, such as technology, clothing and fast food 'superbrands'.

Today, most urban growth is a result of natural increase among those already dwelling in cities. Additionally, formerly small settlements are being reclassified as urban areas as the populace living there grows from within.

Activity 4.2a: Interpreting data tables and column graphs

Use the following data to explore changes in urbanisation.

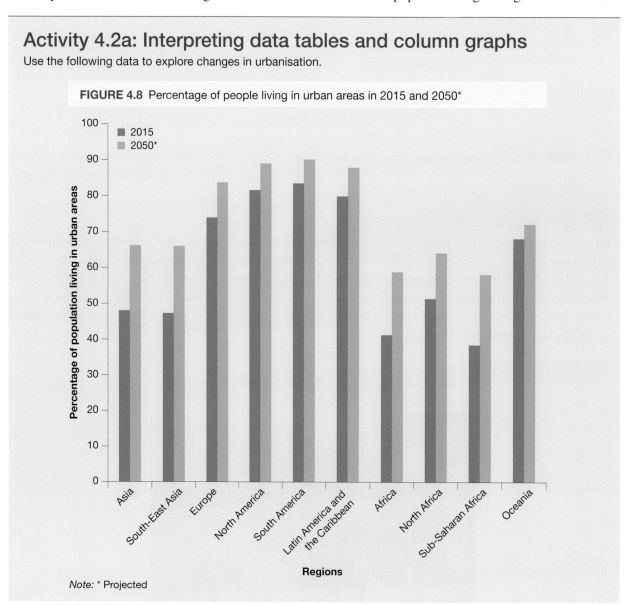

FIGURE 4.8 Percentage of people living in urban areas in 2015 and 2050*

Note: * Projected

TABLE 4.1 Average annual change in urbanisation (%), 1950–2050

	1950–60	1960–70	1970–80	1980–90	1990–2000	2000–10	2010–20*	2020–30*	2030–40*	2040–50*
Asia	1.9	1.2	1.4	1.7	1.5	1.8	1.3	1.0	0.8	0.7
South-East Asia	1.8	1.5	1.7	2.1	1.9	1.6	1.2	1.1	0.9	0.8
Europe	1.0	1.0	0.7	0.4	0.1	0.2	0.3	0.3	0.4	0.4
North America	0.9	0.5	0.0	0.2	0.5	0.2	0.3	0.3	0.3	0.2
South America	1.9	1.5	1.2	0.9	0.7	0.3	0.3	0.2	0.2	0.2
Latin America and the Caribbean	1.8	1.5	1.2	0.9	0.7	0.4	0.3	0.3	0.3	0.2
Africa	2.6	1.9	1.7	1.6	1.0	1.1	1.1	1.1	1.0	0.9
North Africa	2.0	1.6	1.1	1.0	0.6	0.4	0.4	0.5	0.7	0.7
Sub-Saharan Africa	3.3	2.1	2.1	1.9	1.3	1.4	1.4	1.3	1.1	1.0
Oceania	0.7	0.5	0.1	−0.1	−0.3	0	0	0.1	0.2	0.3

Note: * Projected
Source: United Nations, *World Urbanization Prospects: The 2018 Revision*

TABLE 4.2 Average annual urban population growth rates (%), 1950–2050

	1950–60	1960–70	1970–80	1980–90	1990–2000	2000–10	2010–20*	2020–30*	2030–40*	2040–50*
Asia	3.9	3.5	3.5	3.8	3.0	3.0	2.3	1.7	1.3	0.9
South-East Asia	4.4	4.2	4.1	4.3	3.5	2.9	2.3	1.9	1.5	1.3
Europe	2.0	1.8	1.2	0.8	0.2	0.4	0.3	0.3	0.3	0.2
North America	2.7	1.8	1.0	1.2	1.6	1.2	1.4	1.1	0.8	0.5
South America	4.6	4.1	3.5	3.0	2.3	1.6	1.2	0.9	0.7	0.4
Latin America and the Caribbean	4.6	4.2	3.6	3.0	2.4	1.7	1.4	1.1	0.8	0.5
Africa	4.8	4.4	4.4	4.4	3.6	3.6	3.6	3.4	3.1	2.8
North Africa	4.7	4.4	3.6	3.6	2.5	2.1	2.2	2.0	1.8	1.8
Sub-Saharan Africa	5.5	4.6	4.9	4.8	4.0	4.1	4.1	3.7	3.4	3.0
Oceania	2.9	2.7	1.7	1.5	1.1	1.6	1.5	1.3	1.2	1.1

Note: * Projected
Source: United Nations, *World Urbanization Prospects: The 2018 Revision*

TABLE 4.3 Rate of urbanisation and Gross Domestic Product (GDP) per capita

Country	Rate of urbanisation (%) 2015–20*	GDP per capita ($US '000)
Angola	4.3	6.3
Australia	1.4	46.7
Bangladesh	3.2	3.5
Brazil#	1.1	15.1
Burundi	5.6	0.8
Canada	1.0	44.0
Chad	3.8	1.9
China#	2.4	15.5
Denmark	0.5	49.4
Ecuador	1.7	11.2
Germany	0.3	48.7
India#	2.4	6.5
Indonesia	2.3	11.6
Japan	−0.1	41.4
Nepal	3.2	2.4
Niger	4.3	0.9
Nigeria	4.2	5.8
Russian Federation#	0.2	26.9
South Africa#	2.0	13.2
UK	0.9	42.6
USA	1.0	57.4
Vietnam	3.0	6.4

Note: * Projected, # 'BRICS' countries — Brazil, Russia, India, China and South Africa
Source: United Nations, *World Urbanization Prospects: The 2018 Revision*

Explaining and comprehending urbanisation data tables

1. Using table 4.1, construct ten sets or a selection of clustered column graphs for each of the decades to show contrasting patterns of urbanisation. (You could also construct multiple line graphs to show these patterns.)
2. Describe the differences between the regions noting those that have experienced broadly similar or very different rates of urbanisation.

TABLE 4.4 Population growth rates (% change)

	1950–60	1960–70	1970–80	1980–90	1990–2000	2000–10	2010–20*	2020–30*	2030–40*	2040–50*
Asia	2.0	2.3	2.2	2.0	1.5	1.1	1.0	0.7	0.4	0.2
South-East Asia	2.6	2.7	2.4	2.2	1.7	1.3	1.1	0.8	0.6	0.3
Europe	1.0	0.8	0.6	0.4	0.1	0.2	0.0	0.1	−0.1	−0.2
North America	1.8	1.3	1.0	1.0	1.1	0.9	0.7	0.7	0.5	0.4
South America	2.7	2.6	2.2	2.1	1.6	1.2	1.0	0.7	0.5	0.2
Latin America and the Caribbean	2.8	2.7	2.4	2.0	1.7	1.3	1.1	0.6	0.5	0.3
Africa	2.2	2.5	2.7	2.8	2.5	2.5	2.5	2.3	2.1	1.9
North Africa	2.7	2.7	2.5	2.6	1.9	1.7	1.9	1.5	1.3	1.1
Sub-Saharan Africa	2.1	2.5	2.8	2.8	2.7	2.7	2.7	2.5	2.3	2.0
Oceania	2.2	2.2	1.5	1.6	1.4	1.6	1.5	1.2	1.0	0.8

Note: * Projected
Source: United Nations, *World Urbanization Prospects: The 2018 Revision*

Analysing urbanisation data and applying your understanding

3. (a) Semi-logarithmic paper (semi-log paper) is different from common graph paper because the scale on one axis increases at a constant rate of ten, and divides into cycles, the end of each cycle being ten times greater than the end of the previous cycle. Using the data from table 4.3, construct a scattergraph on semi-log graph paper to show the relationship between Gross Domestic Product (GDP) and rates of urbanisation (2010–15). Label Brazil, the Russian Federation, India, China and South Africa (the main five emerging national economies, often referred to as BRICS). Refer to the Resources tab for a worksheet with semi-log graph paper.
   (b) Draw the line of best fit (trend line).
   (c) What does your graph show about the relationship between urbanisation and economic development (GDP per capita)?
4. (a) With reference to figure 4.8, compare the levels of urbanisation in 2015 between Africa and Asia, and the rest of the world.
   (b) How are the differences expected to change by 2050? Consider in particular the gap between their respective levels of urbanisation.
5. (a) Using data from table 4.4 construct histograms to show the estimated population growth for the four decades between 2010 and 2050 for Africa, Asia, North America, South America, Europe and Oceania.
   (b) Does there appear to be a link between population growth rates and levels of urbanisation in Africa and Asia that is different to the rest of the world?
6. Watch the introduction and interactive data visualisation about urbanisation and economic growth in Brazil, Russia, India, China and South Africa from the International Institute for Environment and Development (see the Resources tab). Table 4.3 will also help you to answer these questions.
   (a) Describe the changing global pattern of the relationship between urbanisation and national incomes.
   (b) Describe China's trajectory, and suggest two reasons to explain this.
   (c) Compare India's trajectory with that of China.
   (d) Write a paragraph outlining how the Russian Federation's trajectory is different to that of other countries.
   (e) Explain how the trajectory of South Africa compares with other countries.

> **Resources**
>
> **Digital doc:** Semi-log graph paper (doc-29169)
> **Weblink:** International Institute for Environment and Development

4.2.3 Informal settlement

For decades, the pull factor of a perceived better life in the city has led many people to leave rural areas. Push factors, such as extreme poverty or lack of work opportunities, can lead rural migrants to see the urban lifestyle as their only means of survival. In some countries, urban life becomes attractive because residents benefit from the competition between businesses and from government assistance. For instance, people in Mexico City receive cheaper education, health care and transport than their rural counterparts. It is therefore little wonder that thousands of people from rural settlements have migrated to the fringes of Mexico City to improve their standard of living.

FIGURE 4.9 A favela in Morumbi, on the outskirts of São Paulo

Unfortunately, the hopes of such migrants are not always realised. People are forced to live in slum environments, squatter areas and shanty towns, usually found on the outskirts of the city. They are overcrowded, and lack access to safe drinking water and sanitation. They are subject to insecure **tenure** because the migrants illegally occupy unused land. Globally, about 1 billion people live in slums or squatter settlements and it is estimated that around 2 billion people will live in urban slums by 2030.

Squatter settlements are also often in areas with high environmental risks, such as on mountainsides prone to landslides or in the vicinity of waste dumps. All of this can negatively affect wellbeing and safety, exacerbating mental health disorders, including depression and anxiety, or leading to greater violence and crime. The UN (Habitat) estimates that over half of Africa's urban population lives in slums and faces the aforementioned challenges of poor health and crime. In preparation for the 2016 Olympic Games in Rio de Janeiro, a policy of police occupation of the *favelas* commenced in 2008. The aim was to rid the *favelas* of violent drug traffickers.

Dharavi, India

Mumbai, formerly known as Bombay, is West India's commercial and financial capital. The port city continues to attract migrants from the rural hinterlands of central India. With high birth rates and a large influx of migrants, Mumbai's population continues to grow rapidly. In addition to this, it is estimated that 15 per cent of all urban dwellers in India live in slums.

In the middle of Mumbai is Dharavi, Asia's largest slum. Covering approximately 240 hectares, it is home to more than 1 million people, many of whom are second-generation residents whose parents moved there years ago.

FIGURE 4.10 Dharavi, Asia's second largest slum, is located in central Mumbai and is home to more than 1 million people.

Originally a small fishing village in the mangroves, Dharavi is now a sprawling economic powerhouse of small-scale business and light industry. The 'city within a city' produces embroidered garments, leather goods, pottery and plastic, which are sold both locally and on international markets. Annual turnover is more than US$650 million.

The slums have a severe shortage of living accommodation. Almost none of the people who live in Dharavi own the land, but many own their homes and businesses, some of which they rent out. Rents are very low compared with nearby Mumbai. Most houses have electricity, and some have running water but there is little other infrastructure.

A shortage of facilities means existing ones must be shared. There is, on average, only one toilet for every 1400 people, 78 per cent of which do not have running water. The squalor of open sewer drains and piles of garbage make the slums a haven for disease, particularly during the monsoon season. Despite the poverty, Dharavi is the most literate area of India with a literacy rate of 69 per cent. Ironically, crime is much lower than in other urban parts of the country.

Despite the economic and social benefits to residents, the Indian government has often seen the area as a social embarrassment to progress. Plans to renew and improve the living standards of slum dwellers have been proposed since 2004, but they often fail to materialise because of poor consultation, lack of trust in the government by residents, corruption and a strong desire of residents to live in an unregulated economy.

 Resources

Weblink: 'Slum' is a loaded term. They are homegrown neighbourhoods.

FIGURE 4.11 Shanty houses in Dharavi

Activity 4.2b: The challenges of informal settlements

Consider this hypothetical situation and complete the task below.

The central government has put forward two low-risk options for the development of the Dharavi slums.

Option 1: Under the direction of the Slum Rehabilitation Authority (SRA), the 240-hectare area would be zoned into four types of improved housing with each stage reconstructed one at a time after another has been completed.

- Area 1 (60 per cent) would have old dwellings replaced by many 20 storey units, which could accommodate about 100 residents or 20 families per level, i.e. 2000 per building. Each family would be entitled to a new four room unit with separate kitchen and toilets and would only pay 50 rupees per week. Four high rise blocks would share a common play area for children.
- Area 2 (20 per cent) would have old buildings replaced with 10 storey units identical to Area 1 but would have an extra bedroom. Residents would pay 75 rupees per week.
- Area 3 (10 per cent) would be similar, but buildings would only be 5 storeys, meaning more playground space for children. Payments would be 100 rupees per week.
- Area 4 (10 per cent) would be more up-market units: 5 storeys with two bedrooms, but payments would be 125 rupees per week.

All buildings would be connected to power and would be fitted with running water, sewerage and have garbage collection. All legal residents and squatters would be eligible. Illegal residents would be moved. Public meetings would be held and brochures distributed to explain the plan to residents

Option 2: Under the administration of The Central Utilities Department, people would remain in their existing locations and improvements to services and facilities would be commenced over a three-year period.

Initially, running water and sewerage lines would be installed to improve sanitation and mitigate health risks. Before these projects are fully completed, belowground electricity and internet cables would be added to each group of houses. Where houses are temporary, shared hubs and facilities would be installed with people gaining access through a coded system and low payment.

Included in the installation program is the creation of approximately 500 new jobs where people can either collect garbage and take it to distribution points for recycling or disposal, or be part of a utility maintenance team to ensure breakdowns are minimised.

All legal residents and squatters as well as illegal residents would be eligible.

Public meetings would be held and brochures distributed to explain the plan to residents.

Propose and justify actions

Task: Examine the worth of these two proposals and consider which is better for the residents of Dharavi and the overall future of Mumbai.

You must consider a range of opinions from residents, developers, government officials, engineers and health professionals. Use the decision-making matrix based on economic, social and environmental criteria to help you make your decision.

Redevelop Dharavi proposals	Environmental (+ and –)	Social (+ and –)	Economic (+ and –)
Redevelop four areas with multi-tenement housing units			
Leave the slums as they are and improve facilities and services			

4.3 Air quality in the megacity

While urban living offers many economic opportunities, jobs and services, urbanisation has always raised environmental, social and health concerns. High population densities create concentrated health risks and environmental hazards. Many urban-based activities such as manufacturing also impose external costs on society. These are costs that are paid by someone other than those responsible for the original activity, such as cleaning up after decades of industrial pollution.

4.3.1 Air pollution

As urban populations increase, health conditions worsen, particularly in developing countries that lack the resources to adequately tackle the problem of pollution. A 2016 study of air pollution in megacities by the UN World Health Organization (WHO) and United Nations Environment Programme (UNEP) demonstrated that ambient air pollution concentrations increased the risk of serious health effects. According to WHO data, most of the 3.7 million deaths resulting from outdoor air pollution each year are among urban populations.

Between 2008 and 2013, WHO and UNEP analysed air quality readings from 795 cities for levels of small and fine particulate matter, which can increase the risk of non-communicable diseases, such as stroke, heart disease, lung cancer and asthma, and account for a large proportion of global mortality. The study found that, globally, air pollution levels increased by 8 per cent in urban areas. In high-income countries in Europe and the Americas, however, urban air pollution levels were generally lower than in low income countries in South-East Asia, which recorded increasing levels of urban air pollution during that time.

The Asian Development Bank 2002 Integrated Vehicle Emissions Reduction Strategy for Greater Jakarta specifically measured the air quality in Jakarta's regions, including for levels of nitric oxide. Vehicle emissions contributed 71 per cent of the nitric oxide, industry produced 26 per cent and 3 per cent was generated from domestic sources. Nitric oxide pollution contributes to acid rain, ozone depletion, lower yield crops and potential long-term health effects in people, such as respiratory disease. Table 4.5 shows the normalised vehicle emission of nitric oxide in Jakarta.

WHO data has also shown that unhealthy urban environments, such as slum areas found in the poorest districts of cities in the developing world, are closely linked to communicable diseases. These include diseases such as tuberculosis, which spreads more easily in over-crowded conditions with a lack of adequate ventilation, as well as vector-borne diseases such as dengue, a consequence of unsafe water storage and poor waste management that allow mosquito populations to thrive.

The traditional view of the relationship between economic development and the biophysical environment is that economic growth leads to environmental decline. This means that, as a country develops economically, its natural environment will suffer the effects of land degradation, air and water pollution, resource depletion and loss of biodiversity. Moreover, the faster a country industrialises in order to achieve economic growth, the greater the detrimental effects on the environment.

In recent years, this traditional view of economic development and the environment has been challenged by some economists, who have argued that once a country reaches a certain level of income, the quality of its environment improves rather than declines. The **Environmental Kuznets Curve** (see figure 4.12) was created to demonstrate this view, based on the work of economist Simon Kuznets. Kuznets developed a hypothetical curve that graphs economic inequality against income per capita over the course of time. By studying Latin

TABLE 4.5 Normalised vehicle emission* of nitric oxides in Jakarta

Districts	Nitric oxide
1	0.18
2	0.47
3	0.06
4	0.14
5	0.34
6	0.56
7	0.42
8	0.42
9	0.13
10	0.15
11	0.30
12	0.46
13	0.47
14	0.17
15	0.20
16	0.17
17	0.63
18	0.41
19	0.36
20	0.20
21	0.06
22	0.16
23	0.07

Note: * Normalised vehicle emission load for each pollutant is obtained by dividing the vehicle emission load for each area by the highest total emission load value.
Source: Based on ADB report *Study on air quality in Jakarta, Indonesia*, Figure 4.8

American economies with high levels of income inequality, he suggested a country transitioning from being a primarily rural and agricultural economy to an industrialised urban economy would initially experience increasing levels of inequality, but these levels would eventually peak and then decline. Using various indicators of environmental degradation, the Environmental Kuznets Curve suggests an environment will become increasingly degraded as per capita income increases until a certain point at which the community has a strong enough economy and access to the technology to implement strategies to better manage and protect the quality of their environment.

This means that economic growth in high-income countries results in environmental improvement rather than decline. Supporters of this argument point to examples such as improvements in air and water quality in the world's most developed countries as evidence of their view. Increased wealth provides both the financial and technological means to improve environmental quality, and also leads to an increased demand from people in more developed countries for a cleaner environment.

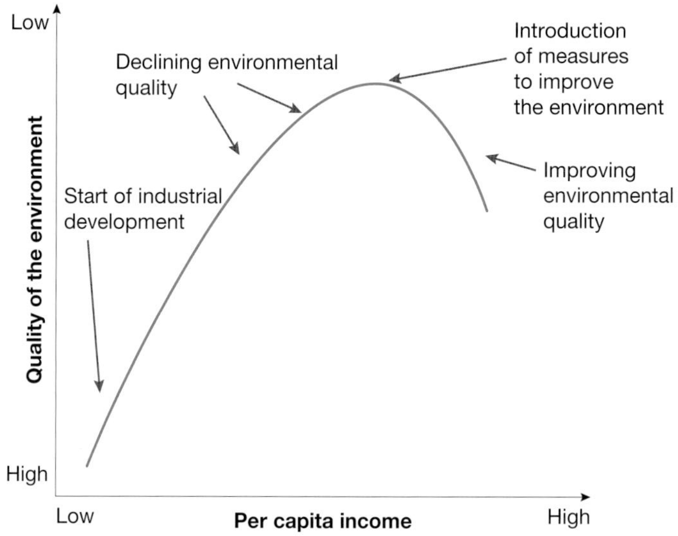

Recent studies in China that examined the relationship between air pollution and industry output in provincial capitals supported the Environmental Kuznets Curve theory for urban areas. Figure 4.13 is a scattergraph that shows the total value of manufactured goods (Secondary Industry Output, which is converting raw materials into products) per capita produced by China's provincial cities measured against air pollution levels of sulfur dioxide, which is produced by burning fossil fuels.

FIGURE 4.12 The Environmental Kuznets Curve

Although these results show a very weak negative correlation between the two variables, they do suggest that provincial cities that have a low value-per-capita output have a very large range in pollution levels. This can be explained by the fact that many will be using old technologies requiring greater use of fossil fuel whereas cities with a higher value output are more likely to use newer technologies, which burn fewer fossil fuels in their manufacturing industries.

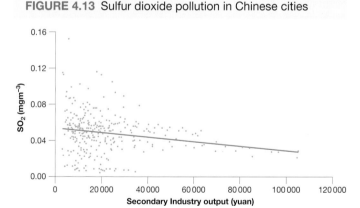

FIGURE 4.13 Sulfur dioxide pollution in Chinese cities

Source: Luo Y, Chen H, Zhu Q, Peng C, Yang G, Yang Y, et al. 2014 Relationship between Air Pollutants and Economic Development of the Provincial Capital Cities in China during the Past Decade. PLoS ONE 98: e104013. https://doi.org/10.1371/journal.pone.0104013

Critics of the Environmental Kuznets Curve point out that the data used to create the curve does not reflect the full range of environmental effects in the world's wealthiest countries. Carbon dioxide emissions, for example, continue to increase with wealth rather than decline once a certain income level has been reached. Similarly, land clearing, loss of biodiversity and overuse of resources also continue to increase in wealthier countries as they develop. The Environmental Kuznets Curve also does not take into account the effects of wealthier countries outsourcing their most environmentally damaging industries to the world's poorer countries or shipping their recyclable waste to developing countries for processing. While development may lead to local pollution levels in rich countries falling, overall global pollution levels might not.

Activity 4.3a: Environmental quality and urbanisation

Based on your understanding of the Environmental Kuznets Curve, answer the following questions.

Explain the information and comprehend the implications

1. In your own words, explain the Environmental Kuznets Curve.
2. Make a list of six ways that environmental quality might be measured (for example rate of deforestation or air quality). Explain whether you think each might demonstrate an Environmental Kuznets Curve, and give reasons for your decision.
3. Explain the limitations of the Environmental Kuznets Curve when considering the impact of urbanisation on a global scale.

Resources

- **Weblink:** Atmospheric chemistry
- **Weblink:** United Nations Population Division
- **Weblink:** Air quality in Jakarta
- **Weblink:** Combatting vehicle pollution

Urbanisation and pollution in Jakarta

Jakarta, in Indonesia, is South-East Asia's largest city and its origins can be traced back to the 4th century when a settlement was built at the mouth of the Ciliwung River. In the early 1500s, Portuguese traders were granted permission to build a fort at Sunda Kelapa, located near the modern city of Bogor. Later that century, the fort was taken over by the Demak Sutanate, a Muslim state from the north, who named the settlement Jayakarta.

A century later, Dutch traders captured the port town and renamed it Batavia. It then became the colonial capital of the Netherlands' Indies and was extensively rebuilt to resemble a Dutch city, with canals for flood protection. As occurred in other colonial cities, the European elite occupied an area detached from that of the indigenous residents. The suburb of Weltevreden was built south of the old city for government officials and wealthy merchants who wanted to escape the commercial hub and unhealthy conditions of old Batavia. As the colony prospered, new residential areas, such as Menteng, were built for the growing middle class and contributed to the rapid expansion of the city in the first half of the 20th century.

These planned suburbs were in sharp contrast to the traditional informal kampungs that existed outside the old city. Kampungs were high-density communities of peasants, mostly migrants from the surrounding rural areas, who participated in semi-subsistence farming on the edge of the city. Kampungs were characterised by an unplanned collection of low-rise wooden or bamboo dwellings, which lacked access to piped water, sewers or electricity.

Japanese occupying forces during World War II renamed Batavia to Jakarta in 1942. The name was retained in 1945 when Indonesia declared its independence. By this time, Jakarta had grown rapidly and had become the most populous city in South-East Asia.

FIGURE 4.14 Artist's impression of Batavia in 1740

Source: Van Ryne, I. The city of Batavia in the island of Java and capital of all the Dutch factories & settlements in the East Indies, London : Robt. Sayer, ca. 1740. Digitised item from: Allport Library and Museum of Fine Arts, Tasmanian Archive and Heritage Office.

In 1950, the first of several master plans designed to guide future growth of the city of Jakarta was produced. Unfortunately, the planners underestimated Jakarta's rate of population growth. They had assumed an annual growth rate of 4 per cent. However, during the 1960s, the city began to develop as a major industrial centre. This was matched by a corresponding growth in its urban population as migrants from throughout Indonesia flocked to the city in search of work. Many of those migrants settled in kampungs situated close to where they could find work. As a result of this excessive, rapid population growth, Jakarta quickly descended into crisis and struggled to provide essential services.

The Master Plan of Jakarta 1965–85 identified four major problems:
- flooding of the low-lying areas
- poor sanitation, especially in respect to the removal of solid waste
- traffic congestion
- poor housing for the rapidly expanding low-income population.

Although the master plan provided an outline for solving the city's problems and catering for its future growth, its achievements were restricted by lack of funds and a reliance on the private sector. The city of Jakarta continued to grow. In 1955, the population was 1 452 000 and had risen to 8 175 000 by 1990. Figure 4.15 shows the rapid increase of built-up area in the Jakarta metropolitan area from 1983–2005.

FIGURE 4.15 The growth of built-up areas in the Jabodetabek metropolitan area

Source: Redrawn by Spatial Vision based on World Bank material 'Indonesia Urban Development Towards Inclusive and Sustainable Economic Growth' by Taimur Samad, 19 September 2012.

Key
— Jabodetabek metropolitan area
---- Jakarta Special Capital Region
▓ Built-up area

The metropolitan area including Jakarta and its surrounding areas is known as Jabodetabek, combining the names of the Jakarta, Bogor, Depok, Tangerang and Bekasi areas, which are home to a combined 25 million people.

Data from 2014, using the Asian Development Bank poverty threshold of $25 US per month, showed that nearly half a million people in Jakarta live in poverty, with North Jakarta and East Jakarta having the highest levels of poverty.

In addition to Jakarta's rapid expansion and high levels of poverty, the city's population density in the major districts is high. In 2016, the population density in central Jakarta was as high as 18 590 pp/sq km. In comparison, Australian Bureau of Statistics data from June 2015 shows that the highest population density of any locality in Queensland is New Farm in Brisbane, which has a population density of 6500 pp/sq km.

FIGURE 4.16 Land use in Jabodetabek, Indonesia

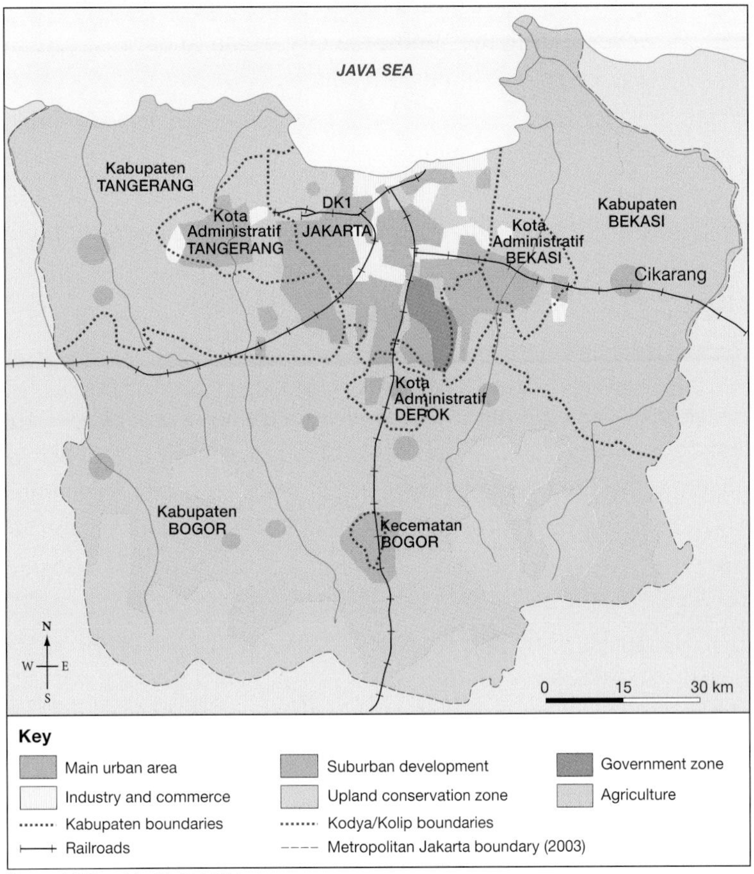

Source: MAPgraphics Pty Ltd, Brisbane

TABLE 4.6 Area of cities in Jakarta and population density

City/Regency	Area (km^2)	Population density (pp/km^2)
South Jakarta (*Jakarta Selatan*)	141.27	15 161
East Jakarta (*Jakarta Timur*)	188.03	12 924
Central Jakarta (*Jakarta Pusat*)	48.13	18 590
West Jakarta (*Jakarta Barat*)	129.54	17 004
North Jakarta (*Jakarta Utara*)	146.66	9 951

Activity 4.3b: Pollution in Jakarta

Use tables 4.5 and 4.6 and figures 4.16, 4.18 and 4.19 to complete the following questions.

Analyse and map pollution data

Go to the Resources tab for a downloadable map that corresponds to the pollution data shown in table 4.5.
1. (a) Using the data from table 4.5, construct a choropleth map to show the pattern of nitric oxide pollution in the districts of Jakarta. (Group the data into five class intervals: <0.10, 0.10–0.20, 0.21–0.30, 0.31–0.40 and >0.40.)
   (b) Describe the pattern of pollution revealed by your map.

FIGURE 4.17 Jakarta's urban districts

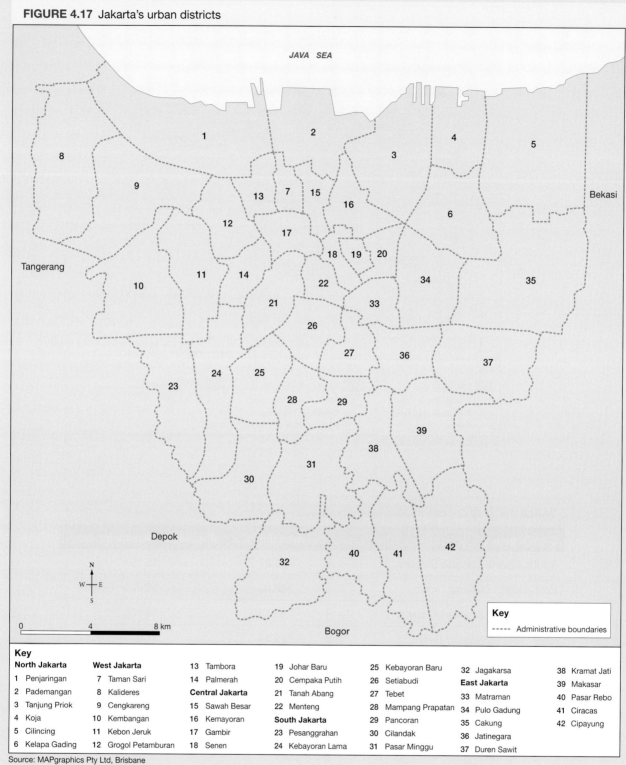

Source: MAPgraphics Pty Ltd, Brisbane

Key

North Jakarta
1 Penjaringan
2 Pademangan
3 Tanjung Priok
4 Koja
5 Cilincing
6 Kelapa Gading

West Jakarta
7 Taman Sari
8 Kalideres
9 Cengkareng
10 Kembangan
11 Kebon Jeruk
12 Grogol Petamburan
13 Tambora
14 Palmerah

Central Jakarta
15 Sawah Besar
16 Kemayoran
17 Gambir
18 Senen
19 Johar Baru
20 Cempaka Putih
21 Tanah Abang
22 Menteng

South Jakarta
23 Pesanggrahan
24 Kebayoran Lama
25 Kebayoran Baru
26 Setiabudi
27 Tebet
28 Mampang Prapatan
29 Pancoran
30 Cilandak
31 Pasar Minggu

32 Jagakarsa

East Jakarta
33 Matraman
34 Pulo Gadung
35 Cakung
36 Jatinegara
37 Duren Sawit
38 Kramat Jati
39 Makasar
40 Pasar Rebo
41 Ciracas
42 Cipayung

2. (a) Using the population density data for Jakarta's districts in table 4.6 and the key showing districts' locations, create a choropleth map to show the patterns of population density in Jakarta.
   (b) Describe the pattern of population density revealed by your map.
   (c) With the aid of your maps, and figures 4.16, 4.18 and 4.19, explain the pattern of air pollution across Jakarta.

(d) Most of Jakarta's nitric oxide pollution is generated from vehicles and industry. Based on Jakarta's population density and areas of higher nitric oxide pollution, can you make any inferences about where the heavy traffic and industrial areas of Jakarta are located? Do your predictions fit with the land use map in figure 4.16?

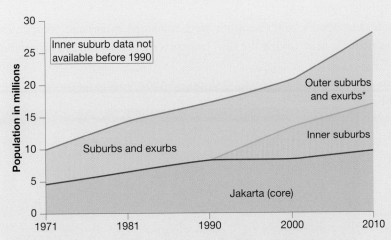

FIGURE 4.18 Jakarta urban area population density, 1971–2010

Note: * An exurb is a settlement outside a city's outer suburbs, often a commuter town.
Source: New Geography, The Evolving Urban Form: Jakarta Jabotabek, by Wendell Cox 05/31/2011, http://www.newgeography.com/content/002255-the-evolving-urban-form-jakarta-jabotabek

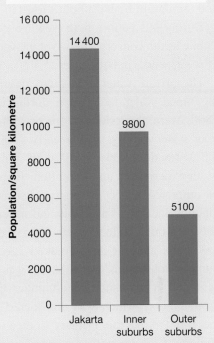

FIGURE 4.19 Jakarta Urban Area: Population Density, 2010

Source: New Geography, The Evolving Urban Form: Jakarta Jabotabek, by Wendell Cox 05/31/2011, http://www.newgeography.com/content/002255-the-evolving-urban-form-jakarta-jabotabek

Resources

Digital doc: Jakarta urban districts (doc-29170)

4.4 Health and sanitation management

4.4.1 Disease in urban areas

It is estimated that 150 million people live in cities with perennial water shortage, which is defined as having less than 100 L per person per day of sustainable surface and groundwater flow within their urban extent. By 2050, demographic growth will increase this figure to almost 1 billion people. With almost 3 billion additional urban dwellers predicted by 2050 an increasing number of cities of all sizes will struggle to meet the increasing demand for water, irrespective of adverse effects associated with possible climate change. Most of the future urban growth will occur in the developing world where the lack of availability of clean water and sanitation infrastructure can easily lead to outbreaks of disease and other health issues.

History very clearly demonstrates what can happen when large numbers of people move to urban areas and live in cramped conditions with poor sanitation. Large cities such as New York and London faced similar problems in the 1800s, when their populations experienced explosive growth. Both cities experienced serious outbreaks of cholera, an infection of the intestinal tract, which is contracted through contaminated food or water. Cholera causes severe diarrhoea and vomiting, which results in chronic dehydration leading to kidney failure and eventually death within days. The disease spreads quickly in areas where people live in crowded conditions and where sanitation and hygiene are poor. New York's 1832 cholera outbreak left over 3000 people dead out of a population of 250 000, which equates to a city of 8 million having more than 100 000 fatalities. The worst hit areas were neighbourhoods such as Five Points, a slum area where African-Americans and immigrant Irish Catholics lived in squalid conditions.

During this period, both New York's and London's water supplies were managed by private companies. Public health concerns, however, played a major role in driving the shift from private to government ownership of water works in both cities.

FIGURE 4.20 Living conditions of working-class residents in east London

The private water companies providing piped water to London during the 1800s were also criticised for providing poor water quality. In the early 1800s, those London households with piped water experienced an intermittent supply of untreated water. In poorer districts, piped water wasn't available and instead people would collect water from shallow wells using a hand-cranked street-pump. Households' 'sanitation' was a cesspool, a pit usually built under the ground floor, which was considered at the time to be the primary source of cholera. Consequently, cesspools were gradually abolished in the 1840s and houses were required to be connected to covered sewers, as opposed to the open form of sewers that had been used previously. These closed sewers, however, flowed directly into the River Thames, resulting in increased sewage pollution of the river.

John Snow, a London physician, believed sewage dumped into the river or into cesspools near wells could contaminate the water supply, leading to a rapid spread of disease. He mapped cholera outbreaks by plotting the distribution of cholera cases. In London's poor Soho district, he noted a distinct cluster of deaths.

Snow examined water samples from various wells and discovered the presence of an unknown bacterium in the Broad Street samples. The pump in Broad Street was disabled and the disease soon abated. Therefore, by identifying the source John Snow was able to demonstrate that cholera was most likely contracted from faecally contaminated drinking water and not vapours. This was a major breakthrough decades before germ theory of disease was generally accepted.

The WHO Global Health Observatory data reports that in 2016 every region of the world had instances of cholera cases. About two-thirds of these cases occurred in developing countries, mostly in sub-Saharan Africa, and the majority (80 per cent) of the 132 121 cases reported in 2016 were in only five countries: Haiti, the Democratic Republic of the Congo, Yemen, Somalia and Tanzania. Despite these reported cases, the WHO estimates that the actual figure of cholera cases each year is far higher: up to 4 million, including up to 143 000 deaths each year.

Cholera is known as a disease of poverty, as it is most prevalent in areas with poor access to clean water and sanitation facilities. This includes city slums and crowded camps for internally displaced persons seeking refuge from civil unrest and natural disasters, such as earthquakes or flooding.

Cholera outbreaks also occur in large cities with ageing infrastructure. In 2008, the Vietnamese capital city of Hanoi experienced a cholera outbreak that produced more than 2490 cases. The outbreak was attributed to sewage from septic tanks contaminating city lakes where people sometimes wash food.

Hanoi's sewers date from the late 1800s, when the city came under French rule as the colony of Indochina. The French were responsible for creating a modern administrative city with broad, tree-lined avenues. However, in the process, many lakes and canals were either removed or reduced in size. Today, although almost eight million people live in Hanoi, the sanitation infrastructure is largely unchanged from the colonial period.

Resources

Weblink: The cholera outbreak in London, 1866

Activity 4.4a: Complete a mean centre analysis

The 'mean centre' is the average x and y coordinate in a map area; it is a useful calculation to determine changes in distribution of events. Complete the following steps to complete a mean centre analysis of figure 4.21, Snow's 1854 cholera map. (A simpler worked example has been provided below to help you follow the instructions.)

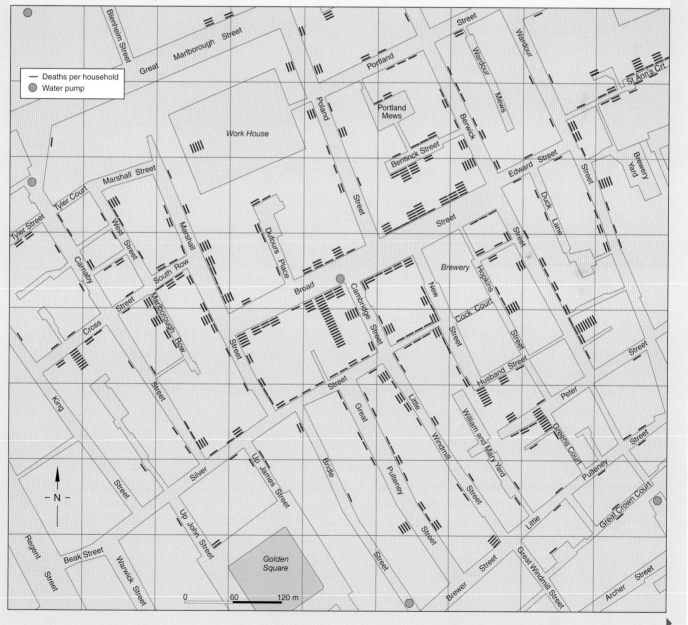

FIGURE 4.21 Snow's map of cholera deaths

CHAPTER 4 Challenges for megacities 235

1. Number the grid rows and columns on the figure 4.21 map. Create two tables: one to record the number of deaths that occurred in each square. Count the number of deaths that occurred in each row of your grid and in each of the columns (see example below). And record the total in one of the corners.
2. Record the total number of deaths shown on the map.
3. Count the number of deaths for each column and multiply the totals by the column number. For example, the first column death total was 21 and this is multiplied by 1. The second column total is multiplied by 2 and so on up to the ninth column.
4. Add up the nine column totals you have calculated and divide by the total number of deaths. This is the mean column value.
5. Repeat the process for the rows. In row 1 at the bottom there were 7 deaths, hence the total for row 1 is 7.
6. Add the total for the eight rows and divide by the total number of deaths. This is the row mean value.
7. These values are then plotted as coordinates as shown in the worked example and the intersection of these coordinates is the mean centre for the distribution.
8. Identify the nearest water pump to the mean centre.

Mean centre analysis worked example

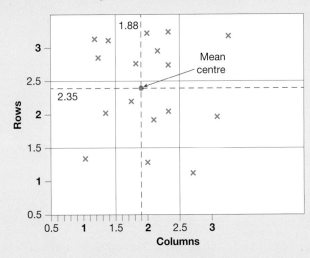

n = total number of points shown in the whole map
$n = 17$

Columns	Frequency	Col x Freq
1	5	5
2	9	18
3	3	9
Totals	17	32

Rows	Frequency	Row x Freq
1	3	3
2	5	10
3	9	27
Totals	17	40

$$\frac{Row \times Freq}{n} = \frac{40}{17} = 2.35 \qquad \frac{Col \times Freq}{n} = \frac{32}{17} = 1.88$$

4.4.2 Access to fresh water

The United Nations has recognised access to water as a human right. Access to clean water sources is expensive in megacities, particularly in developing countries. The UN cites Jakarta and Manila (a city with a wider metropolitan population of more than 12 million) as examples, where residents of urban slums are expected to pay up to ten times more for water than residents of more affluent areas of the same city. A resident of Manila whose income is in the lowest 20 per cent of the population can pay to have the water connected to their home, but the cost is so high and their wages so low, that connection will cost them the sum of their wages for approximately three months.

Three things are required for urban water supplies to meet the needs of their residents.
- A water supply that is ample and sustainable
- Water that is of good quality (potable water, i.e. clean enough for drinking)
- An efficient supply network

The rapid pace of urbanisation has often far exceeded the capacities of many national and local governments to plan and manage the need for additional water.

In any country, developing or developed, failure to invest in adequate maintenance or increasing capacity of water supply that keeps pace with increasing demand can result in critical water shortage. This risk is heightened during drought conditions. Record low rainfall was a key factor in Barcelona's water crisis in 2008, when the Spanish city experienced the worst drought since records began 60 years ago. Reservoirs dropped dangerously low to 25 per cent, only 10 per cent above the threshold for safe drinking water.

Emergency action was taken in the form of fleets of ocean tankers operating for three months at a cost of $25 million. The government was also forced to allow water to be pumped from the River Ebro in the neighbouring region of Aragon. Barcelona's water crisis was mostly due to low rainfall but also to its antiquated, leaky drainage system.

More recently, Cape Town's water crisis of 2018 illustrates the danger of low investment in water supply. Alarm bells began to ring 30 years ago when Cape Town's population grew from 2 million to more than 3.5 million. The city relies almost exclusively on six dams, located in the mountains east of the city, for its water supply. Rather than tackle the supply problem, the focus of action was on curbing demand. Per capita consumption did decline as large water users were required to pay more. In 2014, the dams were full but three years of drought thereafter required the introduction of drastic measures to conserve dwindling water supplies. Residents were restricted to 50 L per person per day and, to heighten people's awareness of the problem, the concept of Day Zero was introduced, marking the day when there would be no water left.

Resources

- **Weblink:** United Nations water and sanitation programs
- **Weblink:** We're (not) running out of water — a better way to measure water scarcity

Beijing: A thirsty giant

With a population of about 22 million, Beijing is struggling to provide an adequate and reliable water supply for its residents. The city's annual per capita water availability is only 120 cubic metres per person, which is well below the United Nations threshold for absolute water scarcity (see table 4.7).

TABLE 4.7 UN Levels of Water Scarcity (UNDESA)

Level of water scarcity	Annual water supplies per person
Water stress	Less than 1700 cubic metres
Water scarcity	Less than 1000 cubic metres
Absolute water scarcity	Less than 500 cubic metres

By 2020, the total annual water consumption in Beijing is expected to reach 4.3 billion cubic metres of water, which is far more than the available, renewable fresh water resources.

Beijing has three water sources:
- surface water captured from five major rivers within the Beijing municipality
- groundwater
- diverted water from the Yangtze River through the South-to-North Water Diversion Project

Beijing is in one of the driest regions of China. As shown in figure 4.22, for much of the year there is little rain. More than 90 per cent of the annual precipitation falls between July and October. Consequently, drought can have a devastating effect on the city's water supply. The drought of 2017 was the worst on record. Less rainfall meant less runoff for sustainable refilling of its surface and groundwater sources.

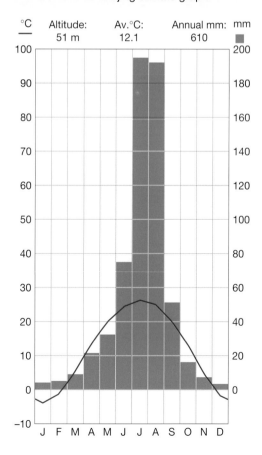

FIGURE 4.22 Beijing climate graph

With 50 000 wells extracting about 1.4 billion cubic metres of water a year, Beijing has a huge groundwater deficit. Between 2000 and 2012, the North China Plain groundwater table, on which Beijing is situated, dropped by 12 metres, a problem that is aggravated by groundwater pollution. In 2014, China's Land and Resources Ministry sampled groundwater in more than 200 cities. They found that 44 per cent of the samples tested 'relatively poor' and requiring treatment, and 16 per cent were rated 'very poor' and deemed unsuitable for drinking. This presents a major concern for Beijing, as groundwater currently contributes more than 70 per cent of the city's total water supply.

Despite being short of water, Chinese cities were accused of wastage by selling it at heavily subsidised prices. In 2010, and for the first time in seven years, Shanghai raised residential water prices by 25 per cent, while Beijing put up the price of water for commercial use by nearly 50 per cent. In 2014, Beijing increased residential water prices by 25 per cent and introduced a separate pricing tier for businesses that saw water prices double.

Beijing's water supply has been supplemented by water diverted from the Yangtze River via the South-to-North Water Diversion project (see figure 4.23). It comprises three routes for diverting water from the Yangtze to the drier north, where other megacities, such as Chongqing and Shenzhen, also face water shortages. However, reduced precipitation, a possible indicator of climate change, has already decreased freshwater reserves in the Yangtze River basin by 17 per cent. The scheme has been heavily criticised for the number of people it has displaced and the potential losses through **evapotranspiration**.

FIGURE 4.23 The South-to-North Water Diversion project in China

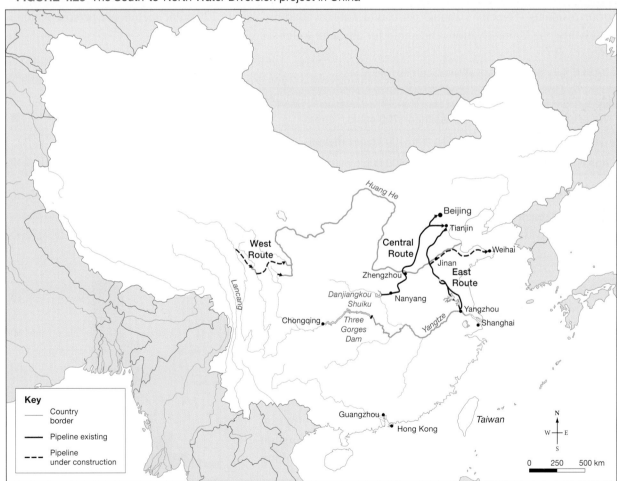

Source: Maximilian Dörrbecker (Chumwa) / Spatial Vision

China's rapid urbanisation and industrialisation over the past few decades have led to large increases in wastewater discharge. Total wastewater discharge is mainly from industrial and domestic sources. More than 80 per cent of wastewater is treated but only 10 per cent is recycled. In Beijing, a third of wastewater is reused for industrial processing or irrigation.

Activity 4.4b: Quenching Beijing's water thirst

With the population of Beijing expected to soar to more than 50 million by 2050, good management of its water resources is critical for the city's future.

The following measures have been identified as possible solutions to the city's long-term water requirements. Critically examine each option, assessing its strengths and weaknesses, and propose a justified set of actions to manage Beijing's water resources.

Option 1: Decentralise industry from Beijing and reduce the amount of agricultural land in the municipality. This is in response to a recent study modelling data on all factors that influenced decreases in groundwater storage (1993–2006). This showed that expansion of Beijing's urban area at the expense of croplands enhanced recharge while reducing water lost to evapotranspiration.

Use figures 4.22 and 4.23, and water supply and use data in the Resources and Environment section of the **China National Bureau of Statistics** weblink in the Resources tab to help you make your decision.

Option 2: Raise water prices for all types of consumer.

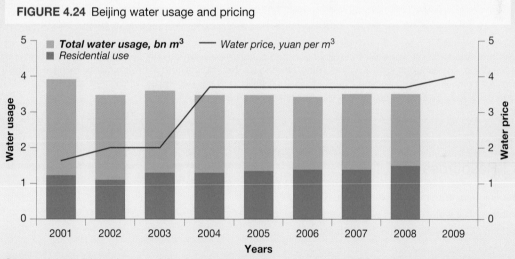

FIGURE 4.24 Beijing water usage and pricing

Source: Adapted from Journal of the American Water Resources and the IOP Conference Series: Earth and Environmental Science.

Option 3: Use recycled water sourced from new types of sewage treatment plants that are cleaner and produce higher water quality. Construction of 17 new processing plants has already begun.

FIGURE 4.25 Beijing water supply and recycling, 2013–20

Note: * Projected
Source: Adapted from Water Scarcity in Beijing and countermeasures to solve the problem at river basins scale, Lixia Wang, Jixi Gao, Changxin Zou, Yan Wang, Naifeng Lin, Nanjing Institute of Environmental Sciences, Ministry of Environmental Protection, Nanjing 210042, China and Beijings Water Resources; Challenges and Solutions JAWRA Journal of the American Water Resources Association 513:614–623, 2015.

Resources

- **Weblink:** China's South-to-North Water Diversion Project
- **Weblink:** The world's largest water-diversion project
- **Weblink:** China National Bureau of Statistics

4.4.3 Groundwater pollution and subsidence

China's Land and Resources Ministry 2014 survey of levels of groundwater contamination showed how much population growth and consequential urbanisation contribute to the groundwater pollution problem. Excessive groundwater extraction in Jakarta has shown the practice can lead to other environmental problems that directly impact on its citizens, two of which are significant land **subsidence** and pollution of groundwater.

Jakarta sprawls across the flat lowland of the northern coast of the West Java province. The city has two sources of water: the Citarum River catchment in the east, which supplies water via the West Tarum Canal, and groundwater extracted from three aquifers located inside a bed of quaternary sediment 50 m thick. The upper **unconfined aquifer** occurs at a depth of less than 40 m, the middle **confined aquifer** between 40 and 140 m and the lower aquifer, also confined, between 140 and 250 m (see figure 4.28). Like the 13 rivers in Jakarta, groundwater flows from south to north.

Only 60 per cent of Jakarta's residents have access to clean, safe treated water. Unequal access is highlighted by the fact that most homes in Jakarta, which are usually in the city's low-income areas, rely on individual shallow wells to supply household needs.

Because there is no efficient municipal sewerage system in Jakarta, 75 per cent of household wastewater is sent to septic tanks. Over 11 per cent of households discharge their wastewater directly into rivers. Off-site sanitation, which includes industry and domestic individual treatment plants, make up the remainder. Septic tanks are a heath concern as many are simply holes dug in the ground, lined with layers of coral rock and fibre at the bottom to filter the water. Often poorly constructed or old, these septic tanks are prone to leak raw sewage, which can then leach into the city's groundwater supply. It is estimated that more than 50 per cent of the shallow wells, those less than 40 m deep, are contaminated by bacteria, such as E. coli, that cause pathogenic diseases.

Deeper extraction is mostly carried out by industry – it is much more localised and the rate of extraction is much greater than that for domestic users. It is estimated that the level of groundwater in north Jakarta has dropped over 50 m since 1910.

FIGURE 4.26 Informal housing in outer Java

Source: New Geography, The Evolving Urban Form, Jakarta Jabotabek, by Wendell Cox 05/31/2011, http://www.newgeography.com/content/002255-the-evolving-urban-form-jakarta-jabotabek

FIGURE 4.27 Kampungs in north Jakarta

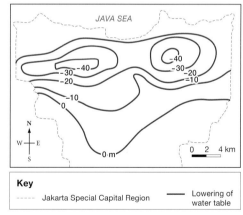

FIGURE 4.28 Head of groundwater level contours (m) of lower aquifer in Jakarta basin

Source: Adapted from Murdohardono and Tirtiomihardjo, 1993.

It is this deep extraction of groundwater that is believed to be one of the main factors responsible for the land subsidence observed in the coastal, western and north-eastern parts of Jakarta. Subsidence is also caused by major building projects on soil of high compressibility. A study of land subsidence in Jakarta from 1982–2010 found spatial and temporal variations, with the annual rates of subsidence between one and fifteen centimetres.

In general, there is a strong connection between land subsidence and urban development activities that require water extraction. Evidence of the problem includes structural damage to buildings around the city centre, expansion of areas prone to flooding and increased inland seawater intrusion. This is because, as the groundwater level drops, the land above compacts and sinks (see figure 4.29).

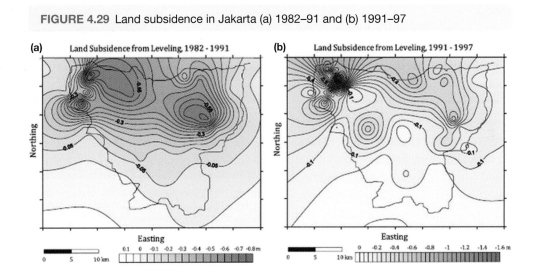

FIGURE 4.29 Land subsidence in Jakarta (a) 1982–91 and (b) 1991–97

Source: © Springer Science+Business Media B.V. 2011 Abidin, H.Z., Andreas, H., Gumilar, I. et al. Nat Hazards 2011 59: 1753. https://doi.org/10.1007/s11069-011-9866-9.

Every year between December and March inclusive, north Jakarta is flooded and residents are moved to evacuation centres. This natural hazard is a result of a combination of factors that include:
- intense wet seasonal rainfall averaging between 200 and 300 mm per month
- low **relief**
- shallow river gradients, which result in flooding when river discharge is high
- absence of flood plains along the riverbanks
- rivers choked by rubbish on river beds
- 8 per cent of the land below sea level aggravated by land subsidence.

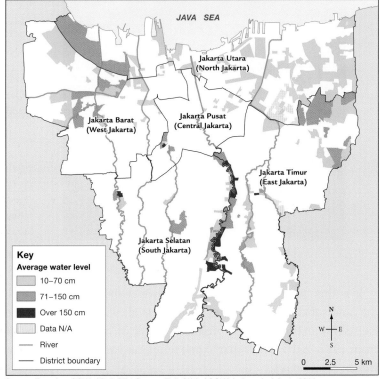

FIGURE 4.30 Areas inundated by severe flooding, which displaced more than 25 000 people in Jakarta in January 2014, after the city was hit by days of torrential rain

Source: Based on OCHA / ReliefWebSource: ReliefWeb / OCHA Indonesia Jakarta 2014
https://reliefweb.int/sites/reliefweb.int/files/resources/Update on Jakarta Flood as of 21Jan2014-R.pdf.

Resources

- **Weblink:** Sewerage and sanitation: Jakarta and Manila
- **Weblink:** Clean water a scarcity in Jakarta
- **Weblink:** Jakarta postpones sewage system building
- **Weblink:** Jakarta floods
- **Weblink:** Jakarta's waste crisis

Activity 4.4c: Sinking Jakarta

Before completing this activity, got to the Resources tab and watch and read the weblinks about Jakarta's subsidence.

Explain groundwater and pollution challenges

1. Explain the following terms in your own words:
   (a) subsidence
   (b) aquifer
2. Why has extracting water from underground made areas of Jakarta more prone to flooding?
3. Explain why reliance on septic tanks in Jakarta has placed people's water supplies in danger of contamination.

Map and analyse the data

TABLE 4.8 Jakarta special capital territory (660 km^2)

Year	Population ('000)	Mean population density ('000/km^2)
1951	2012	3.0
1961	2973	4.5
1971	4579	6.9
1981	6503	9.9
1991	8259	12.5
2000	8389	12.7
2010	9588	14.5
2015	10 323	14.5

4. (a) Using data from table 4.8, construct a composite line (population) and column (population density) graph.
   (b) What impact do you think population and population density have on rates of water extraction?
5. The data points on figure 4.31 represent cumulative subsidence (cm) 1974–2010.
   (a) Draw three **isopleths** of land subsidence at intervals of 50 cm from 50 cm to 150 cm. Some parts of the contours have already been plotted. (A printer-friendly version of this map has been provided in the Resources tab.)
   (b) Describe the pattern of subsidence revealed by your map. Refer to districts and features displayed in figure 4.30.
   (c) How does the pattern of subsidence compare with that which occurred between 1982 and 1991, as shown in figure 4.29a?

Suggest and justify strategies to combat the issue of subsidence

6. How could each of the following measures combat Jakarta's land subsidence and flooding problems?
   - Land use planning
   - Regulation of groundwater extraction
   - Sea wall construction
   - Flood management control
   - Seawater intrusion control (ways to stop seawater entering aquifers)

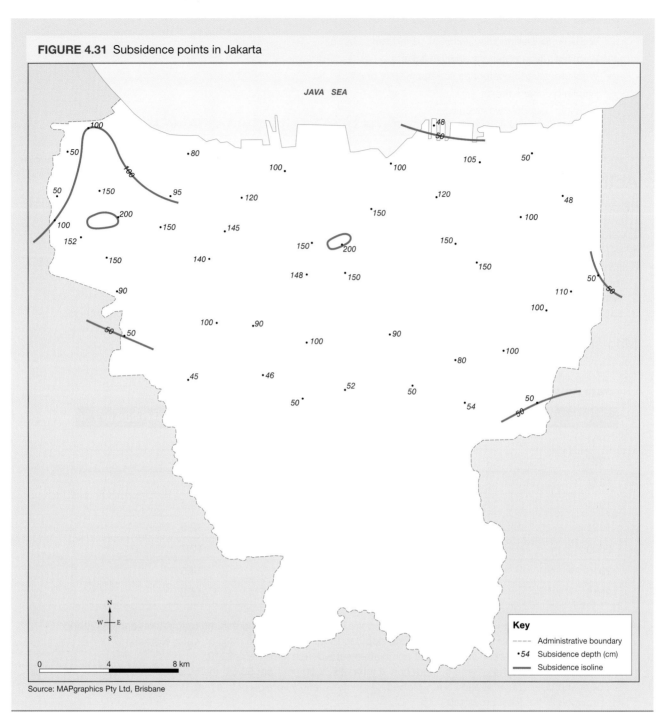

FIGURE 4.31 Subsidence points in Jakarta

Resources

- **Weblink:** Jakarta is sinking!
- **Weblink:** Indonesia crisis: Clean water scarcity in Jakarta
- **Weblink:** Jakarta is sinking so fast, it could end up underwater
- **Weblink:** Jakarta water-related environmental challenges
- **Weblink:** Water, megacities and global change
- **Weblink:** Sinking cities
- **Digital doc:** Jakarta subsidence map (doc-29171)

4.4.4 River pollution

Although cities in the developed world have the infrastructure to provide safe drinking water it does not mean that all their river systems are free of pollution. In New York, for example, Newtown Creek runs through an old working-class area of warehouses and industrial lots in the suburbs of Brooklyn and Queens. It empties into the East River in Manhattan. Measures designed to reduce the amount of pollution date from the 1920s when restrictions were imposed on local oil companies to limit how much waste they could discharge into the creek.

FIGURE 4.32 The location of São Paulo in Brazil

Source: Spatial Vision

As recently as 2014, New York was still failing to meet the 1972 federal Clean Water Act target, which requires the removal of 85 per cent of all pollutants from incoming sewage after treatment. Moreover, the city's treatment plant still experiences frequent overflows when sewage loads are too large for the plant to handle. This is despite New York state's 1992 order for the city to prevent overflows by 2013. The city hopes to be fully compliant by 2022, when its water treatment expansion program is complete. In the meantime, large quantities of untreated sewage enters New York harbour during 'CSO events' – a combined sewer overflow events – which are often triggered by rainfall.

Many rivers worldwide are contaminated by pathogenic bacteria such as E. coli (*Escherischia coli*). This bacteria is capable of causing urinary tract infections, neonatal meningitis and intestinal disorders such as diarrhoea and gastroenteritis. Because E. coli lives in the intestinal tracts of almost all warm-blooded animals, it is present in faeces and therefore untreated sewage. This is an acute problem for some megacities with inadequate water supply and treatment infrastructure. This problem is aggravated where rural migrants have settled in areas lacking any planning and proper sanitation. Raw sewage, including faecal matter, is discharged into rivers, posing a severe danger for public health. This was demonstrated in a study of pathogenic bacteria and their antibiotic resistances in São Paulo, Brazil (see figure 4.32), one of the world's largest megacities.

The Rio Tietê, which flows through the city, is an important water reservoir for São Paulo, but it is heavily contaminated with untreated waste including sewage. A tributary of the Tietê, the Pinheiros River, is notorious for its high level of dissolved methane.

Activity 4.4d: Pollution in the Rio Tietê

Refer to table 4.9 to answer the following questions.

TABLE 4.9 Coliform levels in the Rio Tietê

Site	Levels (Cfu/100ml)	Distance from Biritiba-Mirim (km)
1	0	20
2	115	73
3	1710	104
4	1500	109

(continued)

TABLE 4.9 Coliform levels in the Rio Tietê (continued)

Site	Levels (Cfu/100ml)	Distance from Biritiba-Mirim (km)
5	980	125
6	950	137
7	892	156
8	480	179
9	191	185
10	94	275
11	0	350

Explain the data tables

1. Using data from table 4.9, construct a line graph to show how coliform levels in the Rio Tietê vary with distance from Biritiba-Mirim.
2. Describe the pattern of coliform contamination along the river.

Analyse the data and apply your understanding

3. The level of coliform pollution rises sharply, peaking in inner São Paulo (site 4), but declining in stages in the west near the River Pinheiros and the city of Carapicuíba. What factors might contribute to this pattern? (Referring to a detailed online map of the area may help you to form a conclusion.)
4. Write a paragraph outlining the challenges that these levels of contamination might present for residents of São Paulo. Include details of at least three factors that might influence the severity of the impact on individuals in the community (such as rainfall or personal wealth).

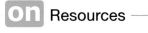

Resources

 Weblink: Megacity drought: São Paulo

4.5 Vulnerability to natural hazards

4.5.1 Factors affecting vulnerability

With the world's population becoming increasingly urbanised, more people are working and living in cities than ever before. The result of this is the concentration of wealth in large urban areas. This excessive concentration makes national and global economies more exposed to anthropogenic and natural disasters. The Lloyd's City Risk Index 2015–25 quantifies this exposure by calculating the potential GDP@Risk in more than 300 cities from these threats. GDP@Risk models the expected loss to GDP from a significant natural or anthropogenic disaster.

Cities, especially in developing economies, struggle to manage rapid population growth. Invariably, it is poor migrants from rural areas who settle illegally in hazardous areas with poorly constructed housing and lack of infrastructure, leaving themselves vulnerable to the impact of natural hazards.

Asia has the largest number of people exposed to natural disasters. Verisk Maplecroft, a UK-based risk management company, has estimated that nearly 1.5 billion people in south Asia are exposed to at least one natural hazard, the most common being flooding. However, they found that African countries are the most vulnerable to risk of poor governance and the lack of preparedness for natural disasters.

The Dartmouth Flood Observatory based at the University of Colorado, USA, estimated that between 2003 and 2008, floods that displaced 100 000 people or more occurred in more than 1800 cities across the world.

Urban researchers are predicting that many of the world's largest cities can expect to be flooded as climate change and rapid urban expansion combine.

One example of the way that the rapid urbanisation can make a place more vulnerable to natural hazards is the Pearl River Delta economic zone in China. The Pearl River Delta economic zone was one of the three major engines in propelling the rapid economic development of China. During the last 30 years, the Pearl River Delta has experienced rapid economic growth and urban expansion. The result was an increase in runoff and abnormal rising water levels during the rainfall season, from June to September. These combined effects overloaded the limited drainage systems and increased the risk of floods.

The Pearl River Delta area is the most highly populated continuous urban area in the world, with more than 55 million people living in the area's cities of Hong Kong, Shenzhen, Dongguan, Guangzhou, Foshan, Zhongshan, Jiangmen, Zhuhai and Macau. The adjacent metropolitan area of Guanhzhou-Foshan is home to an additional 22 million people.

Three quarters of the population in this region now live in urban areas. The city of Foshan is the third most populous city in the region with a population of about 7 million. Since 1990, the municipal government has been spending 0.5 per cent of the GDP each year on flood mitigation infrastructure projects. Figures 4.33 and 4.34 indicate how river discharge at Makou has increased since 1994 but also the success of measures designed to reduce the flood risk because maximum river levels have been reduced.

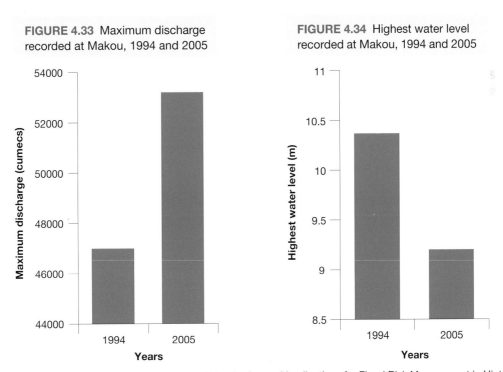

FIGURE 4.33 Maximum discharge recorded at Makou, 1994 and 2005

FIGURE 4.34 Highest water level recorded at Makou, 1994 and 2005

Source: Zhang H, Ma W, Wang X. Rapid Urbanization and Implications for Flood Risk Management in Hinterland of the Pearl River Delta, China: The Foshan Study. Sensors Basel, Switzerland. 2008;84:2223-2239.

Urban land use contributes to flood risk because urban areas contain many impermeable surfaces, such as roads, carparks and roofs. This creates widespread water run-off after heavy rain. Furthermore, urban drainage systems are designed to remove rainwater quickly into neighbouring creeks, but in their natural state these creeks cannot cope with such large volumes of water (see figure 4.35).

In contrast, a catchment in its natural state is more likely to act like a sponge after a rainstorm. Rain is initially intercepted by trees and shrubs before it hits the ground. It can then soak into the soil layer and later percolate down into the rock layer, if it is permeable, before making its way slowly to the river. However, if the rainfall is intense or the catchment is saturated with water from previous rainfall, the water will simply run over the surface and reach the river more quickly. This may cause the river to flood.

FIGURE 4.35 Drainage basin (arrows indicate volume of water)

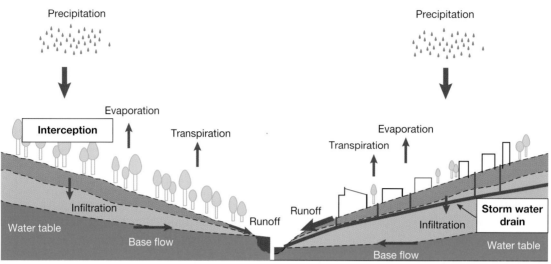

Source: Photo courtesy of Melbourne Water

Storm hydrographs are used to measure the discharge of a creek over time as a result of a rainstorm. The discharge, normally measured in cubic meters per second (**cumecs**), is the volume of water passing a point, usually a gauge station, during a specific period of time. Discharge may also be measured by the height of the creek water above a measuring weir. Data is plotted on a storm hydrograph as a line, whereas rainfall is plotted as a histogram (see figure 4.36).

FIGURE 4.36 Model of a storm hydrograph

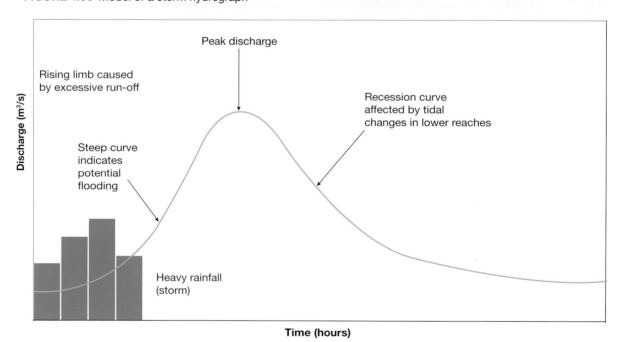

Creeks with a high flood risk have a flash response hydrograph. Discharge rises rapidly after a storm, creating a high peak discharge. Creeks with a low flood risk have a steady or lag response. The peak discharge is much lower and the lag time, the difference in time between peak rainfall and peak discharge, is much longer.

4.5.2 The effects of climate change

As the sizes of cities and the density of population increase, the natural river systems are less able to cope with the levels of run-off from non-permeable surfaces of the growing urban environment, but this is complicated further by the impacts of climate change.

In parts of London, for example, the risk of flash flooding is increasing. British Government scientists now believe that as a result of climate change, longer and more intense rain episodes will become more common and will lead to more flooding. Hammersmith and Fulham has recently been found to be the worst borough for potential flooding, with almost 60 000 homes – 60 per cent of the borough – at risk.

In the north-western London suburb of Brent, however, the flood risk is not from the River Thames. Instead it is from the smaller tributaries, the Silkstream and Brent rivers, and numerous brooks such as the Wealdstone, some of which are underground in conduits but emerge at times of flood, and all of which have been heavily altered from their natural state.

Activity 4.5: Flooding in Brent, London

Refer to table 4.10 and figure 4.37 to answer the following questions.

Explain the data
1. Using data from table 4.10, construct two hydrographs, one for each discharge point.
2. Describe the shape of the hydrographs and explain why they are this shape.

Analyse the data and apply your knowledge
3. Study the location of the gauging stations in figure 4.37.
   (a) What factors might explain the different peaks and lag times at the stations downstream?
   (b) How might people living in the lower catchment be affected by these river heights?

TABLE 4.10 Rainfall and height of Brent River

| Date | Time | Rainfall (mm) | Height above crest of weir (metres) | |
			Wealdstone Brook	Hanwell
16.08.17	18.00	0	0.1	0.5
	21.00	0	0.1	0.5
17.08.17	00.00	20	0.5	0.7
	03.00	31	1.1	1.8
	06.00	24	1.3	2.3
	09.00	10	1.0	3.0
	12.00	0	0.6	3.3
	15.00	0	0.5	3.0
	18.00	0	0.4	2.5
	21.00	0	0.3	1.5
18.08.17	00.00	0	0.2	1.0
	03.00	0	0.2	0.9
	06.00	0	0.2	0.8
	09.00	0	0.2	0.8

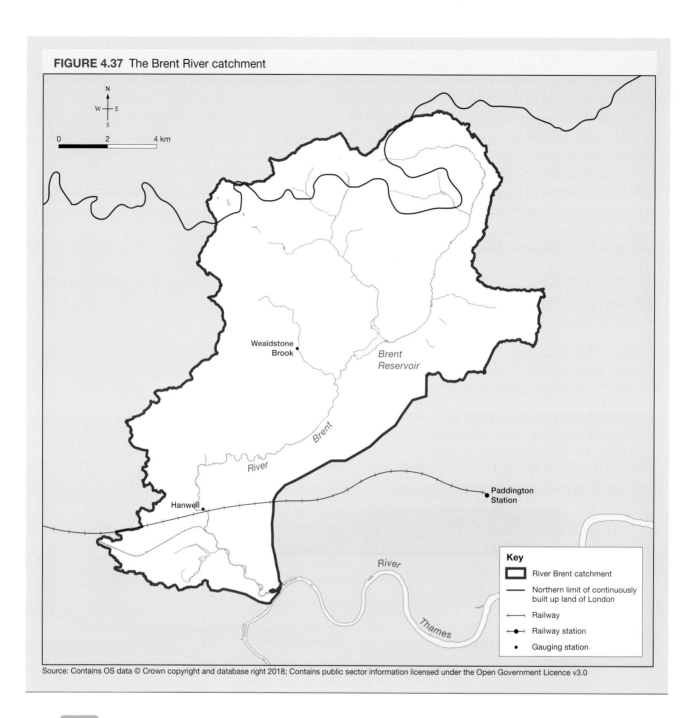

FIGURE 4.37 The Brent River catchment

Source: Contains OS data © Crown copyright and database right 2018; Contains public sector information licensed under the Open Government Licence v3.0

Resources

Weblink: Brent rivers and lakes

Digital doc: Brent River catchment hydrographs (doc-29172)

Manila, Philippines

Metropolitan Manila, with a population of more than 12 million, is one of the megacities most exposed to risk. In a study of megacities at risk, the Economist Intelligence Unit ranked Manila at 55 out of 60 based on 49 risk indicators including digital, health, personal and infrastructure security. Manila was also ranked the highest for rates of death caused by natural disasters, with tropical cyclones (typhoons) and earthquakes accounting for 80 per cent of the share of total GDP at risk according to the Lloyd's City Risk Index. Manila is in the bottom five in infrastructure security, and is second last in digital security. Other south-east Asian cities in the bottom 10 include Vietnam's Ho Chi Minh City, Myanmar's capital, Yangon, and Jakarta.

Manila is located in a highly active cyclone area, has a high risk of flooding, and is adjacent to the Eurasian, Philippines and Sunda plate boundaries, so has high earthquake risk too. The Marikina River, which flows through Manila, runs in a rift valley along two parallel fault lines: the West Valley Fault and the East Valley Fault. The Philippine Institute of Volcanology and Seismology believes that rupturing along the West Valley Fault has a recurrence interval of 200 to 400 years and could result in a 7.2 magnitude earthquake. A seismic event like this in the vicinity of Manila could result in more than 35 000 fatalities. Manila is now inside the recurrence interval for another earthquake.

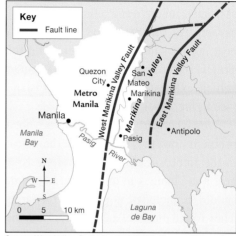

FIGURE 4.38 Location of the West and East Valley Fault lines

Source: West Valley_Fault Line_PHIVOLCS, Bureau of Mines, in coordination with the Board of Technical Surveys and Maps, 1963.

A buffer zone on each side of the fault line, where no building should be established, was recommended but with such rapid urban development it has been impossible to enforce. Poor administration of building standards and the absence of clear planning directions means that local governments have allowed crowding of limited urban space.

Manila is also extremely vulnerable to tropical cyclones, particularly from June to September. On average, the Philippines are struck by 20 tropical cyclones each year. One such example of their devastating effect is the 2013 Super Typhoon Haiyan (see page 36).

Predictably, it is the poorer residents who live in hazardous locations, and are vulnerable to flooding, tropical cyclones and earthquakes without adequate protection.

---

on Resources

🔗 **Weblink:** The Philippines National Disaster Risk Reduction and Management Plan, 2011–2028

🔗 **Weblink:** Philippines: When it rains, it floods

4.6 Heat islands

4.6.1 Urban heat management

In addition to the risks of disease and pollution, large urban environments can also present a direct risk to the health and wellbeing of their inhabitants in other ways. Hot urban environments can be extremely dangerous as heat has the potential to aggravate pre-existing health conditions, including heart disease, lung disease and asthma. Moreover, it is the very old and very young city dwellers who are more susceptible to heat-related illness, especially in poorer neighbourhoods.

An **urban heat island** is a built-up area that is significantly warmer than its surrounding rural areas. The phenomenon was first observed by Luke Howard, a pharmacist working in London during the early 1800s. With a keen interest in meteorology, he recorded and published temperature readings for various neighbourhoods from 1800 to 1830, when the London population grew from 1 to 1.5 million people.

As cities spread, the natural cooling elements of a landscape disappear. The result is that less energy is used up by evaporating water and less of the sun's energy is reflected. More heat is also stored by buildings and the ground in urban areas than in rural areas because materials such as bricks, concrete and bitumen have a high heat capacity, or **thermal mass** — an ability to store heat during sunshine hours and then release it slowly. Heating and cooling equipment, and vehicle exhaust can add to the heat island effect (see figure 4.40).

Even where urban heat islands are not a major problem now, they could exacerbate heatwaves in the future. Researchers in Hong Kong suggest that changes in urban heat island duration should be included in planning for large cities. They found that over a 27-year period, the urban heat island duration in a developing urban area of Hong Kong increased from 13.6 hours to 17.5 hours.

There are a number of measures that can be adopted to mitigate the heat island effect. Los Angeles, for example, was the first city in the USA to make heat reflecting roofs compulsory for new and significantly renovated houses. In 2017 it set itself the goal to cool the city by 1.5°C

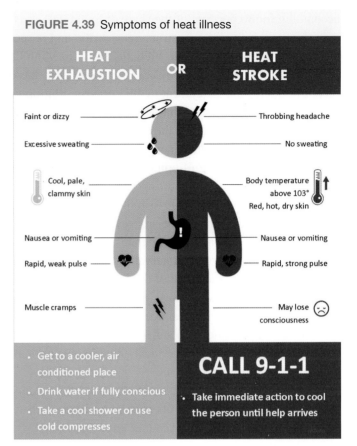

FIGURE 4.39 Symptoms of heat illness

Source: National Weather Service

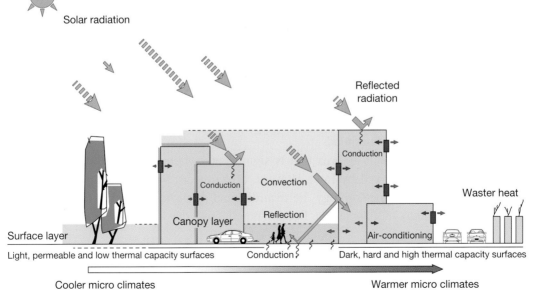

FIGURE 4.40 Urban landcover and the heat island effect

Source: Sharifi, E., & Lehmann, S. 2014. Comparative Analysis of Surface Urban Heat Island Effect in Central Sydney Journal of Sustainable Development, 73, 23–34. doi: 10.5539/jsd.v7n3p23.

by 2035. This involved experimenting with painting streets with a sealant that lowers the surface temperature by 5°C and increasing the city's tree canopy by 40 000 trees by 2019.

The Los Angeles Health Department has also flagged urban heat as a serious and growing threat to public health, particularly as the climate warms. The Los Angeles metropolitan area has a population of around 13 million people, and in October–November 2017 experienced a heatwave that broke many existing heat records for the area. Part of the LA County authority's response to heat in the city is the Emergency Survival Program, which provides tips for avoiding heat illness, but also established 'Cooling Centres', community facilities opened by local authorities on days of extreme heat for people who have no other means of escaping the heat or might be at greater risk of heat illness, such as the elderly, ill or homeless. This may not help to reduce the heat island effect, but it does help to mitigate the impact on the most vulnerable members of the community.

Washington DC has implemented a green area ratio for new developments, while in Seattle a points system is used where property owners select various options, such as tree planting, to meet minimum planning requirements.

Resources

Weblink: Evidence on the effectiveness of interventions during heatwaves
Weblink: Cooling urban heat islands

Activity 4.6a: Understanding and mitigating the heat island effect

Refer to figure 4.41 to answer these questions.

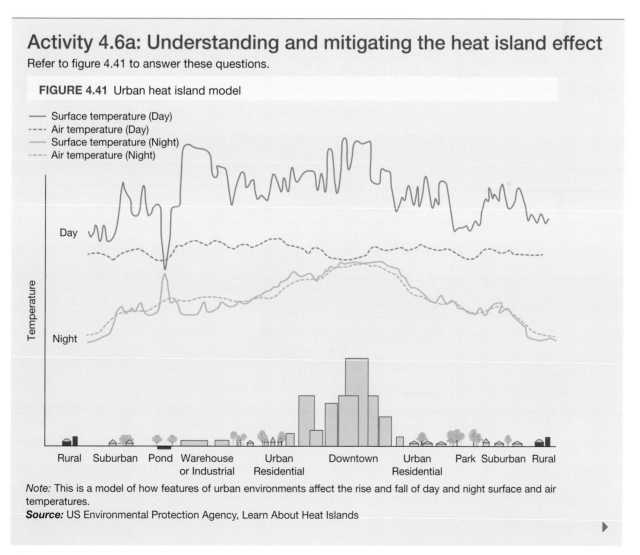

FIGURE 4.41 Urban heat island model

Note: This is a model of how features of urban environments affect the rise and fall of day and night surface and air temperatures.
Source: US Environmental Protection Agency, Learn About Heat Islands

> **Explain the heat island effect and apply your knowledge**
> 1. Describe the contrasting pattern of day and night surface temperatures.
> 2. Compare the day air temperature profiles with the night air temperature profiles.
> 3. Explain why the high-density parts of a city experience the highest temperatures.
> 4. Explain the impact that each of the following measures will have on mitigating the heat island effect.
>   - Tree planting
>   - Vegetated roof tops and wall panels
>   - Light coloured roof tops and pavements

Delhi

With a population of more than 28 million, Delhi, the capital city of India, is one of the world's largest and most extensive megacities. Comprising 11 districts of the National Capital Territory of Delhi, it covers an area of nearly 1500 square kilometres. The average population density is about 11 320 per square kilometre. Delhi's North district has the highest population density at 36 155 persons per square kilometre whereas New Delhi, the least densely populated district, has 4057 persons per square kilometre.

Researchers at the Centre for Science and Environment in New Delhi have identified the heat island phenomenon as a contributing factor to the risk of excessive heat for Delhi's inhabitants. The heat island magnitude in Delhi is high, at around 8°C and, in 2015, was responsible for accentuating heat wave conditions, resulting in melting tarmac roads and a spike in heat-related deaths.

Activity 4.6b: Coping with heat in Delhi
Examine figures 4.42 and 4.43, and then answer the questions that follow.

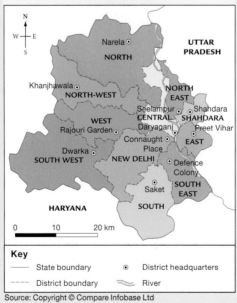

FIGURE 4.42 National Capital Territory of Delhi Administrative Districts

Source: Copyright © Compare Infobase Ltd

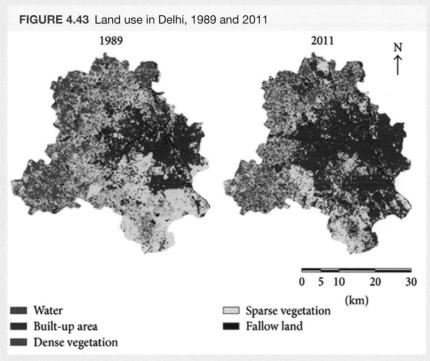

FIGURE 4.43 Land use in Delhi, 1989 and 2011

Source: © 2014 Bablu Kumar et al. Source: Bablu Kumar, Kopal Verma, and Umesh Kulshrestha, "Deposition and Mineralogical Characteristics of Atmospheric Dust in relation to Land Use and Land Cover Change in Delhi India," Geography Journal, vol. 2014, Article ID 325612, 11 pages, 2014. https://doi.org/10.1155/2014/325612.

Explain and analyse the data

1. Describe the changes shown in figure 4.43 by identifying districts where particular types of land use expanded or contracted between 1989 and 2011. Figure 4.42 will help you identify the districts.
2. Refer to table 4.11 to calculate the percentage change in each of the land use categories between 1997 and 2008.

TABLE 4.11 Land use change in Delhi, 1997 and 2008

Land use	1997	2008
Built up land	549	791
Rural land (scrub, forest, etc.)	709	508
Water bodies	68	27

3. Examine table 4.12 to complete this question.
   (a) Using data from table 4.12, construct a spiderchart to show how the built-up density of Delhi's nine administrative districts changed during the period 1977–2014. Built-up density is the ratio of the built-up area to the total area. It indicates how dense the urban area is, irrespective of the location of urban patches within the administrative boundary. (You will find a blank spiderchart template in the Resources tab to complete this task.)
   (b) Describe the pattern of change shown in your graph.
   (c) How do the changes in the distribution of built-up land compare with that of built-up density?
   (d) Explain why the Central district may have remained relatively unchanged.

TABLE 4.12 Built-up density of Delhi's administrative districts*

Administrative district	1977	2014
North	0.28	0.43
North East	0.13	0.62
East	0.27	0.60
Central	0.38	0.40
New Delhi	0.05	0.35
South	0.08	0.37
South West	0.04	0.39
West	0.12	0.60
North West	0.01	0.35
Delhi	0.10	0.40

Note: * There are now 11 districts in Delhi. In 2012, the North East district was divided into North East and Shandara, and the South district was divided into South and South East. These old boundaries have been maintained in this table to allow for comparison.

4. (a) Study figure 4.44, which shows Delhi's changing heat island over a four-day period in 2008. Describe the distribution of relative high and low temperature areas.
   (b) What correlation exists between land use and the pattern of isotherms shown in figure 4.44?
5. What impact might the patterns and trends in Delhi's heat have on the population of the city?
6. Considering the data above, suggest what challenges climate change might pose for the people of Delhi.

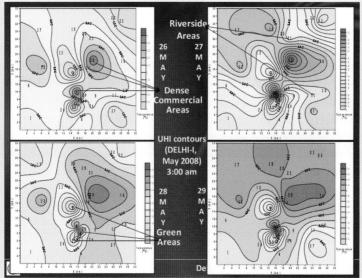

FIGURE 4.44 Delhi's heat island

Source: Assessment of Urban Heat Island Effect in Megacity Delhi by Manju Mohan, 11 April 2013

Resources

Digital doc: Spiderchart template (doc-29667)
Weblink: Water stress in New Delhi

4.7 Food security

4.7.1 Feeding the megacity

The impact of urbanisation on water resources is problematic. Another area of concern is feeding the megacity. Globalisation of food production has revolutionised the food industry but is only part of the solution to the problem of providing food security for all the world's urban populations. As countries become increasingly urbanised, their land use patterns, infrastructure and economies change. Each of these factors then affect the production and availability of food. This challenge presents in different ways in developed and developing nations.

Urban food security in developing countries

Developing countries in particular face two challenges: not having adequate infrastructure to connect producers with consumers efficiently, and the ability to provide food at reasonable prices.

In 2001, the UN Food and Agricultural Organisation (FAO) launched its 'Food for the Cities' initiative with the aim to promote more sustainable and resilient food systems, particularly in developing countries experiencing rapid urbanisation and changing land-use. A key component of this initiative was to secure urban and **periurban** horticulture in local agricultural development plans.

Urban and periurban agriculture have an important role to play in providing access to nutrient-rich food. People living in slum areas inevitably have very limited access to fresh fruit and vegetables. Consequently, their food supply mainly consists of cheap foods with low nutritional value. Where fresh fruit and vegetables are available, they are often unaffordable as the poorer urban households spend more than half their income on food. They are therefore highly vulnerable to increases in food prices.

Urban agriculture has the capacity to overcome this situation by providing a secure source of nutritional food for the urban poor. Its role in developing countries is important. For example, in Hanoi, Vietnam, 80 per cent of fresh vegetables and more than a third of the city's egg supply is produced by urban and periurban farms.

However, rapidly growing cities struggle to sustain agriculture due to increasing urban sprawl, which has made land physically unavailable and expensive. Farming is then relegated to any areas that are available, often near polluted water bodies, drains and in backyards where irrigation is feasible. In Ghana's capital Accra, the Odaw River is used extensively by urban farmers despite the fact that the river has become the receptacle of Accra's rubbish and solid waste. This raises serious health concerns given the city's reliance on the area for its vegetable supply.

FIGURE 4.45 Eagle Street Rooftop Farm covers a 550 m² rooftop space in Greenpoint, Brooklyn.

In Colombo, Sri Lanka, the problem is different. There the existing wholesale and retail markets are poorly located and access is hampered by chronic traffic congestion. This is exacerbated by poorly maintained markets and unhygienic conditions, resulting from increasing amounts of organic and inorganic waste, which then present a health risk.

Urban food security in developed countries

Developed countries, however, can approach this issue in different ways. In 1999, the idea of a 'vertical farm' as a way for densely populated areas to feed themselves was launched in the USA. The idea was that not only could fruit and vegetables be supplied fresher, but also that these vertical farms would have a smaller footprint compared to the current system of importing vegetables into cities. New York City was one of the early test cases and since then the 'skyscraper farm' industry has expanded.

Singapore relies heavily on imported food and only produces 7 per cent of its vegetables. In 2013, Singapore opened a vertical farm on a commercial scale that produces 500 kilograms of vegetables every day for a local supermarket chain.

In the developed world, continued expansion of the city fringe has adversely impacted the supply of fresh vegetables. In Australia, the food bowls adjacent to both Melbourne and Sydney are threatened by urban sprawl. If by 2050 Melbourne's population tops the predicted 7 million mark, it is estimated its fresh food needs will increase by 60 per cent. However, the loss of adjacent farmland to accommodate Melbourne's expansion means that the food bowl's ability to meet the city's food requirements is likely to drop below 20 per cent by 2051. This will increase the cost of perishable food, as having food production and processing close to market has significant cost savings for both producers and consumers. Transport costs are lower and there is less spoilage and waste. There are also environmental benefits if food does not need to be transported over long distances.

The Sydney food bowl produces 20 per cent of the city's needs, including 55 per cent of the supply of meat, 40 per cent of eggs, 38 per cent of dairy products, and 10 per cent of vegetables. Much is produced in small blocks between two and five hectares in size. However, as urban development expands into the city fringe it is estimated that on current trends these farms will be responsible for just 6 per cent of the city's food needs in less than 20 years.

 Resources

🔗 **Weblink:** Improved rural and urban linkages: Building sustainable food systems
🔗 **Weblink:** The world's first commercial vertical farm opens in Singapore

Activity 4.7: Feeding a city
Refer to tables 4.13 and 4.14 to complete the following questions.

TABLE 4.13 Poverty and food insecurity in Sri Lanka

District	Population below poverty line (%) 2010	Food insecure population (%) 2010
Colombo	3	38
Gampaha	3	34
Kalutara	6	27
Kandy	10	20

(continued)

TABLE 4.13 Poverty and food insecurity in Sri Lanka (*continued*)

District	Population below poverty line (%) 2010	Food insecure population (%) 2010
Matale	12	22
Nuwaraelyia	7	9
Gallee	10	31
Matara	11	22
Hambantota	7	17
Kurunegala	12	21
Puttalam	10	29
Anuradhapura	6	20
Pollonnaruwa	6	14
Badulla	13	14
Moneragale	14	11
Ratnapura	10	20
Kegalle	11	22
Batticaloe	20	32
Ampara	12	17
Trincomalee	12	23
Jaffna	16	14
Vavuniya	2	10

Source: Epidemiology Unit Ministry of Health, Sri Lanka, and Geetha Mayadunne and K.Romeshun, Sri Lankan Journal of Applied Statistics, Vol 14-1.

Comprehend and analyse the data to identify patterns

1. Using data from table 4.13, analyse the correlation between the following based on the data provided for Colombo, Sri Lanka.
   (a) the level of urbanisation by district and food insecurity
   (b) poverty and food insecurity.
2. (a) Using table 4.14, describe the pattern of farming in Melbourne's food bowl.
   (b) Study table 4.14 and figure 4.46. Referring to specific commodities, explain the factors that might influence whether food types are well represented in Melbourne's food bowl or are sourced from other areas. Consider what may be required in terms of the amount of land required, water resources or climate needed to produce specific products.
   (c) Based on your conclusions about Melbourne's food bowl, suggest what challenges might affect the production of food in cities with much higher populations or much lower levels of development.

TABLE 4.14 Food types produced in Melbourne's food bowl by percentage

Food type	% of Victoria's production occurring in Melbourne's foodbowl
Dairy	12%
Sugar	0%
Fruit	8%
Oil crops	7%
Cereal grains	3%
Vegetables	47%
Red meat	15%
Chicken meat	81%
Rice	0%
Legumes	1%
Eggs	67%

Source: Sheridan, J., Larsen, K. and Carey, R. 2015 Melbourne's foodbowl: Now and at seven million. Victorian Eco-Innovation Lab, The University of Melbourne

FIGURE 4.46 A snapshot of Melbourne's food bowl

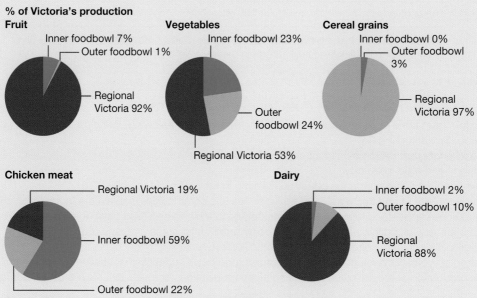

% of Victoria's production

Fruit
- Inner foodbowl 7%
- Outer foodbowl 1%
- Regional Victoria 92%

Vegetables
- Inner foodbowl 23%
- Outer foodbowl 24%
- Regional Victoria 53%

Cereal grains
- Inner foodbowl 0%
- Outer foodbowl 3%
- Regional Victoria 97%

Chicken meat
- Regional Victoria 19%
- Inner foodbowl 59%
- Outer foodbowl 22%

Dairy
- Inner foodbowl 2%
- Outer foodbowl 10%
- Regional Victoria 88%

Source: Sheridan, J., Larsen, K. and Carey, R. 2015 Melbourne's foodbowl: Now and at seven million. Victorian Eco-Innovation Lab, The University of Melbourne

4.8 Waste management

4.8.1 The challenge of waste disposal

As urbanisation increases so does one of its unpleasant by-products: solid waste. The World Bank estimates that by 2025 the world's urban population will be producing more than 2 billion tonnes of waste per year. Urban areas, by virtue of their high population densities, produce an enormous amount of waste and for cities in many developing countries, it is becoming an environmental catastrophe.

Developing nations, urban centres and megacities produce huge amounts of waste, but do not always have the infrastructure to collect or process it. One of the most famous megacity rubbish dumps in the world is the enormous 'Smokey Mountain' in Manila. With approximately 44 per cent of Manila's 10 million residents living in poverty, many of the poorest residents live near the rubbish dump and scavenge there for food and a living.

FIGURE 4.47 The mountainous dumps of rubbish in urban areas in Sri Lanka pose a number of health and safety threats to the workers.

Rubbish-lined streets are also common in cities of developing Asian nations that have not yet reached megacity status. The broader metropolitan area of Colombo, Sri Lanka, has a population of approximately 5.6 million people and has steady population growth. Colombo produces more than 1200 tonnes of waste each day, which ends up in the city's mountainous dumps. On the north-east outskirts of the Colombo is the notorious 90-metre-high Meethotumulla rubbish dump. Such enormous dumps are prone to collapse, burying homes and people working in the area. Considering the projected growth for cities in Asia, many that are already struggling to manage their waste, it is easy to see how waste will become an even bigger management issue in the future.

Resources

- **Weblink:** Life after Smokey Mountain
- **Weblink:** Life scavenging in Manila's garbage mountain
- **Weblink:** Sri Lankan rubbish dump collapse

4.8.2 Garbage and disease

Garbage can also contribute to the spread of disease. Dengue, a vector-spread disease, is particularly endemic in tropical countries like Sri Lanka where the *Aedes aegypti* and *Aedes albopictus* mosquitos are widely adapted to urban and suburban environments. They can lay eggs in as little as 5 to 10 ml of water, which easily collects in discarded rubbish as small as a bottle cap.

With an annual rainfall of 2400 mm, the blocked drains and uncollected rubbish in Sri Lanka's urban areas are ideal breeding grounds for mosquito larvae. This is particularly the case in the south-west, where humid and wet conditions allow the mosquito to live longer. The heavy monsoon rain seasons do, however, provide some respite as they flush out the mosquito breeding grounds.

The south-west is also the most urbanised and densely populated part of the country. The population density of Colombo was approximately 20 000 people per square kilometre in 2017. It is particularly high in the poorer districts of the city. This allows disease to spread relatively quickly. In 2017, Sri Lanka suffered its worst outbreak of dengue in history.

Activity 4.8: Dengue fever

Refer to the **Sri Lanka Ministry of Health Dengue Fever data** and **Sri Lanka Population and Housing data** weblinks in the Resources tab to answer the following questions.

Explain and analyse the data

1. Using data from the most recent complete year listed in Sri Lankan Dengue fever statistics, complete the table below to show the number of reported dengue fever cases in each administrative district in Sri Lanka. Refer to the Sri Lankan population statistics to add the per cent of urban population in each district. You will need to add the year in brackets in the final column.

   TABLE 4.15 Reported cases of dengue fever in Sri Lanka

Administrative district	Urban pop. (%)	Reported cases
Jaffna		
Killinochchi		
Mulativu		
Mannar		
Vavuniya		
Anuradhapura		
Trincomalee		
Polonnaruwa		
Puttalam		
Matara		
Batticalaoa		
Kandy		
Kagalle		
Gampaha		
Nuuwara Ellya		
Badulla		
Monaragala		
Colombo		
Kalutara		
Ratnapura		
Galle		
Matara		
Hambantota		

2. (a) Using data from your table, choose one of the following techniques to examine the hypothesis that there is a link between the level of urbanisation and the incidence of dengue fever.
   - Scattergraph and best fit line
   - Spearman Rank Correlation (see page 264 for how to calculate a Spearman Rank Correlation)
   - Choropleth mapping

(b) What do you notice about the timing of the peaks in both outbreaks and monsoonal rains?
(c) Why do you think there is a time lag between the peaks of rainfall and the spikes in dengue fever?
(d) Write a paragraph comparing the seasonal pattern of dengue outbreaks with Colombo's rainfall pattern and suggest reasons for any relationship you can identify between the two.

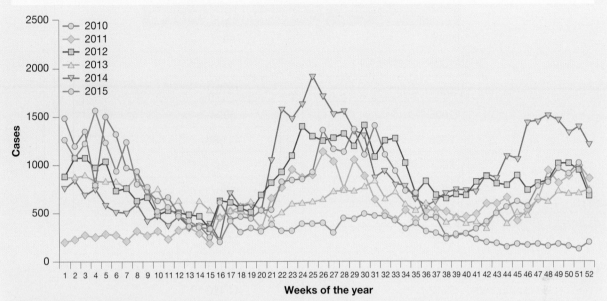

FIGURE 4.48 Distribution of dengue fever cases in Sri Lanka, 2010–15

Source: © 2017 Sirisena et al. Source: Sirisena P, Noordeen F, Kurukulasuriya H, Romesh TA, Fernando L 2017 Effect of Climatic Factors and Population Density on the Distribution of Dengue in Sri Lanka: A GIS Based Evaluation for Prediction of Outbreaks. PLoS ONE 121: e0166806. https://doi.org/10.1371/journal.pone.0166806

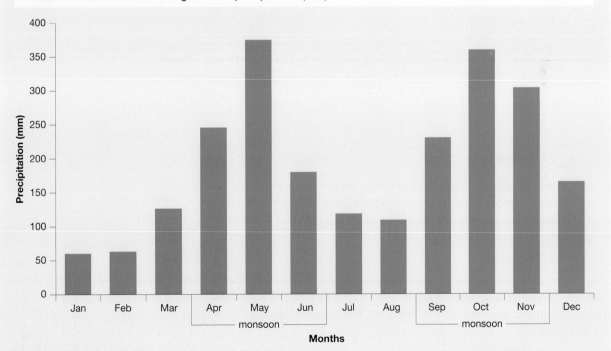

FIGURE 4.49 Colombo average annual precipitation (mm)

Source: www.allmetsat.com

3. (a) Using the data showing the number of cases of dengue reported each year in Sri Lanka, create a line graph to show the change in reported cases of dengue fever.
   (b) Explain the trends during this period.
   (c) Explain two factors that might have contributed to the trends you have identified.

TABLE 4.16 Annual dengue cases, 2002–17

Year	Number of reported cases
2002	8931
2003	4673
2004	15 463
2005	5994
2006	11 980
2007	7332
2008	6607
2009	35 095
2010	34 109
2011	28 473
2012	44 456
2013	32 063
2014	14 700
2015	29 777
2016	54 945
2017	184 442

Source: © 2014 National Dengue Control Unit — Ministry of Health — Sri Lanka

Resources

- **Weblink:** Philippines Department of Health: Dengue Management Plan
- **Weblink:** Dengue factsheet: World Health Organization
- **Weblink:** Sri Lanka Ministry of Health dengue fever data
- **Weblink:** Sri Lanka population and housing data

Calculating correlations: Spearman's Rank correlation coefficient

The scattergraph and best fit line are useful for identifying any anomalies (outliers) that don't fit in with a general pattern of results. An alternative method is to use a statistical test. This establishes whether the correlation is statistically significant or if it could have been the result of chance alone. Spearman's Rank correlation coefficient is a technique which can only yield a result between 1 and minus 1. You must be comparing at least 5 pairs of data for the test to work.

Method — calculating the coefficient

Step 1 Lay out your data in a table.

Step 2 Rank the two data sets. To do this, assign '1' to the largest number, '2' to the second largest and so on. The smallest value in the column will be assigned the lowest ranking. Do this for both sets of measurements. If two or more values are the same, or 'tied', they are assigned the average rank. For example, if there are three tied scores that are second they will all be ranked 3 – i.e. 1, 3, 3, 3 – and then 5, 6 and so on. If there are only two tied ranks, say 7, they become 7.5, 7.5 and then 9, 10 and so on.

Step 3 Find the difference in the ranks (d): This is the difference between the ranks of the two values on each row of the table. Do not worry about the signs +ve or –ve as in the next step they are squared.

Step 4 Square each rank difference you have just calculated.

Step 5 Add up all these squared differences.

Step 6 Substitute your answers into the formula. The formula has two constant values, 1 and 6.

- Σ = sigma (sum of)
- d = difference between ranks
- n = number of pairs in your data (the number of pairs in your table).

$$r_s = 1 - \frac{6\Sigma d^2}{n^3 - n}$$

For example to determine if there is a correlation between the size of a town and the number of petrol stations, you might consider the following data:

Population	Rank (pop.)	Number of petrol stations	Rank (no. of stations)	Difference between ranks	Difference between ranks squared
30 000	4	6]	5	1	1
20 000	5	5	4	1	1
45 000	2	8	2	0	0
39 000	3	7	3	0	0
58 000	1	9	1	0	0]
					$\Sigma d^2 = 2$

$$1 - \left(\frac{6 \times 2}{5^3 - 5}\right)$$

$$1 - \frac{12}{120}$$

$$1 - 0.1 = +0.9$$

The closer the value is to +1 or −1, the stronger the likely correlation (or likelihood that one variable causes another). A perfect positive correlation is +1 and a perfect negative correlation is −1. This value of +0.9 suggests a very strong positive relationship.

FIGURE 4.50 Positive and negative correlation scale

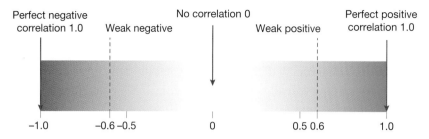

A Spearman Rank significance graph can be used to test the significance of the relationship by looking up the value of +0.9.
- Work out the 'degrees of freedom'. This is the number of pairs in your sample minus 2 (n–2). In the example, it is 3 (5 – 2 = 3 degrees).
- Plot your result on the Spearman's Rank significance graph. The *x*-axis shows the Spearman's Rank correlation coefficient and the *y*-axis shows the degrees of freedom (see figure 4.51).

FIGURE 4.51 The Spearman's Rank significance graph

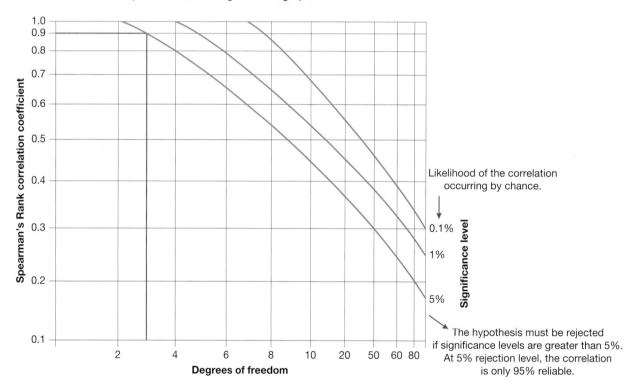

- Although 0.9 is a high coefficient, our result is below the 5 per cent significance level, therefore it is quite possible the result was the product of chance and statistically that is not considered significant. The sample of 5 towns is, after all, exceedingly small.
- If the result had been above the 0.1 per cent significance level, then we could be 99.9 per cent confident the correlation has not occurred by chance.

Remember that just because you get a correlation between two sets of variables does not prove that one variable causes another variable. It can suggest there is a relationship but only further research can actually prove that one thing affects the other. The size of the sample is important. The more data you have, the more reliable your result.

4.9 Economic opportunities and challenges

4.9.1 Urban growth and economic change

India's economic development

India is expected to increase its urban population by more than 400 million by 2050. The annual growth in urban population in India between 2010 and 2015 was the highest among the major economies. As in China, migrants from the rural areas make up a sizeable proportion of the urban population. The last census revealed that over one-third of the urban population comprises migrants from India's rural areas.

India's economic development has followed a different path from many other countries. Structurally, the economy transitioned quickly from agriculture to services and has become the third largest economy in terms of purchasing power parity.

In 1991, the government initiated economic liberalisation reforms. These include the reduction of tariffs and interest rates, and ended many public monopolies but more importantly encouraged foreign investment in many sectors of the economy. Consequently, India has progressed towards a free-market economy, with a much reduced state control of the economy. Per capita incomes have more than doubled since 2005 and by 2015 India outpaced China in terms of GDP growth rate.

China's economic development

The late 1970s marked a major turning point in the direction of China's economic and subsequent urban development. In 1978, Deng Xiaoping became premier and immediately launched a series of reforms aimed at transforming China into a more market-based economy. The open-door policy was instrumental in this change. This liberalised China's foreign trade system and was accompanied by the creation of Special Economic Zones (SEZ) in south-eastern coastal China. Designed to stimulate economic growth, tax incentives were provided in order to attract foreign investment and technology. The first four special economic zones designated in 1980 were based around four small cities in the Guangdong and Fujian provinces. One of these cities, Shenzhen, was a small fishing town but is now a hugely successful trade hub and manufacturing centre with a population of more than 12 million.

FIGURE 4.52 Shanghai, with a population of 23.41 million and a population density of 6100 people per square kilometre, is the world's second largest megacity behind Tokyo.

FIGURE 4.53 Workers inspect television panels in a Shenzhen factory

The success of the SEZs encouraged the government to create new foreign trade areas in cities such as Shanghai. Known as Economic and Technological Development Zones (ETDZ), these zones focus on specific industries, such as high-tech research and development. In 2013, Shanghai was one of four cities given Free Trade Zone status and benefitted from faster foreign investments approvals and more relaxed trade regulations. In 2017, seven more FTZs were approved but unlike the previous ones they are located in underdeveloped areas in China's interior to boost regional economies.

These economic initiatives inevitably had a huge impact on China's rate of urbanisation. With an economy firmly wedded to exports and investment, China witnessed an explosion in the growth of cities in both number and scale. This has been fuelled by a tidal wave of migration of 500 million rural Chinese people into the cities since 1980. With so many people, and therefore potential consumers, now living in urban areas there is now a significant shift towards a consumption-driven economy.

A product of this rapid urbanisation has been the megacity. The UN definition of urban areas gives China a total of 6 megacities. However, the New Urbanisation Programme 2014–2020 aims to create even larger mega-urban regions by merging the nine cities in the Pearl River Delta region, thereby drawing 48 million people together.

Eleven of these mega-urban regions are planned, the largest of which is Jingjinji. This plan is designed to create a more balanced economic structure, partly by moving heavy polluting industries and some services away from the capital of Beijing. This ties in with State Council of China's decree to cap the size of Beijing at 23 million people by 2030.

The backbone of these mega regions are the transportation networks. These will comprise new high-speed and conventional railways, metro systems, light rail and highways. The idea is that with such fast networks people will feel as though they are all part of the same urban entity. The plans also incorporate extensive areas of green space and rural areas within the city clusters.

FIGURE 4.54 Regions of China, including SEZ locations

Source: Spatial Vision

 Resources

🔗 Weblink: China's growing cities
🔗 Weblink: The growth and development of megacities

China's ghost cities

A somewhat curious bi-product of China's rapid urbanisation has been the ghost city. These have sprung up in less-developed provinces such as Qinghai, Sichuan and Gansu in China's interior. As part of a post-GFC (Global Financial Crisis) strategy they were designed to bring prosperity to the region.

Located in inner Mongolia, Ordos Kangbashi was a product of the region's prosperous coal mining industry and its ambitious government. The idea behind the project was that the new city would attract people, especially the new middle class wanting the benefits of a modern city. It was anticipated that many would come from the Dongsheng district, 25 km to the north.

It was also expected that a new city would diversify the economy by attracting new business and relocated government agencies. Unfortunately, the new city got off to a poor start. Demand for housing was initially sluggish and investors pulled out, leaving projects unfinished. The local government tried filling its empty buildings by accepting housing exchange certificates from people elsewhere in China for a property in Ordos. Called a *fangpiao*, a housing exchange certificate is granted when a property is requisitioned by the government. In the first five years, Ordos' population grew to 30 000, by 2018 it had reached 153 000, well short of the original target of 1 million. This has not been helped by a decline of 15 per cent in inner Mongolia's GDP in 2017.

Another economic factor influencing the growth of Chinese megacities is the location of employment opportunities. In 2010, it was estimated that over one-third of rural children were living with close relatives because their parents had moved to the cities for work. This is partly explained by the *hukou* system, which restricts access to education and welfare to the place where the parents were born. A relic of previous government policy, the system was designed to control where people lived.

FIGURE 4.55 The ghost city, Ordos Kangbashi

Activity 4.9: Urbanisation in China

Refer to the **China National Bureau of Statistics** weblink in the Resources tab to answer the following questions.

Explain and analyse the data

1. Using data from table 4.17, construct a multiple line graph to show the changing level of urbanisation in China and India from 1900 to 2015.

TABLE 4.17 Percentage of national population living in urban areas

Year	India	China
1900	11	7
1910	10	8
1920	11	9
1930	12	10
1940	14	11
1950	17	12
1960	18	16
1970	20	17
1980	23	19
1990	26	26
2000	28	36
2010	31	49
2015	33	58

Analysing data and applying your understanding

2. (a) Create a table to show the annual GDP per capita of each of the provinces of China for the last full calendar year. (You will find this data in the Annual National Accounts section of the **China National Bureau of Statistics** weblink in the Resources tab. A template to create your table can also be found in the Resources tab.)
   (b) Using your data table and the blank map of China in the Resources tab, create a choropleth map to show the pattern of annual GDP across China's provinces for the last full calendar year.
   (c) Describe the pattern evident in your map.

3. (a) Create a graph to compare the annual disposable income of urban and rural households in China for the last 20 years. (You will find this data in the Annual People's Living Conditions section of the **China National Bureau of Statistics** weblink in the Resources tab.)
   (b) Why do you think encouraging greater migration of rural people to the cities has assisted China in transitioning to a consumer-based economy?

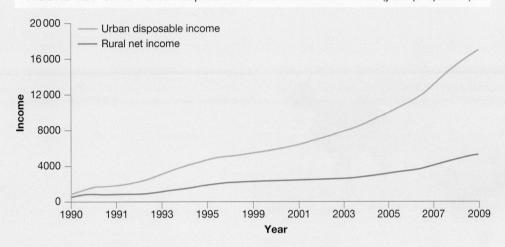

FIGURE 4.56 China – Urban disposable income vs rural net income (yuan per person)

4. (a) Using data from table 4.18, determine whether there is a correlation between wage index, an indication of labour costs, and migration share of total migration in China.
   (b) Explain whether or not you think your result could suggest that migration flows were influenced by wage index differences? Justify your response.
   (c) What other factors do you think could explain the pattern of migration?

TABLE 4.18 Wage index and migration share in China

City	Wage index	Migration share (%)
Jinchang	6.7	0.1
Tianshui	5.7	0.1
Zhuha	6.8	1.2
Guangzhou	6.6	1.2
Shanghai	6.4	0.8
Beijing	6.2	0.7
Quazhu	6.3	1.8
Shenzhen	6.9	2.3
Dongguan	6.9	3.9

Resources

- **Weblink:** China's ghost cities
- **Digital doc:** GDP of provinces in China table template (doc-30358)
- **Digital doc:** Blank map of China (doc-29173)

4.10 Sustainable development in megacities

4.10.1 Challenges and opportunities for developed countries

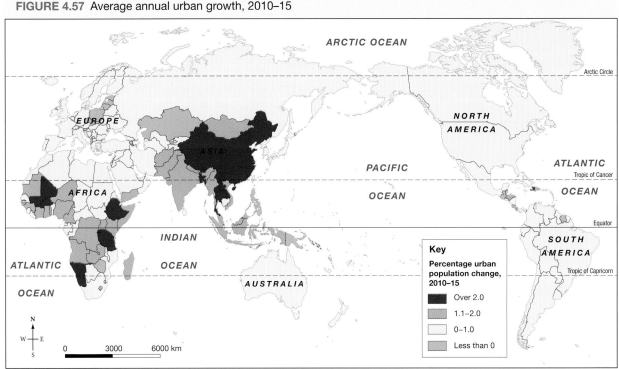

FIGURE 4.57 Average annual urban growth, 2010–15

Source: Reproduced under CC BY-4.0 licence https://creativecommons.org/licenses/by/4.0/ ©2018 The World Bank Group, World Bank Staff estimates based on United Nations, World Urbanization Prospects.

Waste management and sanitation

Large populations produce large amounts of waste that need to be processed and disposed of somewhere, but cities don't have the extensive tracts of land to do this. So where does the waste go? Some developed countries, such as Australia, export some types of waste to other countries. China and India, for example, have at different times received Australian recycling for processing. This may be a 'sustainable' option for the developed nation's environment, but not for the developing nations, which may not have the means of processing the waste in a sustainable or safe way.

The population of New York City, for example, produces around 12 000 tonnes of rubbish a day (with only about 2000 of that recyclables).

The annual budget for waste removal in New York is approximately US$1.5 billion (2016). In addition to this, city authorities need to manage street sweeping, and general cleaning and maintenance of public facilities and streets. Rubbish collected in New York is managed in a number of ways:
- Shipped or trucked about 1000 km to one of the state's landfills (80 per cent of mixed residential waste)
- Incinerated for fuel (20 per cent of mixed residential waste)
- Recycled (either locally – mostly paper and metals – or overseas – mostly plastics)

New York has implemented policies to help manage this volume of waste in a more sustainable way. In 1989, they began a compulsory recycling program, which runs as part of a broader plan that sees wasted separated and managed accordingly.

Other initiatives implemented as part of the city's waste reduction plan include:
- Reduce the percentage of recyclables that are sent to landfill (the approximate capture rate for common recyclables, such as paper and plastic, is currently about half)
- Develop household and community composting and separate organic matter collection
- Reduce or eliminate non-recyclable material, such as single-use plastic bags, Styrofoam food containers, etc
- Implement separate disposal programs for electronics that have reusable and recyclable component parts
- Encourage recycling and reuse of clothing and consumer goods
- Provide incentives for businesses to improve their waste management processes
- Implement school programs to change waste disposal habits of the next generation, including a 'zero waste' scheme.

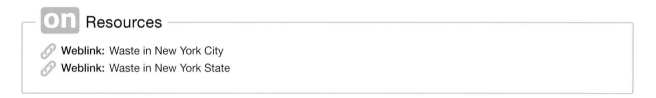

Resources

Weblink: Waste in New York City
Weblink: Waste in New York State

Infrastructure

Poor infrastructure also negatively affects the sustainability and liveability of cities. In some instances, land on the rural-urban fringe of developed nations' megacities is sold off to create new housing and industrial estates. Services such as roads, electricity, water and gas are in place before development begins, but shopping complexes, medical services, schools and recreational facilities that are present in existing suburbs are generally delayed in new suburbs until the area has an established population – generally, a service needs to be economically viable before it will be put in place. This creates difficulties for residents who have to travel long distances for work, school or basic services such as healthcare — often in their own cars, as extensive public transport infrastructure may not yet be in place. This means that they not only travel further, but are less likely to do so in a sustainable way that requires any physical exertion, such as walking to a train station or riding a bike. This contributes to rising obesity levels and health issues.

As cities move outward, they often expand at a faster rate than infrastructure. Traffic congestion inevitably follows as the road network is choked and unable to cope with more and more people making the daily commute to work, as employment opportunities are often limited in new outer suburban fringes.

Resources

Weblink: Health impacts of suburban sprawl

Impact on the natural environment

The expansion of a city's urban footprint also makes significant changes to the natural environment. Clearing vegetation can reduce biodiversity and natural habitats. Natural drainage is impacted as green spaces are lost.

With these changes, water no longer flows over the surface of the land to be taken up by vegetation, into our waterways or infiltrates the soil, but rather runs across concrete and bitumen into stormwater systems. Disrupting the natural waterways also leaves some areas at a higher risk of flooding (see subtopic 4.5).

When an urban grows, food will need to travel vast distances from farms and factories to feed the population of a megacity, adding to each individual's **carbon footprint** (the total greenhouse gas emissions produced as a result of their activities). With so many people living in such a concentrated area, this impact is potentially significant. The cost of transporting this food also creates issues for the urban poor as the cost of transport is added to the cost of the produce at the market.

Expanding cities and populations in developed nations also have greater energy needs to maintain their lifestyle. Generating high levels of electricity also presents challenges for the natural environment, particularly in countries that rely on coal-fired power stations, which emit CO_2. For example, after the forced closure of many of their nuclear power plants after the Fukushima disaster in March 2011, Japan's CO_2 emissions rose to a record high for the year ending March 2014, just three years after the country reverted to relying on coal and gas-generated power. Before this, approximately 26 per cent of Japan's power had been generated by nuclear plants.

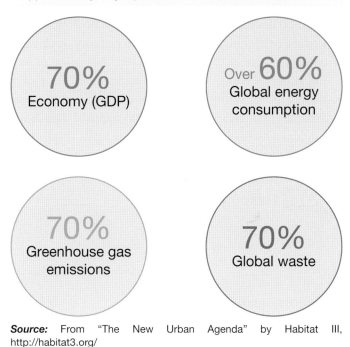

FIGURE 4.58 The Global Context: Cities today occupy approximately only 2 per cent of the total land area.

Source: From "The New Urban Agenda" by Habitat III, http://habitat3.org/

Resources

Weblink: Japan's CO_2 emissions hit second-highest on record

Social impacts

Closer to the city centre, where land is expensive and in short supply, medium- and high-density housing replaces many of the original dwellings, changing the character of these inner-city areas. However, services and infrastructure are already well-established – employment, shopping precincts, educational and health services are already in place and public transport provides a viable alternative to the family car.

The cost of moving into these areas is often out of the reach of people on lower incomes, first home buyers and young families, and so they are forced into the outer areas where housing is more affordable. This urban sprawl places groups with limited income at a greater disadvantage as they often have to travel greater distances for work or services, which is more expensive than travelling shorter distances, meaning they spend more of their limited income. This also has flow-on social effects, as those who have to travel greater distances for work also spend less time at home with their families. These factors combine to create social segregation between low and middle–high income earners. In countries such as the United States, in which a significant proportion of the funding for public education comes from local taxes or funding, school districts with a high proportion of high-income families collect more revenue. This has a direct impact on standards of educational facilities.

The rise in population also leads to an increase in population density and high-density dwellings. High density living may have sustainability advantages over urban sprawl, but it can also have a negative social impacts: less green space, a reduced sense of community and increased crime rates. But the challenges faced by megacities in developed countries also extends beyond the provisions of infrastructure. The development of digital communications, for example the rise of e-commerce, also presents challenges for large urban populations, as the traditional ways of earning an income – manufacturing, trade and service industries – change.

FIGURE 4.59 The impact of urban sprawl on the community

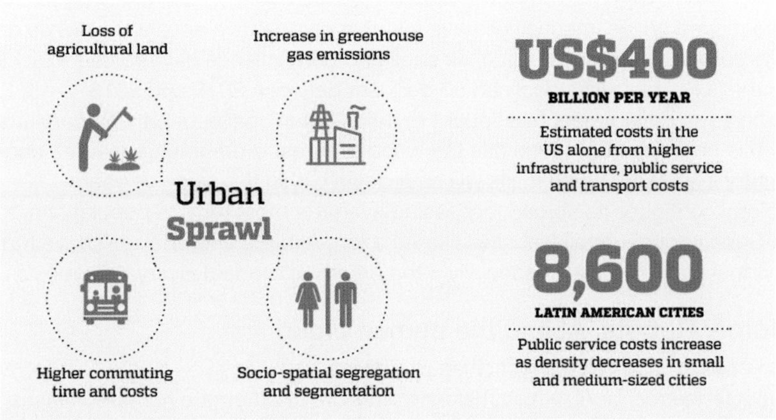

- Loss of agricultural land
- Increase in greenhouse gas emissions
- Higher commuting time and costs
- Socio-spatial segregation and segmentation

US$400 BILLION PER YEAR
Estimated costs in the US alone from higher infrastructure, public service and transport costs

8,600 LATIN AMERICAN CITIES
Public service costs increase as density decreases in small and medium-sized cities

Source: Copyright © United Nations Human Settlements Programme, 2016, World Cities Report 2016

Can the world's megacities survive the digital age?

Today, megacities have become synonymous with economic growth. In both developing and developed countries, cities with populations of 10 million or more account for one-third to one-half of their gross domestic product.

Many analysts and policymakers think this trend is here to stay. The rise of big data analytics and mobile technology should spur development, they assert, transforming metropolises like Shanghai, Nairobi and Mexico City into so-called 'smart cities' that can leverage their huge populations to power their economies and change the power balance in the world.

As technology researchers, however, we see a less rosy urban future. That's because digitization and crowdsourcing will actually undermine the very foundations of the megacity economy, which is typically built on some combination of manufacturing, commerce, retail and professional services.

The exact formula differs from region to region, but all megacities are designed to maximize the productivity of their massive populations. Today, these cities lean heavily on economies of scale, by which increased production brings cost advantages, and on the savings and benefits of co-locating people and firms in neighbourhoods and industrial clusters.

But technological advances are now upending these old business models, threatening future of megacities as we know them.

Manufacturing on the fritz

One classic example of a disruptive new technology is 3-D printing, which enables individuals to 'print' everything from ice cream to machine parts.

As this streamlined technique spreads, it will eliminate some of the many links in the global production process. By taking out the 'middle men,' 3-D printing may ultimately reduce the supply chain to just a designer on one end and a manufacturer on the other, significantly reducing the production costs of manufactured goods.

That's good for the profit margins of transnational companies and consumers, but not for factory cities, where much of their transportation and warehousing infrastructure may soon become redundant. Jobs in manufacturing, logistics and storage, already threatened across many large sites, may soon be endangered globally.

In short, 3-D printing has transformed the economies of scale that emerged from industrialization into economies of one or few. As it spreads, many megacities, particularly Asian manufacturing centers like Dongguan and Tianjin, both in China, can expect to see widespread disruption to their economies and work forces.

Decline of the shopping mall

The retail sector is experiencing a similar transformation. Shopping malls, for example, which once thrived in megacities, are now suffering from the advent of e-commerce.

The value proposition of shopping malls was always that their economies of scale were location-dependent. That is, for malls to be profitable, they had to be sited near a large consumer base. Densely populated megacities were perfect.

But as stores have moved online, megacities have lost this competitive advantage. While online shopping has not completely replaced brick-and-mortar retail, its ease and convenience have forced many shopping malls to close worldwide. In the U.S., mall visits declined 50 per cent between 2010 and 2013.

Cities in China, where the government has sought to build its national economy on consumption, will be hit particularly hard by this phenomenon. China has the world's largest e-commerce market, and it is estimated that one-third of the country's 4000 shopping malls will shut down within the next five years.

As mobile technology continues its spread, accessing even the most remote populations, this process will accelerate globally. Soon enough, retail websites like Amazon, Alibaba and eBay will have turned every smartphone into a virtual shopping mall, especially if the dream of drone delivery becomes a reality.

The new work force: Robots, AI and the human cloud

Changes in the business world will also affect cities worldwide.

Thanks to artificial intelligence, or AI, which makes it possible to automate numerous tasks, both manual and cognitive, these days it's goodbye, human bank tellers and fund managers, hello robots.

Even in jobs that cannot be easily automated, the digitised gig economy is putting people into direct competition with a global supply of freelancers to do tasks both menial and specialised.

There are certainly benefits to crowdsourcing. Using both AI and the crowdsourced knowledge of thousands of medical specialists across 70 countries, the Human Diagnosis Project has built a global diagnosis platform that's free to all patients and doctors – a particular boon to people with limited access to public health services.

But by taking collaboration virtual, the 'human cloud' business model is also making the notion of offices obsolete. In the future, medical professionals from various specialties will no longer need to work near to each other to get the job done. The same holds for other fields.

In a world without office space, traditional business and financial centers like New York and London would feel the pain, as urban planning, zoning and the real estate market struggle to adjust to firms' and workers' changing needs.

Crisis in the making

At some point, all this change may end up meaning that economies of scale matter much, much less. If that happens, population size – currently the motor of the modern metropolis – will become a liability.

Megacities have long struggled with the downsides of density and rapid urbanization, including communicable disease, critical infrastructure shortages, rising inequality, crime and social instability. As their economic base erodes, such challenges are likely to grow more pressing.

The damage will differ from city to city, but we believe that the profound shifts underway in retail, manufacturing and professional services will impact all of the world's seven main types of megacities: global giants (Tokyo, New York), Asian anchors (Singapore, Seoul), emerging gateways (Istanbul, São Paulo), factory China (Tianjin, Guangzhou), knowledge capitals (Boston, Stockholm), American middleweights (Phoenix, Miami) and international middleweights (Tel Aviv, Madrid).

And because 60 per cent of global GDP is generated by just 600 cities, struggle in one city could trigger cascading failures. It's conceivable that in 10 or 20 years, floundering megacities may cause the next global financial meltdown.

If this forecast seems dire, it's also predictable: Places, like industries, must adapt with technological change. For megacities, it's time to start planning for a disrupted future.

Source: This article was originally published on *The Conversation*. Written by Christopher H. Lim and Vincent Mack. Published under a Creative Commons licence.

Activity 4.10a: Megacities in developed nations

Complete the questions below to demonstrate your understanding of the challenges and opportunities of urbanisation in developed nations.

Explain the challenges and opportunities

1. Create a Venn diagram to summarise the potential challenges and opportunities for individuals in a megacity in a developed nation. In your answer consider factors including employment, housing, transport, sanitation and waste management, health and education, food and water security, energy needs and social factors such as community engagement or crime.
2. Explain the ways in which e-commerce and new technology might affect liveability in a megacity in a developed nation.

Identify the challenges and propose a response

3. Technological developments have led to changing job markets and economies in developed nations. Make a list of potential challenges that might arise in a city when there are high levels of unemployment. Propose and justify one strategy that authorities might employ to combat one of the challenges you identified.

4.10.2 Challenges and opportunities in developing countries

Population predictions indicate that over the next 25 to 30 years most urban growth will occur in Africa and Asia — historically the least urbanised continents in the world. It is also in these places, where birth rates remain high, that urban growth will continue. It is predicted that Asia will have 30 megacities by 2025. While Africa will have only three, it has the fastest rate of population increase. By the end of the 21st century, it is estimated that the world's largest cities will be in Africa. Many countries in the developing world experience extreme poverty, civil unrest and famine. These push factors tend to drive people to abandon their rural homes and move to cities, lured by the prospect of safety, employment, shelter and an improved standard of living. It is in these cities that population growth is highest; the UN estimates that 90 per cent of the world's population growth is taking place in the cities of the developing world. The issues faced in these nations mean that the quality of life of their many inhabitants is seriously threatened.

The UN estimates that by 2030 more than 2 billion people worldwide will be living in overcrowded slums. In the African nations of Ethiopia, Malawi and Uganda, 90 per cent of the population already lives in urban slums.

Moving to an urban environment does, however, offer opportunities that are not available in small rural or regional centres in many developing countries. Access to healthcare and education might be sporadic or far more difficult in remote areas, and opportunities for work might be rare. In times of drought or civil unrest, aid agencies are more likely to be accessible in urban centres.

Accommodating rapid population growth

Cities in the developing world lack the infrastructure and capacity to deal with a sustained and large influx of people. The result is the development of shanty towns or slums on the outskirts of major cities, where crime rates are high due to a lack of resources and employment. In Mumbai, for example, where housing is expensive, residents are forced to live in cramped conditions, usually a great distance from their workplace. The long commute may take as long as two hours on clogged motorways or overcrowded public transport. Even though many residents reside close to rail and bus services, the city has tended to spread to the north, meaning that people must commute to the south to access the commercial hub of the city.

The combination of low or no income and a lack of access to education or distance from schools, means that children and young people living in slums are more likely to be required by their families to work at a young age before completing much formal education.

The rapid rise of shanty towns also presents an issue for sanitation, waste disposal and air pollution (see section 4.4). Without access to power grids or clean fuels, low grade fuels that produce high levels of smoke and toxic gasses are used for cooking, heat and light sources. In 2009, the World Health Organization estimated that in developing countries about 56 per cent of the population rely on solid fuels for cooking, with up to 89 per cent of people in the world's least developed countries having no access to modern fuel sources.

FIGURE 4.60 A view of Maharashtra, India, one of the largest slums in the world

Combating health issues

The urban poor also generally experience poorer health due to a lack of access to clean water, fresh healthy food and access to or the income to pay for healthcare and medicine when they fall ill. Crowded living conditions with inefficient or non-existent rubbish collection and sanitation also provide the perfect conditions for the spread of diseases, such as cholera, and parasitic infections, such as worms and diarrhoea. These diseases are treatable, but without access to healthcare or the money to pay for medicine they can be fatal.

Air quality is lower and airborne viruses such as influenza and Zika virus (a serious health concern in the lead-up to the 2016 Rio Olympics) are more easily spread in overcrowded urban areas. Recent studies by the African Population and Health Research Centre found that people who live in urban slums also have higher child mortality and undernutrition rates, meaning babies and children growing up in this environment are more susceptible to other illnesses and are more likely to have stunted growth and development.

 Resources

- Weblink: Slum health is not urban health: why we must distinguish between the two
- Weblink: Ranking the world's megacities is a wake-up call for women's rights
- Weblink: World Health Organization: Energy access

Activity 4.10b: Megacities in developing nations

Complete the activities below to demonstrate your understanding of the challenges and opportunities of urbanisation in developing nations.

Identify the challenges and opportunities

1. Create a Venn diagram to summarise the potential challenges and opportunities for individuals in a megacity in a developing nation. In your answer consider factors including employment, housing, transport, sanitation and waste management, health and education, food and water security, energy needs and social factors such as community engagement or crime.
2. Explain the ways in which rapid population growth might affect liveability in a megacity in a developing nation.

Identify the challenges and propose a response

3. Rapid increases in urban population present challenges for communities and for the individuals who relocate. Make a list of potential challenges that might be faced by a low-income family in a developing country moving from a rural area to an urban area. Suggest and justify one strategy that would minimise the negative impact of such a move for the family. In your answer, outline clearly who would need to implement the strategy: government authorities or other agencies such as non-government support agencies or community groups.

4.10.3 The New Urban Agenda (Habitat III)

Quito, Ecuador, played host to the United Nations global summit on Housing and Sustainable Urban Development in October 2016. It was the third such conference in a series of three, and has been referred to as Habitat III.

The four-day conference brought together around 30 000 people from 167 countries with a single aim: the sustainable development of global cities, no matter what form they take or where they are located. It followed the adoption of the 2030 Agenda for Sustainable Development set in 2015. The result, the New Urban Agenda, is a blueprint for sustainable urban development over the next two decades – cities that are not only prosperous, but also cater for both the cultural and social wellbeing of their inhabitants, while at the same time protecting the environment.

Goal 11, for example, is designed to 'make cities inclusive, safe, resilient and sustainable'. Under this goal, ten or more targets have been set, ranging from the provision of affordable housing, upgrading slum settlements, the provision of green spaces, improving air quality and the provision of transportation services that are not only safe, but also affordable and sustainable.

Goal 13 addresses the issue of climate change, recognising that greenhouse gas emissions continue to rise at unacceptable levels – almost 50 per cent since the 1990s – and that this poses grave consequences for global warming. We have already witnessed melting ice caps, rising sea levels (threatening, in particular, many low-lying island and coastal communities) and an increase in weather-related natural disasters, including the intensity of blizzards, storm activity and cyclones, all of which have intensified.

Goal 6 considers issues related to access to safe drinking water and sanitation for everyone. These two issues alone are responsible for the inflated mortality rate, especially in children, from preventable disease. The targets set under this goal are aimed predominantly at rehabilitating and protecting ecosystems, while at the same time addressing the needs of more than 1.8 billion people who do not have access to an improved water supply free from faecal contamination and the 2.4 billion people who do not have basic sanitation services.

Sustainable megacities

With both the rate of urbanisation and the size of urban areas increasing, it is important that town planners give considerable thought to layout, especially in light of the Sustainable Development Goals as adopted under the New Urban Agenda.

Inner city areas have been transformed through urban renewal, where medium- to high-density housing is more common. Communal green spaces, and vertical and rooftop gardens help address some of the issues associated with land clearing of the past.

Outer suburban areas are developing satellite suburbs, where thriving communities enable individuals to live, work and play in relative proximity to home. Provided that public transport networks also keep pace, cars are also not needed on a daily basis. The biggest challenge is in ensuring that as cities expand in developing countries, the living standards of all inhabitants are transformed, in the context of Sustainable Development Goal 11 (see page 281 for more about sustainable urban planning).

FIGURE 4.61 The 17 Sustainable Development Goals designed to ensure a prosperous future for all while using the environment in a sustainable way.

Source: From Sustainable Development Goals, 2015 United Nations

Activity 4.10c: Sustainable Development Goals

Refer to the **United Nations: The New Urban Agenda** weblink in the Resources tab to complete the following activities.

Explain the Sustainable Development Goals

1. Create a table or chart of the UN Sustainable Development Goals that summarises the aim of each goal and include two targets set under each goal as examples.

Suggest and justify a response

2. Choose one of the targets set in the UN's Sustainable Development that could apply equally to developed and developing nations (for example, Goal 3 targets aiming to reduce the number of traffic fatalities and injuries, or to improve the treatment of substance abuse). Propose and justify two strategies that could be implemented by a government, regardless of their levels of development, to help meet that target.

FIGURE 4.62 Cities play an important role in ensuring sustainable development

Source: Copyright © United Nations Human Settlements Programme, 2016, World Cities Report 2016

Weblink: United Nations: The New Urban Agenda

4.10.4 Sustainable solutions: ecocities

An **ecocity** is built on the principles of sustainable living by eliminating carbon waste and producing energy only from renewable resources. The move is away from low-density housing to more compact medium- and high-density housing with most people living within walking or cycling distance of their workplace. Green spaces and natural habitats meander through and separate settlements, ensuring biodiversity and easy access for residents to both nature and recreation.

In terms of transport, there is little need for cars, as residents live within walking or cycling distance of their workplace and shopping hubs. Public transport connects settlements and runs on a frequent and reliable timetable to meet the needs of those who need to travel longer distances.

Vauban, Germany

Vauban in southern Germany is located in the district of Freiberg and is an example of urban renewal based on the principles of sustainable living in an urban environment. Approximately 5000 people live in the settlement, which also boasts more than 600 jobs. Every home is within easy walking distance of a tram stop, schools, businesses and shopping centres, virtually eliminating the need for private motor vehicles.

The area was planned around green transportation. A network of shared pedestrian and bike paths have been integrated into the development and pass through or past open spaces, making the journey to work or around the neighbourhood more enticing.

Vauban is also well-serviced by public transport, in the form of buses and trams. The development is set out in a linear style along the tram tracks, with all homes being within easy walking distance of a tram stop. The tram network provides an easy and direct connection to the larger township of Freiberg.

The level of private car ownership has fallen dramatically with the passage of time, with fewer than 20 per cent of residences owning a private vehicle. Residential streets do not have parking spaces for cars. Vehicles in this residential area must slow to a walking pace and operate on the idea of pick up and deliver, with strictly no parking. Households must make an annual declaration stating their status of owning or not owning a vehicle. Those with a vehicle must pay a high annual premium, akin to the purchase price of a new vehicle to park their vehicle in a multi-storey complex on the outskirts of the development. Further incentives to remain car free include discounted tram rides and membership in a car-sharing club.

FIGURE 4.63 All buildings in Vauban have solar panels on the roof. Excess energy is fed into the city's power grid, benefiting all residents.

All buildings must meet minimum low energy consumption targets, and solar panels are a noticeable addition to rooftops. The city has its own power grid, which purchases excess energy from households to compensate for any deficiencies in their own energy consumption. Profits from the sale of surplus energy production is split equally between households. Under this arrangement, households that produce more energy than they need make a profit, while at the same time households with an energy deficit have all their energy needs met for a low annual payment of $200–$300.

Household organic waste is treated in a central **anaerobic digester system**. Vacuum pipes suck ecological sewerage into the digester, generating biogas which produces energy that can be used to heat homes or for cooking. Greywater is purified using **biofilm plants**. Harmful bacteria in wastewater is filtered through the naturally occurring bacteria in plants and soil. Once cleaned, the wastewater is returned to the water cycle.

Vauban is part of the much larger city of Freiberg, renowned as being the greenest city in Germany. The residents of Freiberg share a love of cycling and their town has more than 400 kilometres of bike paths, and provides parking for 9000 bicycles. The streets there are also relatively car free. While other cities provide park-and-ride facilities for motor vehicles, in Freiberg, a bike-and-ride service will be found at many transit stations. Pedestrians are also a common site on the streets of Freiburg.

Other transportation options are provided in the form of buses, trains and trams. 70 per cent of Freiburg's population has a mere 500 metre stroll to the nearest tram stop. Trains run at 7.5-minute intervals during peak periods and ticket costs are heavily subsidized to encourage residents to use them.

Shops and offices are located on the ground floor of apartment buildings to provide residents with easy access to commodities they need on a daily basis.

Waste levels have been substantially reduced due to waste minimisation and recycling schemes. An anaerobic digestion waste system, similar to that used in Vauban, is also in use.

On the outskirts of the city, small privately-owned gardens allow inhabitants to grow their own food. Plans are in place to increase this number to allow more residents the same opportunity. The region also boasts more than 600 hectares of parks and 1600 playgrounds to cater for the recreational needs to the population and foster biodiversity.

Activity 4.10d: Sustainability of the megacities

Complete the following tasks to consolidate your understanding.

Explain the information and identify the patterns

1. Describe the features of an ecocity.
2. Future population growth and the growth of megacities will occur more in the developing world than developed countries. Explain whether the building of more ecocities might provide a way to manage this growth.
3. Using a Venn diagram, show the similarities and differences between the problems faced by megacities in developed and developing countries. Include both challenges and opportunities in your diagram.

Apply your knowledge and suggest a response

4. Investigate Sustainable Development Goal 6, 11 or 13 in more detail. Discuss a specific challenge in achieving that goal for one megacity in the developed world, and one in the developing world.
5. Use the information presented in this chapter to suggest ways that megacities can manage the challenges of climate change and global warming.
6. Are the sustainable solutions used in places such as Vauban and Freiburg transferable to cities with populations over 10 million? Justify your answer.

Resources

- **Weblink:** UN Sustainable Development Goals
- **Weblink:** Cities and urbanisation

GLOSSARY

absolute change the change in the number of people living in a specific area

Accessibility/Remoteness Index of Australia (ARIA) a means of classifying a place as 'remote' using GIS data to measure road distance to service centres. This classification system uses a sliding scale to measure remoteness.

adakitic describes volcanoes containing magma formed under island arcs (a chain of volcanic islands) due to sea floor subduction. They are most common in the western Pacific Rim.

anaerobic digester system a system that uses micro-organisms to break down biodegradable objects without oxygen

animal invasion the introduction of a non-indigenous animal to an area, which negatively impacts the environment

anthropogenic processes processes that involve human activity, for example, the burning of fossil fuels to produce electricity

aseismic ridge a mountain ridge or chain of seamounts on the ocean floor usually associated with hot spots (Hawaii or the Galapagos Islands) rather than seafloor spreading

asthenosphere the upper layer of the mantle, below the lithosphere, usually more than 100 km below the surface. It is where rock becomes molten and allows the solid tectonic plates to move over it.

atmosphere the gaseous parcel of air surrounding the Earth, consisting mainly of nitrogen, oxygen, argon, carbon dioxide and water vapour

atmospheric hazard a potentially damaging natural event generated in the troposphere, such as a severe storm, tropical cyclone (typhoons and hurricanes), tornado, blizzard and wind storm

Australian Standard Geographical Classification (ASGC-RA) a means of classifying a place as 'remote' using Census Collection Districts on the basis of the average ARIA score within the district

bioaccumulation the process by which pollutants enter and spread along the food chain by passing from one food source to another by being eaten

biodegrade to break down through natural processes

biofilm plants processing plants or systems that use the naturally occurring bacteria in plants and soil to purify wastewater

biological processes processes that are vital for organisms to live, for example, plants require the process of photosynthesis to survive

biophysical environment both living (biotic) and non-living (abiotic) surroundings of an organism or population. It is made up of the elements of the atmosphere, hydrosphere, lithosphere and biosphere.

biosphere the part of the Earth containing elements of air, water and land where living things can survive. Also called the ecosphere.

carbon footprint the total greenhouse gas emissions produced as a result of a person's activities

city a large urban settlement with clearly defined boundaries and municipal functions

climatological hazard a hazard that occurs due to the climatic conditions of an area, such as bushfires, droughts and heatwaves

closed systems systems that have boundaries but permit an exchange of energy with their surroundings, but not exchanges of matter. For example, the water cycle (hydrological cycle) is a closed system as water cannot be created or destroyed but it has inputs and outputs of energy.

community a system of interacting and interdependent social groups occupying a particular area

confined aquifer groundwater that lies under a layer of impermeable rock or clay, which prevents the water from moving in or out of the aquifer.

conurbation an extensive urban area that forms when there is an amalgamation of several cities, for example, Tokyo–Yokohama in Japan or Gold Coast–Brisbane–Caboolture in Queensland

Coriolis effect a force created by the spin of the Earth, by which free-moving objects are deflected to the left in the Southern Hemisphere and to the right in the Northern Hemisphere

counterurbanisation describes the migration of people from urban to rural areas

cumecs an abbreviation of 'cubic metres per second'

disease a condition that causes harm to, or interferes with, the normal functioning of a living thing

dispersed settlements settlements that tend to be spread out across a large area with very few examples of nucleation (buildings in clusters)

ecocity a city built on the principles of sustainable living by eliminating carbon waste and producing energy only from renewable resources

ecological hazard an interaction between living organisms or between living organisms and their environment that could have a negative effect

ecological hazard zone the specific areas, spaces or places at risk of experiencing an ecological hazard

ecosphere see **biosphere**

endemic located in a specific area

Environmental Kuznets Curve a hypothetical curve that graphs environmental damage against income per capita over the course of time

epicentre the point on the Earth's surface directly above the focus when an earthquake has occurred

epidemic a disease outbreak that affects many people in a specific region

evapotranspiration the process by which water moves into the atmosphere. This occurs through evaporation (from the ground or surfaces) and through the transpiration of plants.

exposure refers to the degree or likelihood of a place, person or thing being affected by a hazard, in terms of risk assessment

faults large cracks in the Earth's crust, often associated with the boundaries of the Earth's tectonic plates

flow movement when rock, soil or sand mix with water and air and move downhill in a flow

focus the exact location in the crust or mantle where an earthquake rupture occurs. Seismic waves radiate away from the focus

gentrification the process of older areas, usually of lower socioeconomic status, being slowly bought out and renovated by wealthier people

geological hazard a potentially damaging natural event that occurs in the Earth's crust, such as a volcanic eruption, earthquake or tsunami

geomorphic hazard a potentially damaging event on the Earth's surface – such as an avalanche, landslide or mudslide – that is often caused by a combination of natural and human processes

globalisation the process by which the world has become increasingly inter-connected through freer movement of capital, goods and services. It is reflected in the value of cross-border world trade expressed as a percentage of total global GDP.

green villages see **ring villages**

Gross Domestic Product also known as GDP; the total value of all goods and services produced within a country, usually over a period of one year

gyres sections of ocean that form smaller, swirling circular currents, usually at the point where two larger ocean currents meet

hamlet a tiny settlement consisting of a small number of residential and work buildings, with possibly some other low-order functions, for example a general store

hazard something that has the potential to cause harm. It may be obvious, for example, a flooded section of road, or not obvious, for example, a damaged hidden electrical wire.

hazard zone an area that may be affected by a natural hazard, for example, areas vulnerable to flooding based on past events or areas likely to be affected by pyroclastic flows from a volcano

hydrological hazard an extreme event with a high-water component, such as flash flooding due to storms, cyclones, ice melt or storm surges and tsunamis

hydrosphere the water portion of the Earth's biophysical environment, including oceans, ice caps, glaciers, rivers and lakes

inclusive planning a style or method of settlement planning based on the idea that it is important to take into consideration the health, wellbeing, needs and wants of current and future residents

indigenous describes something as being native to or originating from a specific place, for example, kangaroos are indigenous to Australia. Note that the word often has a different meaning when capitalised. The term 'Indigenous' refers to people of Aboriginal or Torres Strait Islander descent in some parts of Australia.

infectious diseases contagious or communicable diseases that are spread by being passed from one person to another

inland drainage basin describes an area in which water from sources such as rivers, melting snow or rain collect in a lake or inland sea rather than flowing to the ocean

inter-tropical convergence zone (ITCZ) the zone near the Equator where trade winds of the Northern and Southern Hemispheres meet. The intense heat, warm water and high humidity create what is an almost permanent band of low pressure. The monsoon trough seen on weather charts is part of the ITCZ.

invasive plant a non-indigenous plant species that has been introduced to a specific area by people (either intentionally or accidentally) and has multiplied to an extent that it threatens to damage or is damaging the economic, environmental or social value of a place

inverted barometric effect the change in sea level due to changing air pressure above it. For example, when a low-pressure cell moves over a body of water, its level rises. On the other hand, a high-pressure cell has the effect of pushing sea levels down. For each 10 hPa change in air pressure, water levels may change by about 10 cm.

isolated settlements smaller, individual dwellings, usually not close to settlements, in areas where natural resources may not be plentiful enough to sustain a large population

isopleth the lines on a map that show points of equal measurement, such as depth of ash fall or height above sea level

La Niña the cooling of the water in the equatorial Pacific Ocean. This is associated with changes to weather patterns such as wetter than normal seasons.

landslide the large-scale movement of rock, debris and soil down a slope due to unstable conditions, which may be caused by heavy rainfall, an earthquake or a volcanic eruption. A landslide under the ocean can cause a tsunami.

linear settlements also known as ribbon settlements; settlements in which buildings are strung out along lines of transport or communication, including villages that exist along roads or at intersections, or buildings that congregate along rivers and other waterways

liquefaction when saturated or partially saturated soil loses its firmness and displays the properties of a liquid, such as when an earthquake shakes and loosens wet soil in low-lying areas, and the soil loses rigidity and moves like fluid, covering things in its path

lithosphere the solid, outer part of the Earth, which includes the brittle upper portion of the mantle and crust. It is broken into large slabs called plates.

magma hot molten rock formed below or within the Earth's crust. It reaches the surface through volcanic or plate tectonic activity and becomes lava and eventually igneous rock.

magnitude a measure of size, for example, earthquakes are measured according to magnitude on the Richter Scale.

megacities cities with 10 million or more inhabitants

Mesh Blocks the smallest geographical area defined by the Australian Bureau of Statistics

metropolis the largest single urban settlement in a state (often the capital) or district

microplastics small particles of plastic, measuring less than 5 mm in diameter, the result of larger plastic objects degrading

mitigation strategies and actions to reduce or eliminate a hazard's level of impact if it does occur

monofunctional development developments that have a single function or purpose, such as a development that is solely for residential buildings

mudflow when large amounts of suspended silt and soil move quickly down a slope. They tend to occur mostly on steep slopes but can happen anywhere ground is unstable due to loss of vegetation.

nanoplastics tiny particles of plastic, less than about 1000 nm in diameter (0.001 mm), the result of larger plastic objects and microplastics degrading

natural disaster a large natural event, such as a tornado, flood, earthquake or landslide, that causes considerable loss of life, damage to property and infrastructure, and/or destroys sections of the environment

natural hazard an extreme event that occurs either in the lithosphere, on the Earth's surface or in the atmosphere. They can be highly destructive and cause considerable harm to living things and property. Examples include tropical cyclones, tornadoes, earthquakes and volcanoes.

non-point source broad areas from which a pollution hazard originates, for example, run-off from city streets

nucleated settlements settlements where some buildings have come together in groups or clusters due to economic or social/cultural reasons

open systems systems that allow an exchange of energy and matter across their boundaries. For example, a catchment or drainage basin permits both energy and matter to enter and escape. Water comes in as rain and either runs to the ocean or infiltrates into the ground. Some is lost back to the atmosphere through evaporation.

pandemic a disease outbreak that is prevalent across a wide geographic area including beyond the region in which the outbreak began

periurban located on the edges of a city

periurbanisation the process of urban growth that creates hybrid landscapes with dispersed rural and urban characteristics

photodegradation the process by which objects are broken down or degraded by exposure to light, usually from the sun

point source a particular location from which a pollution hazard originates, for example, an industrial site or an oil tank

pollutants substances introduced into the environment that are potentially harmful to human health and to the natural environment

preparedness actions taken by communities so they can maintain an ability to respond to, and recover from, natural hazards if they occur

pressure gradient a spatial way of explaining the rate of change of pressure between places on a synoptic chart. If isobars are close together, the pressure gradient is tight (winds are strong) whereas if isobars are far apart, the pressure gradient is weak (winds are light).

prevention strategies and actions to stop a hazard from occurring

primary impacts the immediate effects of an event, for example, water washing away roads and buildings in a flood

primary sector industries producing raw materials, such as agriculture or mining

pseudo-urbanisation a type of urbanisation without economic growth that results in large numbers of poverty-stricken residents living in informal settlements

pull factors factors that attract people to an area

push factors factors that encourage people to leave an area

P-waves also known as primary waves; high-frequency seismic waves that travel fastest and are measured first at a seismic station. P-waves can pass through solid rock and liquids.

pyroclastic clouds rapidly moving currents of hot air, gases and ash that run from the crater down the sides of a volcano. They are extremely lethal due to their high speed and lack of sound.

pyroclastic cones steep conical volcanic cones built by a combination of lava flows and cinder/ash from pyroclastic eruptions. They can form quickly and remain active for long periods.

qualitative data data based on personal experience

quantitative data data that can be measured as an exact quantity

relative change the percentage difference in population over a specific time period

relief the difference in height between one point and the surrounding area

ribbon settlements see **linear settlements**

ring villages also known as green villages; settlements found in more remote parts of the world. They are built around a central meeting area and point to the communal aspect of many traditional lifestyles.

risk the potential for something to go wrong. This is a subjective assessment about actions that may be predictable or unforeseen.

risk management strategies and actions to reduce or mitigate risk based on the known consequences of encountering a hazard

Rural Remote and Metropolitan classification (RRMA) a means of classifying a place as 'remote' using population size and distance to the nearest service centre. There are seven categories in the scale.

Saffir-Simpson Hurricane Wind Scale a system of rating tropical cyclones (including hurricanes and typhoons) from 1 to 5, based on sustained wind speed. It is used to estimate potential threat to people and property.

sea floor spreading the divergence of two oceanic crust plates

secondary impacts the effects of an event that occur hours or days after a disaster, often caused by or following on from primary effects, for example, people becoming ill from contact with contaminated water after a flood

secondary sector manufacturing industries such as food processing plants or paper milling

seismic waves waves of energy travelling away from an earthquake. They are like huge vibrations and may travel through the Earth's mantle and crust or along the surface.

slope failure when the pull of gravity causes a hill or mountain slope to collapse

storm surges an abnormal rise of ocean height due to strong onshore winds and/or reduced atmospheric pressure. They are common when tropical cyclones approach a coast.

subduction a geological process where two tectonic plates collide at convergent boundaries

subduction zones the areas of the mantle in which convergent plates collide. Under the ocean, these areas are called trenches.

subsidence when part of the land sinks or collapses

suburbanisation the development and outward spread of new suburban areas

system a dynamic structure with inputs, processes and outputs. If a change occurs in one part of the system, it will affect the whole system. The four major systems of the Earth may be further divided into smaller systems such as biomes, forests, oceans and troposphere.

tectonic plates large sections of crust that make up the lithosphere. Plates are about 100 km thick and are made up of either oceanic crust or continental crust.

tenure the occupancy or lease of an area of land

tertiary impacts the long-term effects of an event, often as consequences of the secondary effects, for example, people permanently relocating to another area because they are too afraid to stay in the hazard zone

tertiary sector service industries such as retail, hospitality, healthcare or education

thermal mass a large capacity to absorb and retain heat energy

thermohaline circulation deep currents in the ocean caused by changes in water density that are created by changes in water salinity and temperature

town an urban settlement that is smaller than a city but larger than a village, and offers low-order functions, such as grocery shops, schools and cafes

troughs long, extended zones of low atmospheric pressure, often associated with fronts and occurring near the surface or in the upper levels of the atmosphere. The monsoon trough is an extended zone of low pressure over northern Australia during summer.

unconfined aquifer groundwater that is open to the atmosphere through holes or pores in the soil or rock

Urban Centre and Locality small settlements with a population of more than 200 people

urban centres areas that have an Urban Mesh Block population greater or equal to 45 per cent of the total population and a dwelling density greater or equal to 45 dwellings per square kilometre (sq km), or a population density greater or equal to 100 persons per sq km and a dwelling density greater or equal to 50 dwellings per sq km, or a population density greater or equal to 200 people per sq km

urban heat islands microclimates created in an urban environment by the hard, dark surfaces that attract and retain heat, such as roads and buildings

Urban Mesh Block a Mesh Block with a population density of 200 persons or more per square kilometre

urban sprawl the process of outward expansion of urban areas from the CBD into the surrounding countryside and invading adjacent towns, regions and undeveloped land

urbanisation growth in the proportion of a population living in urban environments

vector-borne diseases diseases that are carried by organisms, such as mosquitos or fleas, that are capable of transmitting disease-producing bacteria, viruses and parasites from one person to another

village a small settlement with a residential population and a small number of low-order services, such as a grocery shop

vulnerability the degree of risk faced by a place, person or thing, based on an approaching hazard's potential impact, the place's degree of preparedness, and the resources available to respond

wind set-up the vertical increase (rise) in seawater levels due to wind stresses on the ocean surface

wind shear a sudden change in wind speed and/or direction over a relatively short distance in the atmosphere, generally due to altitude

INDEX

A

ability to respond (to a hazard)   12
ABS definitions of settlement sizes   145–6
absolute change   192
Accessibility/Remoteness Index of Australia (ARIA)   149
acid rain   53, 96, 107, 109, 110
acidic volcanoes   53
adakitic volcanoes   56
adaptation (to hazards)   11
aerial imagery *see* satellite and aerial imagery
affordability
  and population distribution   169–72
  and urban sprawl   185, 186
aftershocks   41, 46
Agbogbloshie Dump, Accra, Ghana   119–21
agricultural land, loss of, and urban sprawl   183–4
agricultural runoff, impact on Great Barrier Reef   105–7
air-borne diseases   126
  see also influenza
air pollutants   75, 107–18
  sources and effects   108–9
  types of   107–8
air pollution   75, 94, 95
  Brisbane   112, 113–14, 116–17, 118
  and climate   116
  deaths from   109, 110, 111, 278
  factors affecting   110–16
  hazard mitigation   116–18
  impact of   109–10
  megacities   227–33
  Mexico City   110–11, 112, 113, 114, 116, 117
  and population growth   112
  and topography   113–15
air quality
  in megacities   226–33, 278
  monitoring   108, 117–18
air quality data, analysing   118
air quality index   108, 117
algal blooms   78
alien invasive species   80, 83
anaerobic digester systems   282
analysing spatial and statistical data   70
animal invasions   74, 81
  impact (cost) of pest species   82
  see also invasive plants and animals; rabbits

Anopheles mosquito   131
anthropogenic pollution
  marine environment   97–103
  sources   94
anthropogenic processes   73
anthropogenic wastes   95
aquatic weeds removal   85
Aral Sea   121–5
  anthropogenic activities affecting   123
  changes, 1960–2017   121, 122
  as closed inland drainage basin   121
  contaminant hazards   124
  environmental, social and economic impacts   124
  reducing the level of risk   124–5
artificial intelligence (AI)   276
aseismic ridge   56
assessing and responding to hazards   5, 12, 15–19
asthenosphere   40
atmosphere   6
atmospheric composition   21
atmospheric hazards   3, 6, 20–1
  Australia   9
  preparedness strategies   29–30
  processes that create   21–2
  responding to   28
  thunderstorms   22–3
  tropical cyclones   14–15, 24–8
atmospheric pollutants   75, 107–18
Australia Plate   47, 48
Australian Bureau of Meteorology tropical cyclone category system   26
Australian places   139
  defining 'place'   140–1
  FiFo communities, central Qld   179–81
  historic urbanisation   141–3
  Indigenous places   156–8
  key questions   139
  land use   149–52
  most populous urban areas   159, 160–1
  natural and cultural places   162–3
  remote and rural places   146–8, 149, 175–8
  responding to challenges facing   203–7
  urban places/urban areas   146–8, 151, 182–203
  see also settlements

Australian Standard Geographical Classification (ASGC-RA)   149
Australian Statistical Geography Standard (ASGS)   145, 146
Australia's population distribution   158–62
  factors affecting   163–75
avalanches   63

B

Bali, Indonesia
  marine pollution   100
  volcanic eruptions   61–2
Barcelona, Spain, water supplies   236
Bardon Park sustainable use, Brisbane (fieldwork example)   209–14
basic volcanoes   53
bathroom chemicals, as marine hazard   103
Beijing, China
  groundwater use   237–8
  recycled water use   239, 240
  water supply   237–40
  water usage and pricing   239
bioaccumulation   95, 103
biodegradable wastes   95
biodiversity
  invasive species impact on   82
  loss of, and urban sprawl   183–4
biofilm plants   282
biohazards   73
biological control of invasive species   86
biological processes   73
biomagnification   95
biophysical environment   4
Biosecurity Act 2014   85, 92
biosphere   6
  environmental pollutants   75
Black Death   126
Black Saturday bushfires   17
  factors influencing   17–18
  how the fires changed response strategies   18–19
Bounded Locality   146
Brent, London, flooding   249–50
Brisbane
  early settlement   141
  motor vehicle dependency   187
  Paddington   191
  population change   174, 192–3
  population characteristics   192–5

West End   195–203
zoning   188
Brisbane's air pollution
   air quality monitoring   118
   and climate   116
   hazard mitigation   116–17
   and population growth   112
   sources   112
   and topography   113–14
Broad Street pump, London, and cholera deaths   234, 235
bushfires   17
   'Prepare. Act. Survive' response framework   18

C

cane toads   74
Cape Town, South Africa, water crisis   237
carbon dioxide   53, 75, 108
carbon footprint   274
carbon monoxide   109
Carnegie Ridge   56
cause of hazard   11
central Queensland, FiFo communities   179–81
chemical control of invasive weeds   85
China
   economic development   267–70
   ghost cities   269–70
   Pearl River Delta, risk of floods   247
   regions   269
   South-to-North Water Diversion project   238
   Special Economic Zones (SEZ)   267–9
   urbanisation   270–1
   see also Beijing
cholera   76, 127–31
   cases reported by year and continent   129
   deaths, and London water supplies, 19th century   234, 235
   impact   130–1
   modern cities   234–5
   oral rehydration therapy   130
   WHO prevention strategies   129–30
choropleth maps   163–4
Christchurch, New Zealand, earthquakes   45–6, 47–50
   deaths and destruction   48
   economic cost   48
   liquefaction   48
   social effects   49
cities   155
   inputs and outputs   183
   the rise of   216–18

   as urban heat islands   251–6
   vulnerability to floods   246–8
   see also ecocities; megacities
climate, and population distribution   167
climate change
   and ecological hazards   75
   and glacier melting   68
   impact on Murray–Darling Basin Plan   79
   and infectious diseases   131–5
   and invasive species   77, 84
   and tropical cyclones   28
   and UN Sustainable Development Goals   279
   and urban flood risk   249
   and vulnerability to natural hazards   13
climate zones, Australia   167
climatological hazards   6
   Australia   9, 17
closed systems   6
Colombo, Sri Lanka
   garbage and dengue fever   261, 262
   urban agriculture   257
   waste disposal   261
colonial settlements in Australia   141
column graphs, interpreting   220
communicating your ideas   209
communication technology, rural and remote areas   178
communities   140
   rural and regional Australia   181
   see also Fly In-Fly out (FiFo) communities
confined aquifers   240
continental crust   41
continental plate   40
continental transform fault   47
control of invasive species   85–7
conurbation   143
convection currents   40
converging faults   41
converging plate boundaries   41, 43
coral reefs
   at risk from chemicals   103
   bleaching of   103, 105
Coriolis effect   21, 24, 98
correlations, calculating   264–7
cost of living   19
Cotopaxi, Ecuador   56–7
counterurbanisation   173–4
Crichton's Risk model   15, 16
crust   40, 41
cultural and natural places, Australia   162–3
cultural and religious backgrounds, and population   168
cyanobacteria   78

D

damage potential from hazard   11
data tables, interpreting   221–3
deaths
   causes of, low- and high-income countries   128
   from air pollution   110, 111, 278
   from cholera, 19th century   234, 235
   from infectious diseases, malnutrition and birth problems   126, 129
   from malaria   131
debris   64, 65
Delhi, India
   heat island phenomenon   254–6
   land use   255
demographic factors
   affecting population distribution   172–5
   urban and rural places   147–8
demographics, West End, Brisbane   197–200
dengue fever   131, 263
developed countries
   sustainable development, megacities   272–7
   urban food security   258
developing countries
   slum environments   217–18, 224–5, 277–8
   sustainable development, megacities   277–9
   urban food security   257
development density, urban and rural places   147
Dharavi, Mumbai, India, slum environments   224–5
disaster management   16
diseases   74, 76, 77, 125
   caused by air pollution   109, 278
   and garbage   261–4
   hazard zones   216–31
   in urban areas   233–6
   see also air-borne diseases; deaths; infectious diseases; vector-borne diseases; water-borne diseases
dispersed settlements   159
diverging faults   41
diverging plate boundaries   41, 43
dot density maps   163
drainage basins   248
drought, primary, secondary and tertiary impacts   19
duration of hazard   11, 76

E

E. coli (*Escherichia coli*)   245
e-waste recycling   119–21

early Australian settlements, site and situation   141
Earth, structure   40
earthflows   65
earthquakes   40, 41, 43–53
   Australia   8
   Christchurch, New Zealand   45–6, 47–50
   comparing effects of   52–3
   magnitude   43–4
   Manila, Philippines   251
   in Mexico   50–2
   Nepal   69
   Papua New Guinea   46
   responding to   45–7
   seismic waves   41, 43–4
   South America   56–7
   strength measurement   43, 44
Earth's core   40
Earth's physical systems   6
East African Rift Zone   41
Ebola virus   74, 126
ecocities   281
   Freiberg, Germany   281, 282–3
   principles   281
   Vauban, Germany   281–2, 283
ecological hazards   72
   atmospheric pollutants   75, 107–18
   definition   72
   factors affecting severity of impact   76–80
   infectious and vector-borne diseases   76–7, 125–36, 234–5, 245
   key questions   72
   lithospheric pollutants   75, 119–25
   marine hazard zones   97–107
   plant and animal invasions   74, 80–94
   pollutants   74, 75, 77, 94–7
   as result of biological and anthropogenic processes   73
   risk management   75–6
   types of   73–5
   vulnerability to   76
   water extraction from Murray–Darling Basin   77–80
economic activity, urban and rural places   146–7
economic development
   China   267–70
   India   267
   and urban growth   267–71
   and urbanisation   217–18
economic factors
   affecting population distribution   168–72
   affecting vulnerability   13
economic growth
   and environmental quality   227–8
   and megacities in the digital age   275–6
   and urbanisation   219–20, 222

economic impact
   on FiFo communities   180
   of invasive species   82, 90
ecosphere   6
Ecuador
   volcano alert codes   57
   volcanoes   56–7
educational services, and population distribution   167
electricity generation   274
Emerald, Qld   178
employment
   and population distribution   168–9
   rural and remote areas   175
endemic cholera   130
energy consumption   274
energy use, and urban sprawl   184
environmental degradation
   and economic growth   227–8
   through urban sprawl   183–4
environmental impact, of invasive species   82, 90
Environmental Kuznets Curve   227–8
environmental pollutants   75
environmental quality, and urbanisation   227–9
epicentre   41, 43
epidemics   126
evapotranspiration   239
ex-tropical cyclone Linda, and dangerous surf   32–3
exotic species   80
exposure (to hazard)   15–16
extinct species   82

F

faults   40, 41
favelas   217–18, 224
feral cats   81, 82
feral goats   82
feral pigs   81, 82
feral plants and animals   81
field study preparation
   example - sustainable use of public space in Bowman Park, Brisbane   209–14
   geographic inquiry model   207–9
fire ants in Australia   91–4
   eradication program biosecurity areas   92
   impacts   91–2
   locations   91
   as noxious species   92
   responding to the hazard   92–3
fire seasons in Australia   10
fishing nets, lost   99, 101
flooding, and urban sprawl   184
floods   5
   Brent, London   249–50
   and climate change   249
   Jakarta   242

   Manila, Philippines   251
   Nepal   67, 68, 69
   preparedness   29
   urban areas   184, 246–9, 273
flow movement   63
   classification   64
   types of   65
Fly In-Fly Out (FiFo)
   definition   179
   Parliamentary Inquiry into   179, 181
Fly In-Fly Out (FiFo) communities   165, 175, 179–81
   factors driving FiFo work   179
   impacts on   180–1
   in Queensland   179
focus   41
fold mountains   41, 67
food chains   95, 103
food security, megacities   257–60
foxes   81, 82
Freiberg, Germany, as green city   281, 282–3
frequency of hazard   11, 76
fresh water
   access to
      megacities   236–40
      and UN Sustainable Development Goals   279
   and population distribution   165
   quality and availability, rural and remote areas   178
freshwater runoff, impact on Great Barrier Reef   105–6

G

garbage, and disease   261–4
garden cities   203–4
gentrification   191, 197, 201
Geographic Information System (GIS)   149
geographic inquiry model   207
   analyse the data   209
   communicate your ideas   209
   evaluating the options   209
   gathering information   208
   planning   207–8
   proposing actions   209
geography, and population distribution   165
geological hazards   3, 6, 40
   Australia   8
   earthquakes   40, 41, 43–53, 69
   processes that create   41–3
   volcanic eruptions   40, 53–63
geomorphic hazards   3, 6, 63–4
   Australia   8
   and flow movement   63, 64, 65
   mass wasting   63, 64–5
   responding to   65–7
   and slope failure   63, 65

Ghana, urban agriculture   257
ghost cities, China   269–70
glacier melting, and climate change   68
Global Context   274
Global Ocean Conveyor Belt   98
Global Seismographic Network (GSN)   43
global warming   13, 36
globalisation   220
graphing data, using ICT   199–200
Great Barrier Reef
   coral bleaching   105
   human impact on   104–7
   impact of deteriorating water quality from freshwater runoff   105–7
Great Pacific Garbage Patch hazard zone   97, 99–100, 102
green belts   203–4
green villages   159
greenhouse gas emissions   13, 274
Gross Domestic Product (GDP)   219, 222, 274
groundwater pollution, Jakarta   240–1
groundwater use, Beijing   237–8
gyres   98

H

Habitat III   279–81
hamlets   145
Hawaiian volcanoes   53
hazard zones   4–5
   in Australia   9–10
hazardous diseases   74
hazards   73
   in backyard gardens   119
   see also specific types, e.g. ecological hazards
health issues, urban areas, developing countries   278
health and sanitation management, megacities   233–46
health services, and population distribution   167
heat islands   184, 251–6
heat-related illnesses   251, 252
heatwaves affecting eastern Australia, 2017   10
herbicides   85
homesteads   145
hot spots   40, 53, 56
housing affordability and availability
   rural and remote areas   177
   Sydney   170–2
   urban areas   189–90
housing policy, and urban sprawl   186
hurricanes   24, 25, 35, 36
hydrocarbons   109

hydrological hazards   6
   Australia   9
hydrosphere   6
   environmental pollutants   75

I

Icelandic volcanoes   41, 53
impact of ecological hazards, factors affecting   76–7
impact of natural hazards
   factors affecting   11–12
   factors affecting vulnerability   12–13
   managing   16
   primary, secondary and tertiary impacts   14
inclusive planning   204
India
   economic development   267
   heat island phenomenon, Delhi   254–6
   slum environments   224–5, 278
Indigenous Australians
   language groups prior to British colonisation   156
   population by state and territory   157
   population characteristics   156–7
   population distribution   156, 157–8
Indigenous people, Country/Place to   141
Indigenous places, in Australia   156–8
indigenous plants and animals   80
industrialisation
   and environmental decline   227
   and urbanisation   216–17
industry, rural and remote areas   176
infectious diseases   74, 125, 126
   cholera   76, 127–31
   and climate change   131–5
   deaths from   129
   hazard zones   126–31
   location of recent serious outbreaks   127
   and poverty   127
   reducing the risk of   135–6
   see also air-borne diseases; vector-borne diseases; water-borne diseases
infill development   204, 205–6
influenza   74, 76, 77, 126, 278
informal settlements   217–18, 224–6
infrastructure, and sustainability   273
inland drainage basin   121
inter-tropical convergence zone (ITCZ)   24
introduced species   81

invasive plants and animals   80–94
   in Australia   74
   biosecurity measures to reduce risk of   84–5, 92
   characteristics   80–1
   climate change effects   77, 84
   control methods   85–7
   countries with highest number of alien species   83
   definition   80
   fire ants in Australia   91–4
   global hazard zones   83–4
   impact of   81–2
   lantana   81, 87–8
   management   85–7
   mouse plagues   82, 83, 88–90
   reducing the risk of   84–94
inverted barometric effect   27
isolated settlements   159
isopach maps   55

J

Jakarta, Indonesia
   floods   242
   groundwater pollution   240–1
   land use and population density   231–2
   nitric oxide pollution from vehicles   227, 233
   poverty   230
   problems identified in 1965–85 Master Plan   230–1
   sanitation   241
   subsidence   240–4
   urbanisation and pollution in   229–33

L

La Niña   31
lahars   53
land degradation   82
land–rent relationship   185, 186
land subsidence   63, 240–4
land use
   Australia   149–52
   Delhi, India   255
   mapping   151–2
   and mortgage stress   189–90
land use zoning, urban areas   184–7
landslides   41, 63, 65
   events leading up to   67
   in Nepal   67–9
   responding to   65–6
lantana invasion   81, 87–8
lava flows   53
lead contamination in Australia   119
lead poisoning   119
linear settlements   159
liquefaction   41, 48
lithosphere   6, 56, 63

lithospheric pollutants   75, 119–25
   Agbogbloshie Dump, Accra, Ghana   119–21
   contaminant hazards in the Aral Sea   121–5
   hazards in backyard gardens   119
litter in marine environments   97, 98–100, 102–3
liveability, West End, Brisbane   200–3
liveable places   207
London
   cholera deaths, 19th century   234, 235
   climate change and flash flood risk   249–50

M

magma   41
magnitude
   earthquakes   43–4
   ecological and natural hazards   4, 11, 76
Major Urban Areas   145, 146
malaria   74, 131–5
   causative organisms and transmission   131
   constraints on complete eradication   134
   global distribution   133
   hazard zones   132
   prevalence and deaths   131
   prevention and treatment   132–4
malaria parasites, life cycle   132
managing impact (of hazards)   16
managing invasive species   85–7
Manila, Philippines
   vulnerability to natural hazards   250–1
   waste disposal   261
mantle   40, 41, 43
mapping
   comparing aerial images with topographical maps   154–5
   interpreting aerial and satellite imagery   152–3
   land use   151–2
marine creatures, plastics as hazards for   98–9, 100
marine debris
   sources by region   99
   types of   97, 98–9, 100, 101
marine ecological hazards
   factors affecting severity of impact   98–102
   Great Pacific Garbage Patch   97, 99–100, 102
   human impact on the Great Barrier Reef   104–7
   microfibres and medication residues   103–4
   monitoring   102–7
   plastics/microplastics/nanoplastics   75, 97, 98–9, 100, 102–3
   types of   97
marine hazard zones   97–107
marine pollution
   Bali, Indonesia   100
   Senegal   100–1
mass wasting   63
   Nepal   68
   types of   64–5
mean centre analysis   235–6
measles   77
mechanical control of invasive weeds   85
medium household income per week in south-east Qld   170
megacities   215
   air quality   226–33
   challenges/opportunities for developed countries   272–7
   challenges/opportunities for developing countries   277–9
   economic opportunities and challenges   267–71
   food security   257–60
   health and sanitation management   233–46
   heat islands   251–6
   informal settlements   217–18, 224–5
   key questions   215
   survival in the digital age   275–6
   sustainable development   272–83
   and urbanisation   216
   vulnerability to natural hazards   246–51
   waste management and sanitation   261–7, 272–3
   see also cities
Melbourne's food bowl   258, 260
Mercalli Scale   44
Mesh Blocks (MBs)   145
metropolis   144
Mexico City's air pollution   110–11
   air quality monitoring   117
   and climate   116
   hazard mitigation   116, 117
   'no drive days'   117
   and population growth   112
   sources   111
   and topography   113, 114–15
Mexico City's slum environments   224
Mexico's earthquakes   50
   social and economic impacts   50–2
microfibres, as marine hazard   103
microplastics   98, 99
   in the Pacific, monitoring   101–2
mid-oceanic ridges   40, 41
Mindanao Trench   41
mining boom   176, 177
mitigation (hazards)   5
   air pollution   116–18
   mouse plagues   90
mobility and location of hazard   77
Modified Mercalli Scale   44
monofunctional development   182
mortgage stress, and land use   189–90
mosquito-borne diseases   77, 125, 131, 261–4
motor vehicle dependency
   Brisbane suburbs   187
   and urban sprawl   184
Mount Agung, Bali, Indonesia   61–3
Mount Etna, Italy   58
Mount Isa   165–6
Mount Vesuvius, Italy   58–61
   79 CE eruption   58
   emergency planning   60
   evacuation plans   59
mouse plagues   82, 83, 88–90
   economic, environmental and social impacts   90
   factors affecting severity of impact   88–9
   hazard prevention and mitigation strategies   90
mudflows   63, 64, 65
mudslides   63, 64
multicell thunderstorms   22
Mumbai, India
   accommodating rapid population growth   277
   slum environments   224–5
Murray–Darling Basin, water extraction   77–80
Murray–Darling Basin Plan (2014–2026)   79

N

nanoplastics   98
National Broadband Network (NBN)   178
natural and cultural places, Australia   162–3
natural beauty, and tourism   167
natural disasters   4, 70
   by fatalities   7
   risk assessment   16
natural environment, impact of sustainability on   273–4
natural hazards   3
   analysing spatial and statistical data   70–1
   assessing and responding to   5, 12, 15–19
   in Australia   8–10
   cities vulnerability to   246–51
   definition   4
   impact of   10–15
   key questions   3
   and natural disasters   4

risk management   5, 15
systems approach   6
types of   6
see also specific types, e.g. geological hazards
Neolithic revolution   141
Nepal
earthquakes   69
floods   67, 68, 69
landscape profile   68
landslides   67–9
reducing the risks of hazards   69
New Urban Agenda (Habitat III)   279–81
New York
cholera outbreak, 1830s   234
river pollution   245
vertical agriculture   258
waste management   272–3
New Zealand earthquakes   45–6, 47–50
Newcastle earthquake   8
nitric oxide pollution from vehicles, Jakarta   227, 233
nitrogen oxides (NO$_x$)   75, 95, 107, 108
non-point sources of pollution   94, 95
north (aerial and satellite images)   153
northern Australia, tropical cyclones   31–5
noxious weeds   81, 87
nucleated settlements   159

O

ocean circulation system   98
oceanic crust   41
oceanic plate   40
oceanic trench   40
oil pollution   95
open systems   6
Other Urban Areas   145, 146
ozone   107, 109

P

P-waves   43, 44
Pacific Plate   47, 48
Pacific Ring of Fire   53
Paddington, Brisbane   191
pandemics   126
Papua New Guinea, earthquakes   46
particulate matter   75, 107, 108, 109
pathogenic bacteria   245
patterns, shapes and textures (aerial and satellite images)   153
periurban horticulture   257
periurbanisation   174–5
pesticides   85
pharmaceuticals, as marine hazard   103–4
Philippines
Super Typhoon Haiyan   36–7
Typhoon Hato   36
see also Manila, Philippines
photodegradation   98
physical control of invasive species   85
physical factors
affecting population distribution   165–7
affecting vulnerability   12
places
definition   140–1
and population, Australia   141–3
urban and rural   146–8
see also settlements
planning
fieldwork   207–8
garden cities   203–4
inclusive   204
urban villages   204–5
plant and animal hazards, types of   80–1
plant and animal invasions   74, 77, 80–94
Plasmodium falciparum   131, 134
Plasmodium vivax   131
plastic rubbish in the marine environment   75, 97, 98–9, 100
microplastics in the Pacific   102–3
plate tectonic movement   40, 41
point sources of pollution   94, 95
political factors affecting vulnerability   13
pollutants   74, 75, 77, 94–7
definition   94
reducing the risk of   95
see also specific types, e.g. air pollutants
polluted drinking water, and infectious diseases   125, 128–9
pollution   75
groundwater, Jakarta   240–1
prevention and control   95
types of   94
and urbanisation in Jakarta   229–33
world map   96
see also specific types, e.g. air pollution
population change
Australia   158
Brisbane   174, 192–3
demographic factors affecting   172–5
formula   172
Queensland   158, 172, 173, 176
population characteristics, Brisbane   192–5
population density   163–4
population distribution
Australia   158–62
factors affecting   163–75
Indigenous Australians   156, 157–8
methods   163–4
Queensland   163–4, 175
population growth
accommodating, developing countries   277–8
and air pollution   112
Australia and Queensland   158
Australian towns   142
global   218
rates   222
and urban sprawl   186
and urbanisation   218–19, 220, 221
population size
Australian states   159
urban and rural places   146
poverty
and food insecurity, Sri Lanka   258–9
and infectious diseases   127
Jakarta, Indonesia   230
Manila, Philippines   261
and Mexico's earthquakes   50, 51
predictability of hazard   11
preparedness (natural hazards)   5, 11
Nepal   69
tropical cyclones   29–30
volcanic eruptions, Italy   59–60
prescribed medications, as marine hazard   103
pressure gradient   21
prevention (hazards)   5, 11
prickly pear cactus   86
primary air pollutants   107
primary data   208, 209
primary impacts   14, 19
primary sector   146
primary waves   43, 44
prior knowledge, for interpreting satellite and aerial imagery   153
pseudo-urbanisation   217, 218
pull factors   165, 219, 224
push factors   164, 224, 277
pyroclastic clouds   5, 54
pyroclastic cones   53

Q

qualitative data   208
quantitative data   208
quarantine regulations   85

Queensland
  FiFo communities   179–81
  mortgage stress   190
  population change   158, 172, 173, 176
  population distribution   163–4, 175
  resources industry   180
  rural and remote area challenges   175–8
  urban sprawl   182

R

rabbits   74, 77, 81, 82, 83, 86
rapid onset hazards   76
recycled water, Beijing, China   240
recycling e-waste   119–21
relative change   192
relative socioeconomic disadvantage in Australia   169
remote places   149
  challenges   175–8
  measurement scales   149
  see also Fly In-Fly Out (FiFo) communities
resource exploitation and trade, affect on population distribution   168
resources, and population distribution   165
resources industry, Queensland   180
responding
  to atmospheric hazards   28
  to earthquakes   45–7
  to geomorphic hazards   65–7
  to natural hazards   5, 12, 15–19
  to tropical cyclones   29–30
  to volcanic eruptions   56–8, 59–60, 61–3
ribbon settlements   159
Richter Scale   43, 44
ridges (high pressure)   22
ring villages   159
risk   4, 75
risk assessment   15–16
risk management   5, 15
  ecological hazards   75–6
risk management triangle   15, 16
river pollution   245–6
robots   276
rockfalls   65
rockslides   65
rubber vine   74, 81
Rural Balance   146
rural places   146–8, 151
  challenges   175–8
  see also Fly In-Fly Out (FiFo) communities
rural population growth, global   219
Rural Remote and Metropolitan Areas classification (RRMA)   149
rural–urban fringe, and urban sprawl   186

S

S-waves   44
safety, and urban sprawl   185
safety evacuation plans   17
Saffir–Simpson Hurricane Wind Scale   25, 26
salinity levels, Aral Sea   124
San Andreas fault, California   41, 47
sanitation
  poor, in cities   233–6, 241, 245
  and waste management   272–3
São Paulo, Brazil, Rio Tietê river pollution   245–6
satellite and aerial imagery
  comparison with topographical maps   154–5
  scale   152–3
  showing land use   150
  strategies for interpreting   151–3
scattergraphs   70–1, 264–7
sea floor spreading   53
secondary air pollutants   107
secondary data   208, 209
secondary impacts   14, 19
secondary pollutants   95
secondary sector   146
secondary waves   44
seismic waves   41, 43–4
Senegal, marine pollution   100–1
service provision
  rural and remote areas   167
  urban areas   192
services available, urban and rural places   147
settlement size, ABS definitions   145–6
settlements   141
  broad categories   143–5
  early Australian   141
  hierarchy of   143–5
  patterns of   159
shanty towns see slum environments
shield volcanoes   53
shopping malls, decline in   276
Singapore, urban agriculture   258
single-cell storms   22
slope failure   63, 65
slow onset hazards   76
slum environments   217–18, 224–5, 277–8
slump   65
smart cities   206
Snow, John, cholera map, London   234, 235

social demographics, urban and rural places   147–8
social factors
  affecting population distribution   167–8
  affecting vulnerability   13
social impact
  of invasive species   82, 90
  of urban sprawl   274–5
soil creep   65
soil pollution   94, 119
soil quality, and population distribution   166
solar radiation   21
solid waste   257, 261
South America, earthquakes and volcanoes   56–7
South-to-North Water Diversion project, China   238
spatial technologies   208
Spearman's Rank correlation coefficient   264–7
Special Economic Zones (SEZ), China   267–9
speed of onset (hazards)   11, 76
squatter settlements   224
Sri Lanka
  dengue fever and garbage   261–4
  poverty and food insecurity   258–9
  urban agriculture   257
storm hydrographs   248
storm surges   27
storms, preparedness for   29
stratovolcanoes   58, 61
strike-slip faults   41
strip development   182, 184
Stromboli   58
subduction zones   40, 41, 53, 56
subsidence   63
  Jakarta   240–4
suburbanisation   174
sulfur dioxide   53, 75, 95, 107, 109
sunscreen chemicals   103
Super Typhoon Haiyan   36–7, 39, 251
supercells   23
sustainable development
  developed countries   272–7
  developing countries   277–9
  ecocities   281–3
  impact on the natural environment   273–4
  infrastructure   273
  in megacities   272–83
  New Urban Agenda (Habitat III)   279–81
  waste management and sanitation   272–3
Sustainable Development Goals (UN)   279, 280–1
sustainable megacities (UN Development Goal)   280–1

sustainable use of public space in Bowman Park, Brisbane (fieldwork example) 209–14
   collecting data 211–13
   developing criteria 211
   key questions 210–11
   making recommendations 214
   writing the report 214
Sydney
   food bowl 258
   housing affordability 170–2
synoptic maps and weather warnings, reading 32–3
systems, and approach 6

T

tectonic plate boundaries 41, 43, 53, 58, 67
tectonic plate movement 40, 41, 43
tectonic plates, world map 42
temporal spacing 77
tenure 224
tertiary impacts 14, 19
tertiary sector 146
thermohaline circulation 98
3-D printing 275
three-sector theory 146
thunderstorms 22–3
   formation 22
   impacts 23
   types of 22–3
Tokyo, Japan, as megacity 215
topographical maps, comparing with aerial images 154–5
topography, and air pollution 113–15
tourism 167
towns 144
transform fault boundary 41
transform faults 41
transport options, urban areas 191–2
transportation connections, rural and remote areas 177
Tropical Cyclone Debbie 2017 33–5
   rainfall and other effects 33–4
Tropical Cyclone Warning Advice 28
tropical cyclones 24–8
   categories 25–6
   and climate change 28
   effects of 14–15
   formation 24–5
   frequency 36
   northern Australia 31–5
   paths and intensity 26
   preparedness and response 29–30
   and storm surges 27
   world distribution 24
   see also hurricanes; typhoons
tropical depressions 26
tropical storms 26
troughs (low pressure) 22
tsunamis 40, 41
Typhoon Hato 36, 39
typhoons 24
   frequency 36
   Philippines 36–7, 251
   in the western Pacific 35–9

U

UN Intergovernmental Panel on Climate Change 104
UN Sustainable Development Goals 279, 280–1
unconfined aquifers 240
upper mantle 40, 41
urban agriculture 257–8
urban areas/places 146, 151
   access to fresh water 236–40
   demography – social 147–8
   development density 147
   disease in 233–6
   economic activity 146–7, 267–71
   flood risks 184, 246–50, 273
   gentrification 191
   housing availability and affordability 189–90
   land use zoning 187–8
   most populous places, Australia 159, 160–1
   population growth 218–19, 220, 221, 267–71, 272
   population size 146
   service provision and management 192
   services available 147
   transport options 191–2
   as urban heat islands 151–6, 184
   urban sprawl 174, 184–7
   waste management 192, 261–7
   see also Brisbane; cities; megacities; West End
urban Australia, challenges facing 182–95
Urban Centre and Locality (UCL) 146
urban centres 145
urban development, historic 141–2
urban food security
   developed countries 258
   developing countries 257
urban growth see urban areas/places, population growth
urban heat islands 184, 251
   management 251–6
Urban Mesh Blocks 145
urban renewal and regeneration 205–6
urban sprawl 174, 182–7
   characteristics 182, 184
   factors affecting 184–7
   impacts of 183–5
   reasons for 182
   social impacts of 274–5
urban villages 204–5
urbanisation 172–3, 215
   and air pollution 227–33
   annual change in 221
   China 270–1
   and economic growth 219–20, 222
   and environmental quality 227–9
   global patterns of 216–26
   and globalisation 220
   interpreting data tables and column graphs 220–3
   and megacities 216
   and pollution in Jakarta 229–33
   and population growth 218–19
   and the rise of the cities 216–18

V

Vauban, Germany, as ecocity 281–2, 283
vector-borne diseases 74, 125, 126
   dengue fever 261–4
   malaria 74, 131–5
   reducing the risk of 135–6
vertical farming 258
Vesuvius see Mount Vesuvius, Italy
Victorian Bushfires Royal Commission (2009), recommendations 18
villages 144, 159
volcanic ash 53–4, 57
   impacts of 54–5
volcanic eruptions 40, 53–63
   features 54
   impacts 53–5
   Mt Agung, Bali, Indonesia 61–3
   Mt Vesuvius, Italy 58–61
   responding to 56–8, 59–60, 61–3
   South America 56–7
volcanic gases 53
volcanoes 40, 43
   acidic lava cones 53
   basic lava cones 41, 53
   hazard zones 5
   location 53
vulnerability to ecological hazards 76
vulnerability to natural hazards 16, 75
   cities 246–51
   factors affecting 12–13

W

wages 169, 170
waste disposal 274
   challenge of 261
waste management
   garbage and spread of disease 261–4
   megacities 261–7
   rural and remote areas 177–8

sustainable 272–3
urban areas 192
water availability
   access to fresh water, megacities 236–40
   and population distribution 165
   rural and remote areas 178
water-borne diseases 125, 126
   cholera 76, 127–31, 234, 235
   E. coli (*Escherischia coli*) 245
water degradation 82
water extraction from Murray–Darling Basin 77–80
   factors affecting severity of impact 78
   impact of climate change 79
   reducing the risk from 79
   size of Basin 78

water pollution 75, 94, 95, 101
   see also groundwater pollution; river pollution
water quality, impact on the Great Barrier Reef 105
waterways, and urban sprawl 184
West End, Brisbane
   cultural precinct 197
   demographics 197–200
   development history 195–7
   dwelling types over time 200
   gentrification 197, 201
   liveability 200–3
   location 195
   Right to the City movement role 201
   tensions over developments 200–1
   West Village development 201, 202

wildfires 17
wind set-up 27
wind shear 24
World Health Organization
   air pollution deaths 109
   cholera cases, 2016 234
   cholera prevention strategies 129
   ecological hazard definition 72
   malaria prevalence and control 131, 132, 134

Z

Zika virus 278
zoning 204
   urban areas 187–8
zooxanthellae 103